HANDBOOK OF PLANT CELL CULTURE

Published

Volume 1, *Techniques for Propagation and Breeding*
Editors: David A. Evans, William R. Sharp, Philip V. Ammirato, Yasuyuki Yamada

Volume 2, *Crop Species*
Editors: William R. Sharp, David A. Evans, Philip V. Ammirato, Yasuyuki Yamada

Volume 3, *Crop Species*
Editors: Philip V. Ammirato, David A. Evans, William R. Sharp, Yasuyuki Yamada

Volume 4, *Techniques and Applications*
Editors: David A. Evans, William R. Sharp, Philip V. Ammirato

Volume 5, *Ornamental Species*
Editors: Philip V. Ammirato, David A. Evans, William R. Sharp, Yashpal S. Bajaj

Volume 6, *Perennial Crops*
Editors: Zhenghua Chen, David A. Evans, William R. Sharp, Philip V. Ammirato, Maro R. Söndahl

HANDBOOK OF PLANT CELL CULTURE,

Volume 6

Perennial Crops

Editors

Zhenghua Chen
Chinese Academy of Sciences

David A. Evans
DNA Plant Technology Corporation

William R. Sharp
DNA Plant Technology Corporation

Philip V. Ammirato
Barnard College, Columbia University
DNA Plant Technology Corporation

Maro R. Söndahl
DNA Plant Technology Corporation

McGraw-Hill Publishing Company

New York St. Louis San Francisco Auckland Bogotá
Caracas Hamburg Lisbon London Madrid Mexico
Milan Montreal New Delhi Oklahoma City
Paris San Juan São Paulo Singapore
Sydney Tokyo Toronto

Library of Congress Cataloging-in-Publication Data
(Revised for volume 6)

Handbook of plant cell culture.

Includes bibliographies and indexes.
Contents: v. 1. Techniques for propagation and breeding / David Evans . . . [et al.] — v. 2. Crop species / editors, William R. Sharp . . . [et al.] — v. 3. Crop species / editors, Philip V. Ammirato . . . [et al.] — v. 6. Perennial crops / editors, Zhenghua Chen . . . [et al.].
1. Plant cell culture—Handbooks, manuals, etc. I. Evans, David, date. II. Sharp, William R., date. III. Ammirato, Philip V., date.
SB123.6.H36 1983 631.5′23 82-73774
ISBN 0-02949230-0 (v. 1)

234567890 DOC/DOC 943210

ISBN 0-07-010848-X

The editor for this book was Jennifer Mitchell, the designer was Naomi Auerbach, and the production supervisor was Dianne L. Walber. It was set in Times Roman by the McGraw-Hill Publishing Company Professional Reference Division composition unit. Project supervision by The Total Book.

Printed and bound by R. R. Donnelley & Sons Company.

Volume 6 is a translation and adaptation of a work published originally in China by Higher Education Press.

Contents

Preface vii

Contributors x

Part A. Basic and Specialized Techniques 1

1 Potential of Biotechnology in Perennial Crop Improvement, *Zhenghua Chen and William R. Sharp* 3
2 General Techniques of Tissue Culture in Perennial Crops, *Zhenghua Chen and David A. Evans* 22
3 Haploid Induction in Perennial Crops, *Zhenghua Chen* 62
4 Cell Suspension Culture and Mutant Screening in Perennial Crops, *Zhenghua Chen, Yuhua Song, and Wenbin Li* 76
5 Protoplast Culture and Fusion in Perennial Crops, *Wenbin Li and Zhenghua Chen* 92
6 Medicinal and Aromatic Perennial Crops, *Wenbin Li and Da-wei Zhang* 116
7 Tissue Culture of Perennial Ornamental Plants and Virus Elimination, *Wenda Qiu* 127

Part B. Timber and Cork Trees 143

8 Poplar and Willow, *Jingfang Lin* 145
9 Poplar: Anther Culture, *Zhihua Lu and Yuxi Liu* 161
10 Poplar: Ovary Culture, *Kexian Wu* 183
11 Poplar: Rapid Propagation, *Yanlei Ba and Jianshong Guo* 191
12 Eucalyptus, *Quan Ouyang and Haizhong Peng* 199

13 Cork Tree, *Kexian Wu* 216
14 Staghorn Sumac, *Zhiqing Zhu* 225

Part C. Fruit Trees 231

15 Apple: Anther Culture, *Guangrong Xue and Jianzhe Niu* 233
16 Apple: Shoot Tip and Embryo Culture, *Jixuan Wang* 245
17 Crab Apple: Anther Culture, *Jiangyun Wu* 256
18 Pear, *Huixiang Zhao and Nailiang Gu* 264
19 Peach, *Niyun Hu, Zenghai Yang, and Guangming Lu* 278
20 Grape: Anther Culture, *Ziyi Cao* 300
21 Grape: Micropropagation, *Ziyi Cao* 312
22 Hawthorn, *Jingshan Wang and Guang Yu* 329
23 Hawthorn: Embryo Culture, *Yuying Wang, Xinyi Gao, and Kai Fu* 346
24 Black Current: Shoot Tip Culture, *Jiangyun Wu and Dingqiu Huang* 351
25 Citrus: Anther Culture, *Zhenguang Chen* 365
26 Litchi: Anther Culture, *Lianfang Fu* 376
27 Longan: Anther and Embryo Culture, *Wenxiong Wei* 385
28 Loquat: Shoot Tip Culture, *Yongging Yang* 397
29 Kiwi Fruit, *Zhenguang Huang and Suying Tan* 407

Part D. Extractable Products 419

30 Oleaster, *Jingli Pan, Puxuan Wang, Rulan Gao, Hui Fan, and Lizhu Tu* 421
31 Walnut, *Biwen Han and Shulan Liu* 431
32 Oil Palm, *Yuanfang Cui and Zheng Gong* 440
33 Rubber Tree: Anther and Ovule Culture, *Zhenghua Chen, Xuen Xu, and Rensheng Pan* 453
34 Guayule, *Huizhu Xu and Minzhi Qian* 468
35 Tea: Anther Culture, *Zhenguang Chen and Huihua Liao* 475
36 Deodara Cedar, *Min Liu* 480
37 Xanthoceras Sorbifolia, *Yongming Wang* 484

Species Index 493
Subject Index 498

Preface

Volume 6 continues a treatise begun in 1983. Volume 1 detailed the many different techniques and approaches available in the area of plant cell culture that were beginning to play significant roles in basic research, plant breeding, and crop improvement. Volumes 2 and 3, published in 1984, and Volume 4, published in 1986, explored the further development of aseptic culture techniques and their application to specific agricultural crops. Volume 5 summarized applications to ornamental species, especially floricultural crops, a group of plants that were the first commercial beneficiaries of these new and powerful techniques.

This volume explores the development and application of plant cell culture techniques to perennial crops, plants that have proven particularly difficult to manipulate in aseptic culture. Problems that have challenged researchers include the juvenility or maturity and strong seasonal variation within stock plants; the long life cycles that affect explant availability and the analysis of results; and the difficulties in promoting plantlet growth from somatic or microspore embryos and in the rooting of micropropagated or regenerated shoots.

However, the last five years have witnessed extraordinary advances in our ability to manipulate and exploit perennials in cell culture and it is appropriate that a volume in this series be devoted to this group. Woody trees are grown or harvested for a wide range of products, including lumber, cork, fruits, pharmaceuticals, flowers, spices, oils, nuts, and beverages and we are pleased to have a large number of these crops represented here.

The path to publication of Volume 6 differed from previous volumes. Several years ago, Z. H. Chen of the Institute of Genetics of the Chinese Academy of Sciences in Beijing assembled a volume on woody plants patterned after the *Handbook of Plant Cell Culture*. Shortly thereafter, Dr. Chen was a visiting scientist at DNA Plant Technology Corporation along with her husband Professor Ruofu Du, also of the Chinese Academy of Sciences. During this period the decision was made to publish an English-language adaptation as part of this treatise.

For a good number of years, Chinese scientists have been at the forefront in research on perennial crops and particularly the micropropagation, anther and microspore (pollen) culture, and embryo and ovule culture of a wide variety of woody plants. However, much of this pioneering research is published in local journals or only in Chinese and is generally unavailable to the West. We envisioned this volume as providing increased access to research in China in the context of a general synthesis of advances in the cell culture of perennial crops. We are pleased that our contributions are primarily from researchers in China and that a large percentage of the literature cited derives from Chinese publications.

In developing this volume, Dr. Chen decided that it would facilitate comprehension and citation if all personal names, including those of Chinese scientists, were given in the Western fashion, i.e., with the family name placed last. For example, her name in the more traditional form would be Chen Zhenghua but is listed here as Zhenghua Chen or Z. H. Chen. This has been done with all names.

The primary translation took place in China under Dr. Chen's direction and the individual translators are footnoted at the beginning of each chapter. Professor Du substantially reworked and readied the manuscripts for transmittal. Further revisions and editing were carried out in the United States.

The organization of the volume follows the same format as previous ones. The first section summarizes basic and applied techniques as they pertain especially to perennial crops, including haploid induction, cell suspension cultures and mutant screening, protoplast culture and fusion, medicinal and aromatic plants, and virus elimination. Subsequent sections discuss specific crops, i.e., timber and cork trees, fruit trees, and those with extractable products, including nuts, oils, rubber, and beverages. Within each chapter is a review of the literature that highlights key contributions and advances. Protocols are provided; future prospects are discussed and full citations are given for references.

To realize this unique effort, many people were mobilized and we offer our sincere thanks to all of them—our Chinese collaborators and translators, our readers and translators in the United States, and our editors at McGraw-Hill. We especially thank Dr. and Mrs. Alan Kinnersley, who perfected the translation, JoAnn Roadside, our typist, and Laura Patterson McDonald, our editorial assistant who also assembled the Species Index. Special recognition and thanks go to Jennifer Mitchell at McGraw-Hill and Jennifer Carey for their extraordinary patience, perspicacity, and professionalism, which they provided with a full measure of good cheer. Finally, we thank our fellow scientists who have supported us in our efforts and who continue to develop and exploit the tools of plant cell culture in ever more imaginative and productive ways.

The preparation of this volume presented a unique opportunity for cooperation between the United States and China. We very much enjoyed and benefitted from working with scientists from many of the major institutions in China. We are delighted at the outcome and look forward to further collaborations.

Contributors*

Yanlei Ba	Gimsar Forest Seed Station, Xinjiang, China
Ziyi Cao	Plant Physiology Division, Agronomy Department, Gansu Agricultural University, Gansu Province, China
Zhenghua Chen	Institute of Genetics, Chinese Academy of Sciences, Beijing, China
Zhenguang Chen	Fujian Agricultural College, Fujian Province, China
Yuanfang Cui	Institute of Tropical Crops, South China Academy, China
David A. Evans	DNA Plant Technology Corporation, Cinnaminson, New Jersey
Hui Fan	Northwest Institute of Botany, Wugong, Shanxi Province, China
Kai Fu	Institute of Botany, Chinese Academy of Sciences, Beijing, China
Lianfang Fu	Institute of Plant Physiology and Genetics, Fujian Academy of Agricultural Sciences, Fuzhou, China
Rulan Gao	Northwest Institute of Botany, Wugong, Shanxi Province, China
Xinyi Gao	Institute of Botany, Chinese Academy of Sciences, Beijing, China

*For all names, the family name is listed last, as discussed in the Preface.

Zheng Gong — Institute of Tropical Crops, South China Academy, China
Nailiang Gu — Department of Horticulture, Tianjin Agricultural College, Tianjin, China
Jianshong Guo — Gimsar Forest Seed Station, Xinjiang, China
Biwen Han — Beijing Agricultural University, Beijing, China
Niyun Hu — Department of Horticulture, Northwest Agricultural University, China
Dingqiu Huang — Northeast Agricultural College, Harbin, Jian Heilongjiang, China
Zhenguang Huang — Institute of Pomology, Chinese Academy of Agricultural Sciences, Zhengzhou, China
Wenbin Li — Institute of Genetics, Chinese Academy of Sciences, Beijing, China
Huihua Liao — Fujian Agricultural College, Fujian Province, China
Jingfang Lin — Chinese Academy of Forestry Sciences, Beijing, China
Min Liu — Lushan Botanical Garden, Jisngxi Province, China
Shulan Liu — Beijing Agricultural University, Beijing, China
Yuxi Liu — Forestry College of Northeast China, China
Guangming Lu — Department of Horticulture, Northwest Agricultural University, China
Zhihua Lu — Forestry College of Northeast China, China
Jianzhe Niu — Institute of Pomology, Chinese Academy of Agricultural Sciences, Xingcheng, Liaoning, China
Quan Ouyang — Guangxi Forestry Company, Guangxi, China
Jingli Pan — Xian Botanical Garden, Xian, Shanxi Province, China
Rensheng Pan — Institute of Genetics, Chinese Academy of Sciences, Beijing, China
Haizhong Peng — Institute of Beihai Forestry Sciences, Guangxi, China
Wenda Qiu — Department of Horticulture, Agricultural University of Zhejiang, Hangzhou, China
Minzhi Qian — Wuhan Institute of Botany, Wuhan, China
William R. Sharp — DNA Plant Technology Corporation, Cinnaminson, New Jersey
Yuhua Song — Institute of Genetics, Chinese Academy of Sciences, Beijing, China
Suying Tan — Institute of Pomology, Chinese Academy of Agricultural Sciences, Zhengzhou, China
Lizhu Tu — Department of Biology, Inner Mongolia University, Huhehaote, Inner Mongolia, China
Jingshan Wang — Chifeng Institute of Forestry, Inner Mongolia, China
Jixuan Wang — Liaoning Institute of Pomology, Xiongyue, Liaoning Province, China
Puxuan Wang — Northwest Institute of Botany, Wugong, Shanxi Province, China
Yongming Wang — Anshan Institute of Forestry Sciences, Anshan City, Liaoning, China
Yuying Wang — Institute of Botany, Chinese Academy of Sciences, Beijing, China
Wenxiong Wei — Fujian Academy of Agricultural Sciences, Fuzhou, Fujian Province, China
Jiangyun Wu — Northeast Agricultural College, Harbin, Jian Heilongjiang, China

Kexian Wu	Institute of Forestry Sciences, Heilongjiang Province, China
Huizhu Xu	Wuhan Institute of Botany, Wuhan, China
Xuen Xu	Institute of Genetics, Chinese Academy of Sciences, Beijing, China
Guangrong Xue	Institute of Pomology, Chinese Academy of Sciences, Xingcheng, Liaoning Province, China
Yongging Yang	Institute of Fruit Trees, Academy of Agriculture, Fuzhou, Fujian Province, China
Zenghai Yang	Department of Horticulture, Northwest Agricultural University, China
Guang Yu	Chifeng Institute of Forestry, Inner Mongolia, China
Da-wei Zhang	Institute of Genetics, Chinese Academy of Sciences, Beijing, China
Huixiang Zhao	Department of Horticulture, Tianjin Agricultural College, Tianjin, China
Zhiqing Zhu	Institute of Botany, Chinese Academy of Sciences, Beijing, China

Abbreviations

Growth Regulators

IAA	Indole–3–acetic acid
IBA	Indole–3–butyric acid
NAA	1–Naphthaleneacetic acid
2,4–D	(2,4–Dichlorophenoxy)acetic acid
2,4,5–T	(2,4,5–Trichlorophenoxy)acetic acid
CPA	(4–Chlorophenoxy)acetic acid
PIC	Picloram (4–amino –3,5,6–trichloropicolinic acid)
NOA	2–Naphthoxyacetic acid
BTOA	2–Benzothiazoleacetic acid
BA	6–Benzylaminopurine
ZEA	Zeatin
KIN	Kinetin
2iP	(2–Isopentenyl)adenine

Additives

CH	Casein hydrolysate

CW	Coconut water
EDTA	(Ethylenedinitrilo)tetraacetic acid
GA	Gibberellic acid (Gibberellin A_3)
ABA	Abscisic acid
ADE	Adenine

Macro– and Micronutrient Formulations

MS	Murashige and Skoog (1962)
B5	Gamborg et al. (1968)
ER	Erikkson (1965)
WH	White (1963)
SH	Schenk and Hildebrandt (1972)

PART A
Basic and Specialized Techniques

CHAPTER 1

Potential of Biotechnology in Perennial Crop Improvement

Zhenghua Chen and *William R. Sharp**

INTRODUCTION

Economic Importance of Perennial Crops

Many perennial crop species are critical for the economy. The need for wood and wood products is continuously increasing, the storage of energy is a universal problem, and cut wood is needed for firewood and charcoal. Wood is also a raw material for the textile, papermaking, and plastics industries. Wood species include many important oil, beverage, perfume, medicinal, and flowering trees. Perennial crops play an important role in improving sandy soil, re-forming natural environments, recovering good ecological conditions, and beautifying city environments.

*English translation by Kaiwen Yuan.

Tasks in Perennial Crop Improvement

YIELD

Timber Trees. Fast-growing species should be selected so that timber can be harvested and profits gained rapidly. Indices of high yield vary with species, tree age, and region. In general, diameter and height of tree and volume of timber are crucial. Trees with good trunk form and thin bark often have high yields. Good structures for the plant body and crown are required. For example, timber trees should have dense and narrow crowns to favor close plantings.

Fruit Trees. The yield of fruits should be high and stable. Different yield targets should be set for varieties ripening in different periods. The trees should have expanding crowns so that the area of fruit set will be large and the fruit distribution will be even.

Other. In some species (e.g., rubber tree), the products of which are used as raw materials in industry, the trunks should enlarge quickly so that the latex can be tapped early and with high yield. Tea plants must have luxuriant foliage and strong regenerating ability.

QUALITY

Timber. The specific gravity and hardness of timber for building material are important. The timber for papermaking should have long fibers with high fiber content. High-yielding varieties are also needed for resin producing pine trees.

Fruit. Important fruit qualities include size, form, color, taste, juiciness, sweetness or sourness, fragrance, nutrient value, and behavior during storage and transportation. Fruits used for making wine, vinegar, etc., should be bred for qualities such as color of juice, sugar, and acid content. Varieties for canning should have compact flesh; dried fruits should be rich in fat, protein, and vitamins.

Other. Different breeding indices should be considered in selected species, such as latex quality of rubber tree, fragrance of tea, caffeine content of coffee, and active component content of medicinal plants.

STRESS TOLERANCE. Stress tolerance is important for every tree species. Stresses include extremes of temperature (low or high); soil factors (salinity, alkalinity, and metal content of soil); and dry or barren environment.

RESISTANCE TO DISEASES AND PESTS AND TOLERANCE TO PEST CONTROL. Specific requirements are necessary for different tree species (Durzan, 1985).

Necessity of Reforming Perennial Crop Breeding Techniques

The basis of perennial crop breeding can be summarized as follows:

1. Increase sources of genetic variation and introduce new genes.
2. Combine useful genes.
3. Select superior genotypes.
4. Obtain genetically stable and uniform populations.
5. Accelerate propagation and release of seedlings.
6. Preserve germ plasm effectively.

For centuries, conventional breeding techniques have contributed much to the improvement of tree species. However, the resources of useful genetic variation are nearing exhaustion. Selection efficiencies are low, and the breeding cycle is very long. Reform of breeding techniques is, therefore, imperative.

Biotechnology as it has developed in recent years has great potential. If traditional breeding techniques can be combined with biotechnology, new and remarkable progress will be achieved in improvement of tree species. Mutagenesis and screening at the cellular level with physical and chemical factors can greatly increase sources of genetic variation. Cell fusion, introduction of foreign genetic materials, and in vitro culture can be used, and an intact plantlet can be regenerated from a hybrid cell or a cell carrying foreign genetic materials. Such methods enable us to introduce new genes and recombine desirable genes. With recombinant deoxyribonucleic acid (DNA) techniques, we can also add, remove, or modify any specific gene of a tree. Haploid and homozygous diploid plants obtained from anther or unpollinated ovary culture can promote efficient selection of superior genotypes and accelerate the development of genetically stable and uniform populations. Plant regeneration via somatic cell culture will provide new and effective means for fast propagation and release of seedlings. Low-temperature storage has opened additional prospects for germ plasm preservation.

Disadvantages and Advantages of Biotechnology in Perennial Crops

As compared with herbaceous plants, perennial crops present some difficulties for using biotechniques for their improvement.

1. Perennial crops have a long reproductive period, so substantial time and effort are needed to improve a variety.
2. The individual size of perennial crops is big. For some species, only

a few dozen trees can be planted on 1 hectare (ha) of land. If numerous trees are required for a study, a large area of land is needed.
3. Many characteristics of perennial crops are complex and unstable because different parts of the tree are in various physiological states, as they are in different locations and have developed at different times. This physiological variation is always mixed with genetic variation because of different climates of the year.
4. Since many species are cross-pollinated and highly heterozygous, tree explants vary significantly in in vitro culture.
5. Trees grown for years in the field have accumulated various fungi and bacteria, so explants can easily be contaminated during in vitro culture (Bonga, 1982a).

Advantages of biotechnology for perennial crop improvement include the following:

1. Perennial crops have a long reproductive period which provides adequate time for investigations. Some karyotypic variants in annual crops usually do not reproduce, and their life cycle is short, so they cannot be fully investigated. In perennial crops, there is sufficient time not only for studying the variant itself but also for investigating the progeny.
2. A superior F_1 hybrid can be released through vegetative propagation without seed production every year.
3. An improved clonal variety of a perennial crop can be utilized longer in production than an annual crop, in which superior characteristics are more easily lost through interbreeding and short generation time.
4. Rapid propagation of seedlings by tissue culture in perennial crops is much more valuable than that in herbaceous plants, for a greater profit can be obtained from the plant as it grows for several decades in the field.

CELL TOTIPOTENCY AND ITS APPLICATION

Concept

Plant cells are totipotent; that is, they contain all the genetic information required to regenerate a whole plant. The possibility of regenerating a complete plant from a single cell was first proposed by Haberlandt, a German botanist, in 1902. More than 50 years later complete plants were regenerated from cells cultured in vitro. Since then, many investigators have studied plant regeneration under in vitro conditions and have obtained further understanding of cell totipotency. Steward once summarized the function of totipotency in plant development, and he considered that totipotency was a fundamental characteristic of the somatic cell in its cycle (Steward, 1968, 1983). Cell totipotency can be accomplished by nucleocytoplasmic interaction during the cell cycle (Durzan, 1984a).

Totipotent cells can be roughly divided into three types.

FERTILIZED EGGS (ZYGOTES). Gametes segregate during meiosis and form zygotes after pollination. They have great developmental potential and can develop further into seeds. Their material has undergone gene recombination through crossing-over, random segregation of chromosomes during meiosis, and gametogamy and is not yet genetically identified as a sporophyte. So their phenotypes often remain unknown. In vitro culturing of zygotes or embryos from plants with desirable characteristics can give improved plant lines for use in breeding. The zygotes cannot generally be used for rapid clonal propagation unless they are derived from hybridization of two superior pure lines identified through complementation.

DEVELOPING MERISTEMATIC CELLS. Totipotency is most complete in developing meristematic cells, such as those in roots, flowers, and young stems and leaves. These materials can be collected after the phenotype of plants has been identified, so we can objectively select superior materials as explants and regenerate large numbers of plantlets very similar to the donors (Durzan, 1984b). Totipotency in such materials is monitored by observing mitosis. Variations such as somatic chromosomal aberrations that cannot be maintained by sexual reproduction may thus be maintained through meristem culture in vitro.

MALE AND FEMALE GAMETES AND HAPLOID CELLS. Because only one of each pair of alleles remains in a gamete genome, both dominant and recessive genes can be fully expressed; this characteristic is favorable for selection and elimination in breeding. Cell clones derived from haploids are also the most desirable receptor material for use in gene engineering. Only after chromosome doubling can the plants regenerated from male and female gametes or haploid cells undergo reproduction.

Accomplishment of Cell Totipotency

The generation of totipotent cells is shown in Fig. 1. There are three cycles in the figure. Cycle A indicates the life cycle, including alternation of sporophyte and gametophyte generations. In perennial crops, genetic stability is usually maintained by vegetative propagation, and younger trees are generally easier to propagate than older ones. Cycle B shows the

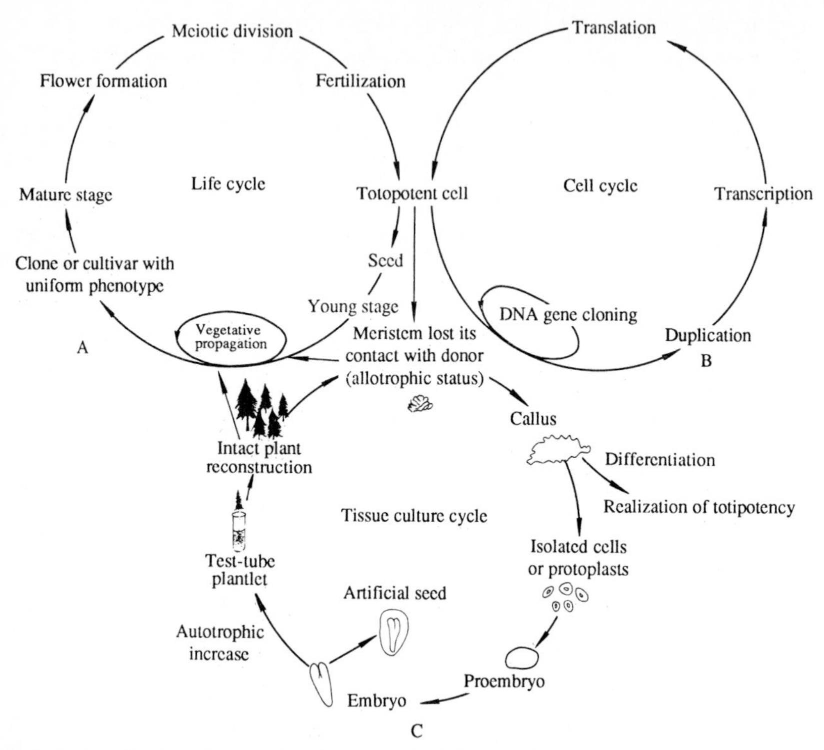

Figure 1. Realization and utilization of cellular totipotency (cf. Durzan, 1980, 1984a, 1984b; Watada et al., 1984; Chaleff and Roy, 1984).

nucleocytoplasmic cycle determined by the cell. Cellular totipotency is accomplished and maintained by DNA replication, ribonucleic acid (RNA) transcription, and protein translation through nucleocytoplasmic interaction. Cycle C is the tissue culture cycle. Tissues or cells in vitro have lost their contact with donors. They depend on artificial nutrition and hormones to undergo metabolism under sterile condition. The cells are then allotrophic. Cellular totipotency in a meristem can be realized via three approaches. First, buds may be directly differentiated from the meristem: very few somaclonal variations arise in this case. Second, callus may be induced from the meristem, and then cellular totipotency is expressed through callus differentiation. Third, embryos can be regenerated from free cells or protoplasts, and intact plants regenerated from the somatic embryos, or they may be made into artificial seeds from which plants can grow. Autotrophic ability is obviously increased at this phase. Cycle B can also be combined with cycle C to produce individual plants with specifically desirable characteristics which may then be put into the life cycle A. With recombinant DNA technology, foreign DNA can also be introduced into cultured cells or protoplasts and expressed in regenerated plants.

With further exploration and utilization of cellular totipotency, more new varieties will be created, and both space and time will be saved for the improvement of existing varieties.

GREAT POTENTIALS IN OBTAINING NEW GENETIC VARIANTS

There are five biotechnological methods for obtaining genetic variants for use in hybridization and breeding. They are (1) gametoclonal variation, (2) somaclonal variation, (3) controlled mutagenesis, (4) somatic hybridization, and (5) recombinant DNA. Genetic variants obtained by these methods have been extensively studied in herbaceous plants and are suitable for use with woody species.

Somaclonal and Gametoclonal Variation

Tissue culture itself is usually a rich resource for producing genetic variation. Variants produced in vitro have been reviewed (Morel, 1971; Murashige, 1974; Green, 1977; Skirvin, 1978). The terms used for variants in some publications refer to the origin of the new regenerated plants. For instance, the plants derived from calli are called calliclones (Skirvin and Janik, 1976), and the plants from protoplasts are named protoclones (Shepard et al., 1980). Larkin and Scowcroft (1981) detailed various sources of somaclonal variation and extensively appraised their significance. They considered that the plants derived from any form of cell culture should be called somaclones and the genetic variation they exhibit should be termed somaclonal variation. However, the sources of explants for culture vary. For instance, gametes or other haploid cells are different from diploid sporophytes. Segregation takes place in the gametes because of synapsis and crossing-over during meiotic division which lead to variation in addition to that occurring during in vitro culture. Cultured gametes thus have two sources of variation in the process of in vitro culture, and it is hard to make a distinction between the two. Some investigators have, therefore, termed this type of variation gametoclonal variation (Evans et al., 1984), to distinguish it from somaclonal variation originating from cultured sporophytic materials.

Somaclonal and gametoclonal variation may be derived from the following sources.

EXPRESSION OF RECESSIVE VARIATION. The genotypes of haploid plants obtained from anther and ovary culture are various. Since a haploid contains only one genome, both dominant and recessive genes may fully show their phenotypes at the cell level and the whole plant level. Especially in perennial crops, recessive genes accumulated for generations under the heterozygous state cannot be fully expressed in diploids, although they can be evident in haploids.

KARYOTYPIC VARIANTS. Gametoclonal variants containing foreign chromosomes can be obtained from anther or ovary culture of interspecific or intergeneric hybrids. For instance, nine new types of plants containing foreign chromosomes were obtained from anther culture of F_1 hybrids of hexaploid *Triticum aestivum* × hexaploid *Secale cereale*. Mixoploids were observed. In the same experiments, self-crossing of the F_1 produced only four types in the F_2 generation, and most of them reverted to the wheat chromosome number. This experiment revealed that foreign addition, substitution, and translocation lines can be directly produced by anther culture. The types of regenerated plants were in accordance with the chromosome constitutions of donor pollen (Hu et al., 1982).

Large numbers of somaclonal variants can be obtained from callus induced from cultured embryos of interspecific and intergeneric hybrids and regenerated to whole plants. For example, triploid hybrids were produced from a cross between diploid *Lolium perenne* and tetraploid *L. multiflorum*, then plants were regenerated from their cultured seed embryos via callus development. Among the 2000 regenerated plants, substantial variations were observed for many important economic traits. Cytological studies indicated that there were many aneuploids and plants with chromosomal inversions, translocations, and deletions. Such variation did not occur in hybrids which were not put into tissue culture (Ahloowalia, 1976, 1978).

In another example, mixoploid plants containing 28 and 56 chromosomes were regenerated from young spike cultures of *Hordeum vulgare* × *Triticum aestivum* (Chu et al., 1984).

CROSSING-OVER DURING MITOSIS. Banding studies have shown that chromosomal synapsis and crossing-over occur during mitosis as well as meiosis. At the diploid level, this somatic synapsis mostly occurs between homologous chromosomes and only rarely between heterochromosomes. In haploids the crossing-over takes place mostly between heterochromosomes. Such crossing-over caused variation in anthocyanin formation in haploid petunia.

CHANGES IN GENE NUMBER. In higher plants, some specific genes can be amplified during differentiation or under specific environmental pressure. With the number of copies of genes, messenger RNA (mRNA) and the relevant protein increase accordingly. It has been reported that DNA in flax (*Linum usitatissimum*) varieties changes under different growth conditions. DNA segments may be amplified in larger plants, although in smaller and normal ones, no duplication occurs (Cullis, 1973, 1975). Increases and decreases in number of copies of ribosomal genes have also been found in a wide variety of crops. In breeding for disease resistance, variant plants resistant to high toxin concentrations could be obtained when the concentration of toxin extracted from a certain fungus was gradually increased in the medium (Gengenbach et al., 1977). Tobacco (*Nicotiana tabacum*) plants resistant to high saline concentrations were obtained from cultures after 11 rounds of selection in medium with increasing salt. This increased saline resistance might be related to an increase in the corresponding DNA.

VARIATION IN CHROMOSOME NUMBERS. Variation in chromosome numbers is a common occurrence in culture. Such variations are particularly evident in plants regenerated via callus differentiation. Perennial crops have a long period of individual development, and variation of chromosome number occurs throughout the developmental process. Multipolar mitosis, karyomixis, and other activities can induce variation of chromosome number and ploidy. Incomplete segregation of chromosomes produces aneuploid cells, which are commonly found in the H_1 generation of pollen plants. Many pollen-derived poplar plants contain aneuploid cells, and in rubber trees, chromosome numbers in somatic cells tend to increase ninefold. These results will be discussed in Chapter 7 and Chapter 32, respectively.

SPONTANEOUS MUTATION. Gene mutation often occurs during culture and results in somaclonal variation. The mutation rate is higher in plants regenerated from calli or protoplasts than those from proliferated buds. The hormone concentration in the medium, genotype, and origin of explants can all affect mutation rate. Evans and Sharp (1983) described somaclonal variation in tomato and demonstrated that some somaclonal variant plants resulted from single gene mutations. The variation observed included fruit color and plant structure, and some variations were favorable for mechanical harvesting. Evans and Sharp suggested that if the best varieties were used for tissue culture, superior variants could be achieved. The variant plants could basically maintain all the superior characteristics of the original variety with only a few, modified undesirable traits or with desirable traits increased so they could soon be used in production.

Induction and Selection of Mutations

The spontaneous mutation rate is usually very low in nature. The rate can be increased by certain physical and chemical factors. An important factor in the great progress of microbial mutation breeding (so that, for instance, the penicillin content of *Penicillium notatum* mutants is 400 times higher than that of the wild type) is that microbes contain only one set of genes, and mutagenesis and selection are conducted at the cellular level. In higher plants the selection efficiency is greatly promoted by using haploid cells in mutagenesis and selection (Bao Wenkui et al., personal communication). Methods for establishing a screening system of mutant cells under in vitro culture are described in Chapter 4.

Protoplast Fusion

Protoplasts contain both nuclear and cytoplasmic genomes. Many experiments with herbaceous plants have shown that protoplasts from different species, genera, and even families can be used. Such fusions have been achieved for *Solanum tuberosum* + *Lycopersicum esculentum* (Melchers et al., 1978), *Arabidopsis thaliana* + *Brassica* sp. (Gleba and Hoffmann, 1980), and *Nicotiana tabacum* + *Atropa belladonna* (Gleba et al., 1982). Qian et al. (1983) have obtained *Glycine max* + *Nicotiana tabacum* fusions and established a hybrid cell line.

Protoplast fusion may result in a nuclear hybrid or a cytoplasmic hybrid. When cells of distant species are fused, chromosome elimination may occur or a whole chromosome set of one parent may disappear. Many variant plants can thus be obtained. Most such plants are morphologically abnormal and produce sterile gametes. However, there are a few hybrid plants that express normal phenotypes and contain genetic information from both parent species (Dudits, 1980). Protoplast fusions within a species or between closely related species form hybrid cells more easily than those between distantly related species.

Protoplast fusion is very useful in perennial crops. Intraspecific recombination in somatic hybrids can shorten the breeding cycle greatly. Since plants are difficult to regenerate from protoplast cultures and fusion products, the work in this field is still in an exploratory stage. Very little work has been done in woody species (for details see Chapter 5).

Recombinant DNA Techniques

Much work has been done on the development of recombinant DNA techniques for crop improvement. However, to establish this technique, some problems must be solved: cloning genes useful for agricultural crops, understanding the structure and function of the genes, modifying and altering test tube genes, transferring genes into most organisms, and expressing the genes in foreign cells. Techniques need to be developed at the molecular, cellular, and whole plant levels.

At the molecular level, the first task is to identify and evaluate a useful gene, then isolate and clone it by using a microbial host. For instance, seed protein genes are some of the most widely studied plant genes. These genes can be isolated from mRNAs produced during seed ripening after reverse transcription and hybridization with complete DNA. Seed storage protein genes have been isolated and cloned in maize, wheat, soybean, and other major crops (Hu et al., 1984). Another important molecular technique is to select and develop vectors for genetic engineering. At present, a promising vector is the Ti plasmid from *Agrobacterium tumefacieus* (Willmitzer et al., 1980). Plant viruses are another group of vectors. Cauliflower mosaic virus is a virus with double-stranded DNA. Its drawbacks for genetic engineering are that it cannot be integrated into the host chromosome and that its carrying capacity is too small. Other vectors should be further investigated, and the existing vectors need to be refined (Sharp et al., 1984).

At the cellular level, recombinant DNA can be employed in an experimental system using stable single cells obtained by protoplast culture or pollen culture. Recombinant DNA may be integrated into cells by infecting protoplasts with modified plasmids containing selective markers. Transformed complete plants can be regenerated by cell selection or by fusing of the transformed protoplasts with normal protoplasts (Sharp et al., 1984).

The final stage of this work is conducted at the whole plant level. This includes transferring the transformed plants into soil, controlling gene expression during their development, evaluating and selecting them, and propagating the transformed plants.

Genetic engineering now is only at an exploratory stage. Genes that control characteristics of higher plants are complicated. It is difficult to master the techniques of gene isolation, purification, and cloning, and of protoplast transformation. Even then, inserted genes are not always expressed. Clearly, genetic engineering technology cannot be applied to plant breeding of woody species in the near future. However, from a long-term point of view, it has broad prospects.

PROPAGATION OF IMPROVED TREE VARIETIES WITH BIOTECHNOLOGY

In conventional tree breeding, long periods are necessary to develop new varieties of woody species and also to replace varieties. Particularly in cross-pollinated species propagated with seeds, it is difficult to maintain the superior characteristics of a new variety. However, by using biotechnology a large number of individual plantlets (including hybrids and polyploids) with improved characteristics may be propagated in a short time.

From a genetic point of view, rapid propagation has three advantages.

PRODUCTION OF POPULATIONS WITH UNIFORMLY SUPERIOR PHENOTYPES. Bonga (1982b) compared genetic gains of sexual propagation and vegetative propagation. He considered that a greater genetic gain could be obtained with vegetative propagation. A normal curve in Fig. 2 shows the height distribution of a forest under natural conditions. Most trees are of medium height with both the highest and lowest ones fewer in number. If the highest trees are selected for both sexual and vegetative propagation simultaneously, the genetic gain from the latter method will be much greater than for the former because some superior characteristics will be lost to recombination. But conventional vegetative propagation is slow, and it is impossible to collect large quantities of explants from the best trees. Explants can be taken from a few trees with the best traits, and large quantities of superior plantlets can be regenerated rapidly.

MAINTENANCE OF CHARACTERISTICS AND COMBINATIONS THAT CANNOT BE MAINTAINED BY SEXUAL PROPAGATION. It is hard to maintain distant hybrids, polyploids, aneuploids, and other karotypic callus variant materials by sexual propagation, but they can be maintained through vegetative propagation. In annual crops, some beneficial epigenetic gene interactions are usually broken by sexual recombination, but they can be kept together with vegetative propagation. Resistance to unfavorable environments and to certain drugs is epigenetically controlled (Carlson, 1982). The inclination of trunks of some coniferous plants regenerated from explants taken from upper lateral branches of donor trees also appears to be an epigenetic phenomenon (Bonga, 1982b).

ARTIFICIAL SEEDS. In order to release improved varieties for production rapidly and effectively, an artificial seed engineering technique has been established for carrot (Zhan, 1984). The "artificial seeds" are tissue culture–derived seed embryos packed in synthetic plastic seed

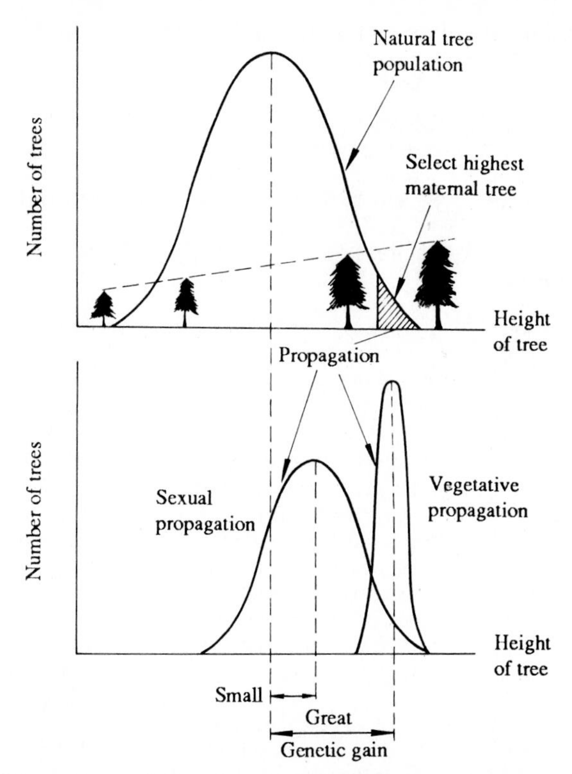

Figure 2. Comparison of genetic gains between sexual and vegetative propagation (Bonga, 1982b). The highest tree from a natural population (top) is selected for further breeding (below). Superior characteristics are maintained in vegetative propagation, giving a population with a much greater average height. In sexual propagation, genetic recombination leads to a normal distribution of the genes controlling tree height. This results in a population with a much smaller average improvement in height.

coats (Durzan, 1980; Sharp et al., 1984). This technique is suitable for seed storage and shipment and is also useful for rapid multiplication of a superior variety or hybrid.

TENTATIVE PROGRAM FOR TREE IMPROVEMENT VIA BIOTECHNOLOGY

COMPARISON OF TWO BREEDING SYSTEMS IN IMPROVEMENT OF TREE SPECIES. With the rapid development of biotechnology, appropriate modification of breeding techniques has begun. Table 1 shows that better results can be achieved with much less space, time, staff, and

TABLE 1. Comparison of Conventional and In Vitro Breeding Systems

Factors	Conventional Breeding	In Vitro Breeding
Growth Cycle	Several decades	25–26 hr
Size	One to several dozen meters	50–100 μm in diameter
Space (10^6 offspring per year)	Several hectares	10 l of suspended cells
Quantities required for mutation rate investigation	10^7 trees	10^7 cells
Quantities required for character investigation	10^3 trees	10^3 cells
Time required for seed production	5–15 years	One month produces 10^6 somatic embryos for use as artificial seeds
Predictability of seed production	Seed setting has on- and off-years and is affected by natural factors	Seed production is completely under artificial control
Genetic uniformity	Much variation occurs during sexual propagation and pollination and selection should be controlled	Less variation in somatic embryos
Multiplication	Cutting or grafting at slow speed	Propagation coefficient of in vitro culture can increase by millions
Ploidy	Difficult to obtain haploid and homozygous lines; diploid materials	Easy to get haploid and pure lines
Breeding methods	Selection is most common; hybridization and mutagenization are available in some species; self-cross and backcross are difficult or impossible	Combination of hybridization and anther culture can accelerate breeding process and increase selection efficiency; mutation and mutant screening are conducted at cellular level with high efficiency; pure lines can be achieved through haploid culture and chromosome doubling; genetic engineering methods can be used

Durzan, D. J. 1980. Progress and promise in forest genetics. *Proc. 50th Anniv. Conf. Paper Sci. Technol.: The Cutting Edge*. Appleton, Wisc., May 8–10, 1978. pp. 31–60. Institute of Paper Chemistry, Appleton, Wisc.

material when biotechnology is used in breeding. Only the evaluation of quality, yield, and uniformity is conducted in the field. Propagation efficiency is also markedly higher than with conventional breeding methods (Durzan, 1980).

Schemes for Tree Improvement

A SCHEME FEASIBLE FOR TREE IMPROVEMENT IN THE NEAR FUTURE. From Scheme 1 (Fig. 3), four types of breeding materials may be obtained for selection:

1. Utilization of heterosis: A pure line can be obtained in a short time through the use of haploids. After evaluation of the pure line, the performance of various crosses between different lines can be evaluated. When the best cross is selected, it can be proliferated by using in vitro rapid propagation to regenerate large numbers of self-rooted plantlets or produce artificial seeds.
2. Direct selection of superior gametoclonal variants: After field identi-

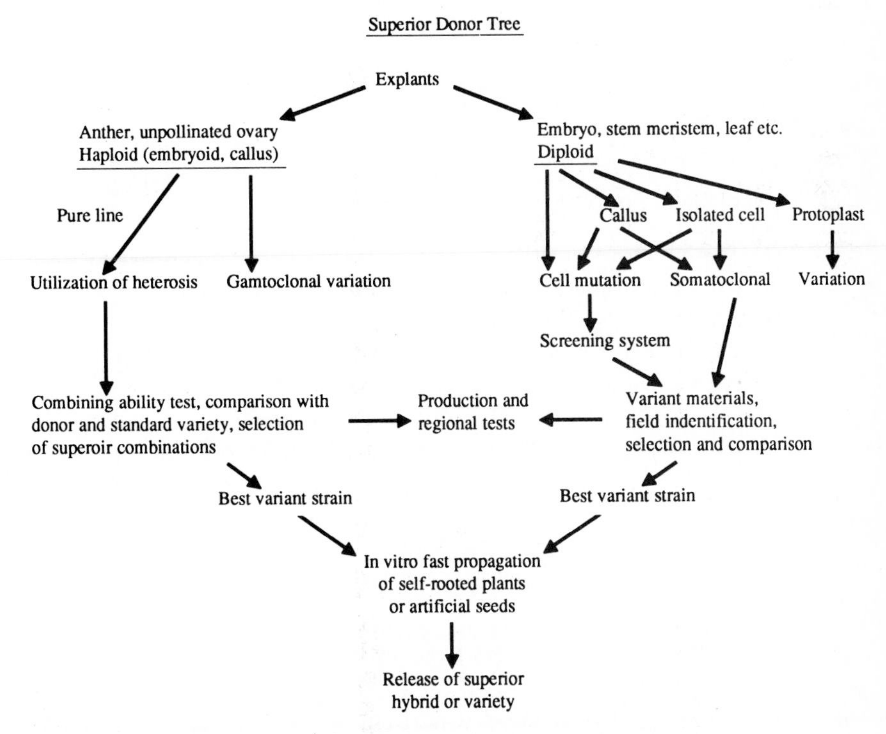

Figure 3. Tree improvement by biotechnology, Scheme 1.

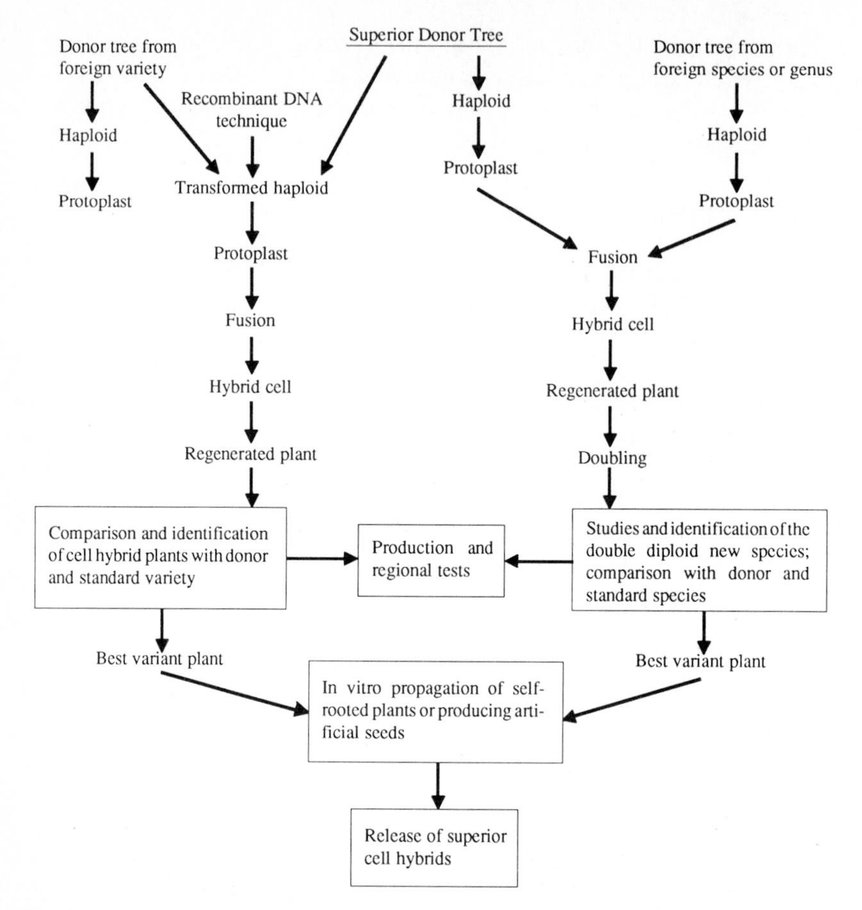

Figure 4. Tree improvement by biotechnology, Scheme 2.

fication, regional trials, and production tests, they are multiplied by in vitro propagation.

3. Screening of mutants: Mutants are physically and chemically screened at the cellular level by using haploid or diploid materials. The best materials are planted in the field for identification and release.
4. Somaclonal variants are selected from haploid or diploid materials. The best variant is selected and released after comparison and identification.

EVENTUAL SCHEME FOR VARIETY IMPROVEMENT. The woody species can be improved efficiently with the techniques of somatic cell hybridization and DNA recombination from haploids. Improved cell lines then can be multiplied rapidly via in vitro propagation. The procedure is shown in Fig. 4. From this procedure, three types of recombinant materials may be obtained for breeding:

1. Intervarietal somatic recombination: When protoplasts from haploid cells of two superior varieties are used for somatic recombination (by protoplast fusion), cell hybrids can be obtained within 1 year. They can be proliferated after selection and identification. The breeding cycle is thus shortened.
2. Interspecific or intergeneric somatic recombination: Distant nuclear hybrids may be obtained by fusion of haploid protoplasts. New species with double diploid cells can be achieved after chromosome doubling of the hybrid cells. After identification, they can be released or used as primitive material for breeding.
3. Introduction of foreign genes by recombinant DNA techniques: A transformed material containing foreign DNA is used for somatic recombination with an improved donor material. Then superior materials containing foreign genes can be obtained and used in breeding.

The techniques in Scheme 2 are rather difficult. There have been very few reports on plant regeneration from protoplasts in perennial crops and even fewer on somatic hybridization. These techniques are still under investigation.

REFERENCES

Ahloowalia, B. S. 1976. Chromosomal changes in parasexually produced rye grass. In: *Current Chromosome Research*. (K. Jones and P. E. Brandham, eds.), pp. 115–122. Elsevier, Amsterdam.

———. 1978. Novel rye grass genotypes regenerated from embryo callus culture. *Fourth Int. Congr. Plant Tissue Cell Cult.*, Calgary, Canada, p. 162 (abstract).

Bonga, J. M. 1982a. Introduction. In: *Tissue Culture in Forestry* (J. M. Bonga and D. J. Durzan, eds.), pp. 1–3. Martinus Nijhoff/Dr. Junk, Dordrecht.

———. 1982b. Vegetative propagation in relation to juvenility, maturity, and rejuvenation. In: *Tissue Culture in Forestry* (J. M. Bonga, and D. J. Durzan, eds.), pp. 387–412. Martinus Nijhoff/Dr. Junk, Dordrecht.

Carlson, P. S. 1982. Plant genetic engineering:III. The experimental system of genetic engineering in higher plants. *Hereditas* **4**(5):41–43 (in Chinese).

Chaleff, R. S. and Rice, T. B. 1984. Herbicide-resistant mutants from tobacco cell cultures. *Science* **223**:1148–1151.

Chu, C. C., Sun, C. S., Chen, X., Zhang, W. X., and Du, Z. H. 1984. Somatic embryogenesis and plant regeneration in callus from inflorescences of *Hordeum vulgare* × *Triticum aestivum* hybrids. *Theor. Appl. Genet.* **68**:375–379.

Cullis, C. A. 1973. DNA differences between flax genotrophs. *Nature* **243**:515–516.

———. 1975. Environmentally induced DNA changes in flax. In: *Modification of the Information Content of Plant Cells*. (R. Markham, D. R. Davies, D. A. Hopwood, and R. Horne, eds.), pp. 27–36. Elsevier, North-Holland.

Dudits, D., Fejer, O., Hadlaczky, G., Koncz, C., Lazar, G. B., and Horvath, G. 1980. Intergeneric gene transfer mediated by plant protoplast fusion. *Mol. Gen. Genet*. **179**:283–288.

Durzan, D. J. 1980. Progress and promise in forest genetics. *Proc. 50th Anniv. Conf. Paper Sci. Technol.: The Cutting Edge*. Appleton, Wisc., May 8–10, 1978. pp. 31–60. Institute of Paper Chemistry, Appleton, Wisc.

———. 1984a. Potential for genetic manipulation of forest trees: Totipotency, somaclonal aberration, and trueness to type. In: *International Symposium of Recent Advances in Forest Biotechnology*, Sponsored by Michigan Biotechnology Institute, 104–125.

———. 1984b. Explant choice. In: *Handbook of Plant Cell Culture*. Vol. 2 (D. A. Evans, W. R. Sharp, P. V. Ammirato, and Y. Yamada, eds.), pp. 471–503. Macmillan, New York.

———. 1985. Tissue culture and improvement of woody perennials: An overview. In: *Tissue Culture in Forestry and Agricultural*. (R. R. Henke, K. W. Hughes, M. J. Constantin, A. Hollander, eds.), pp. 233–256. Plenum Press, New York.

Evans, D. A. and Sharp, W. R. 1983. Single gene mutation in tomato plant regenerated from tissue culture. *Science* **221**:949–951.

———, Sharp, W. R., and Medina-Filho, H. P. 1984. Somaclonal and gametoclonal variation. *Am. J. Bot*. **71**:759–774.

Gengenbach, B. G., Green, C. E., and Donovan, C. M. 1978. Inheritance of selected pathotoxin resistance in maize plants regenerated from cell cultures. *Proc. Natl. Acad. Sci. USA* **74**:5113–5117.

Gleba, Y. Y. and Hoffmann, F. 1980. *Arabidobrassica*: A novel plant obtained by protoplast fusion. *Planta* **149**:112–117.

———, Momot, V. P., Cherep, N. N., and Skarzynskaya, M. V. 1982. Intertribal hybrid lines of *Atropa belladonna* × *N. chinensis* obtained by cloning individual protoplast fusion products. *Theor. Appl. Genet*. **62**:75–79.

Green, M. M. 1977. The case for DNA insertion mutations in *Drosophila*. In: *DNA Insertion Elements, Plasmids, and Episomes*. (A. I. Bukhari, J. A. Shapiro, and S. L. Adhya, eds.), pp. 437–455. Cold Spring Harbor Laboratory, New York.

Hu, H., Xi, Z., Jing, J., and Wang, X. 1982. Production of aneuploid and heteroploid of pollen derived plants. *Proc. 5th Int. Congr. Plant Tissue Cell Cult.*, The Japanese Association for Plant Tissue and Cell Culture, pp. 421–424. Tokyo, July 11–16, 1982.

Hu, Z. 1984. Agrogenetic engineering:IV. Prospects of agrogenetic engineering. Beijing Agric. Sci. **7**:38–41. (in Chinese).

Larkin, P. J. and Scowcroft, W. R. 1981. Somaclonal variation—a source of variability for plant improvement. *Theor. Appl. Genet*. **60**:197–214.

Melchers, G., Sacristan, M. D., and Holder, A. A. 1978. Somatic hybrid plants of potato and tomato regenerated from fused protoplasts. *Carlsburg Res. Commun*. **43**:203.

Morel, G. 1971. The impact of plant tissue culture on plant breeding. In: *The Way Ahead in Plant Breeding*. (F. G. H. Lupton, G. Jenkins, and R. Johnson, eds.), pp. 185–194. 6 Congr. Eucarpia, Univ. Press, Cambridge.

Murashige, T. 1974. Plant propagation through tissue culture. *Annu. Rev. Plant Physiol.* **25**:135–166.

Qian, Y., Fowke, P. C., Rennie, and Wate, L. R. 1983. Ultrastructure and isozyme analysis of cultured soybean-*Nicotiana* fusion products. *Acta Bot. Sin.* **25**(5):397–400. (in Chinese, English abstract).

Rieger, R., Michaelis, A., and Green, M. M. 1976. Epigenetics. In: *Glossary of Genetics and Cytogenetics*. p. 187. Springer Verlag, Berlin.

Sharp, W. R., Evans, D. A., and Ammirato, P. V. 1984. Plant genetic engineering: Designing crops to meet food industry specifications. *Food Technology: Overview* **2**:112–119.

Shepard, J. F., Bidney, D., and Shahin, E. 1980. Potato protoplasts in crop improvement. *Science* **28**:17–24.

Skirvin, R. M. and Janick, J. 1976. 'Velvet Rose' *Pelargonium*, a scented geranium. *Hortic. Sci.* **11**:61–62.

Steward, F. C. 1968. Growth and organization in plants. p. 564. Addison-Wesley, Reading, Mass.

———. 1983. Reflections on aspetic culture. In: *Handbook of Plant Cell Culture* (D. A. Evans, W. R. Sharp, P. V. Ammirato, and Y. Yamada, eds.), pp. 1–10. Macmillan, New York.

Watada, A. E., Herner, R. C., Kader, A. A., Romani, R. J., and Staby, G. L. 1984. Terminology for the description of developmental stages of horticultural crops. *Hortic. Sci.* **19**:20–21.

Willmitzer, L., DeBencckeleer, M., Lemmers, M., Van Montagu, M., and Schell, J. 1980. DNA from Ti-plasmid present in nucleus and absent from plastids of crown gall plant cells. *Nature* **287**:359.

Zhan, X. 1984. Superseeds—embryoids packed in plastics. *Plant Physiol. Commun.* **2**:74 (in Chinese).

CHAPTER 2
General Techniques of Tissue Culture in Perennial Crops

Zhenghua Chen and *David A. Evans*

INTRODUCTION

Tissue culture of perennial crops includes culture of shoot tips; stem segments; embryos; plumular axes; cotyledons; unpollinated ovaries, or ovules; petals; anthers; leaves; isolated cells; and protoplasts. To this date, embryos, shoots, and intact plantlets have been regenerated from certain cells, tissues, and organs of a wide variety of woody species. The further perfection and utilization of these techniques will lead to significant improvement. However, there are many problems yet to be solved. For instance, genetic and physiological differences of explants often reduce cultural repeatability; severe contamination and oxidation of phenolic substances in primary cultures often impede culture development; optimum conditions of different culture steps required for different tree species need to be perfected; and, especially the problems of root initiation from mature explants have not been solved. This chapter will briefly review these problems and will also describe culture procedures and media required for each step in different woody species as well as applications of tissue culture in germ plasm preservation.

EXPLANTS USED FOR TISSUE CULTURE

Juvenility, Maturity, and Rejuvenation of Perennial Crops

CHARACTERISTICS OF JUVENILITY AND MATURITY. All perennial crops go through the two stages of juvenility and maturity during the course of normal development. There are significant differences in tree form, anatomical structure, cell biological characteristics, etc., between the two stages. Therefore, the two stages should be distinguished so that we can learn the nature and characteristics of various explants.

The characteristics of juvenility and maturity in trees are listed in Table 1. Juvenility and maturity are localized in different regions in a seedling plant. In general, the area of juvenility is located at the lower part of a tree: the nearer to the roots, the closer to juvenility. The cells in this area maintain the characteristics of juvenility for a long time. Root suckers developed from an injured tree still have juvenile characteristics. When a juvenile tree reaches a certain age, it becomes mature. The time required for maturation varies with the species, taking, for example, 5–6 years for apple and 7–8 years for pear. The cells in a mature area on a plant will not return to juvenility under natural conditions.

In some perennial crops branches of a mature tree are often used for cutting and bud grafting. The tree used for such a clone can be very old,

TABLE 1 . Characteristics of Juvenility and Maturity*

	Juvenility	Maturity
Nucleus	Small†; surrounded by a thin layer of cytoplasm	Long, rectangular shape; surrounded by endoplasmic reticulum membrane
Chromatin	Noncondensed	Condensed methylated DNA; often polyploid
Ribosome	Free	Attached to membrane
Apical meristem	Large cells	With increased RNA
Leaf	Simple form; large cortical cells; maintained in winter	Complex form; condensed leaf veins; thick
Branch	Obtuse angle without branchlet	Acute angle with branchlet
Main shaft	Short tubule; smooth bark	Thick bark
Cutting	Easy to root	Hard to root
In vitro culture of cells and tissues	Able to induce organogenesis	Hard to induce organogenesis

*The characteristics listed in the table are based on observations of one or several species or tissues.

†The zygote is an exception because its nucleus is rather large.

(BORGER, 1985)

and it possesses typical cell characteristics of a mature tree. Some workers have assumed that continuous division of meristematic cells will ultimately result in senescence; however, senescence is not found in many trees which have undergone generations of asexual propagation (Bonga, 1982).

FACTORS DETERMINING JUVENILITY AND MATURITY. Many authors consider that more and more genes are inactivated in the course of development. One mechanism of gene inactivation is methylation of chromosomal DNA. Other factors may include euchromatinization or heterochromatinization and formation of histone. The maturation process is not determined by the nucleus alone. Organelle DNA also may be a key factor in determining the maturity level of cells. Cytoplasmic organelle DNA can replicate continuously. In the course of perennial crop development, the amount of organelle DNA tends to increase as the cells mature. Organelles differentiate as the tree grows older. Such changes could affect the expression of organelle DNA. The cytoplasm also plays an important role in the maturation process. When the nucleus of a mature cell is transplanted to the cytoplasm of a juvenile cell, the whole cell becomes juvenile (Bonga, 1982).

CELLS WITH LOW MITOTIC ACTIVITY. In plant meristems, cells in different locations mature at different rates. A few cells in some areas in the meristem have very low mitotic activity and mature very slowly. The meristem is formed by slowly dividing cells and surrounded by rapidly dividing cells. Highly active meristematic cells make the plant grow and mature; cells with low meristematic activity maintain genetic stability.

The cells with low mitotic activity are similar to embryonic cells. In apical meristems, such cells can be traced from germination to flowering. There is another important region of mitotic activity: initial cambium cells. These cells divide much more slowly than peripheral cells. Maintenance of the genetic stability of cambium in the course of plant development of an individual tree depends mainly on these initial cambium cells with low mitotic activity (Bonga, 1982).

REJUVENATION. Rejuvenating factors exist in the cytoplasm. As mentioned above, the cytoplasm plays an important role in gene expression, as has been shown by nuclear transplantation experiments. The mechanism of rejuvenation is not yet clear. Some reports consider that sexual rejuvenation processes occur mainly during the prophase of meiosis, because this is a time when cytoplasmic ribosomes and plastids are greatly decreased and mitochondria dedifferentiate (Bonga, 1982). Investigation of sexual rejuvenation is very important in revealing the mechanism of rejuvenation in general. It appears that cells in an undifferentiated state help rejuvenation because the cells in meiosis are in an undifferentiated state. But many important questions are unanswered: how the old information is removed from the cytoplasm, how the cytoplasm gets new information, and when nuclear genetic information is reprogrammed. Answers to these problems will require further research on somatic embryogenesis. Somatic embryogenesis has been used to regenerate trees of a number of species. The following work is being conducted to rejuvenate somatic cells:

1. Inducing vacuoles, cell walls, and membranes to release hydrolases, thereby increasing hydrolase in the cytoplasm. Some cell organelles will thereby be damaged, and the total number of organelles will thus decrease (Beevers, 1976).
2. Using starvation, cold treatment, centrifugation, etc., to partially damage cell organelles and their DNA (Bonga, 1982).
3. Reducing the number of cell organelles with chemical agents. It has been reported that secondary cells containing terpene substances were easier to dedifferentiate than other cells (Lorbber and Muller, 1976), and hemiterpene could stimulate formation of adventitious roots. In addition, antibiotics and ethidium bromide can temporarily block the function of organelle DNA. Ethidium bromide can inhibit aging of fungi, and some antibiotics can stimulate bud morphogenesis (Owens, 1979).

Cell rejuvenation is fundamentally complex. It is not simply the reduction of organelle numbers and their DNA or simplification of organelle structure. Techniques for rejuvenating somatic cells in in vitro culture require further investigation.

MAINTAINING GENETIC STABILITY. Genetic stability is a prerequisite for asexual propagation. Maintenance of superior characteristics is extremely important, particularly when asexual propagation is to be conducted over many generations. The hereditary features of long-lived trees can remain unchanged for a thousand years while growing in complicated environments, which also necessitates considerable genetic stability. Genetic stability is closely associated with the embryonic cells with low mitotic activity existing in meristematic regions discussed previously. These cells can maintain genetic stability because they are located in positions where internal environmental factors are very stable. Even more important, daughter cells containing unaltered DNA can remain in the original position after each division. As the speed of mitosis of these cells is slow, there are very few chances for genetic variation to occur and generate variant cell lineages. The production of sexual and asexual offspring depends on these slowly dividing cells (Bonga, 1982).

Types of Explant

Different types of explants have different responses to in vitro culture. The following types of explants have been used in perennial crop tissue culture:

EXPLANTS WITH BUDS. Explants with buds include the terminal bud and axillary bud of the shoot tip, the nodal bud, and the meristem at the root stem transition zone. Explants with buds are suitable for rapid in vitro propagation of trees. When shoot tips are used for propagation, the apical dominance of the dominant bud should be inhibited to accelerate massive growth of axillary buds, thereby increasing the propagation coefficient. Juvenile buds have greater potential for propagation than mature buds and are generally easier to root.

EMBRYOS. The embryonic explants used for tissue culture include the zygote developed after fertilization under natural conditions and embryos at different phases. These may also include embryos developed after test tube fertilization. These materials contain extremely young meristematic cells. Since these cells have undergone complete rejuvenation through sexual processes, successful tissue culture can easily be achieved. The plantlets produced by embryo culture are all juvenile.

DIFFERENTIATED ORGAN TISSUES. Differentiated organ tissues include diploid tissues of the young stem and leaf, cambium, root, flower, petal, calyx, nucellus, etc. These differentiated tissues can become dedifferentiated after induction of mitosis. Some organs can develop directly into budlike structures or embryos without callus formation. For example, budlike structures can be induced directly from young leaves of coniferous trees (Bonga, 1982). The diploid cells in floral organs are sometimes considered very similar to rejuvenated sex cells, because rejuvenation of these meristematic cells takes place before or slightly after the formation of the floral organs. Therefore, it is easy to obtain adventitious embryos by using young inflorescences at the premeiotic phase and nucellus at the postmeiotic phase as explants, e.g., in *Malus* and *Citrus* (Eichholtz et al., 1979; Button and Kochba, 1977).

POLLEN AND HAPLOID CELLS IN FEMALE GAMETOPHYTES. Pollen and haploid cells contain only half the chromosome number of somatic cells, as they are the result of meiosis. However, in practice whole anthers and unpollinated ovaries or ovules instead of isolated haploid cells are usually inoculated for culture (see Chapter 4).

Selection of Explants

Appropriate selection of explants is very important for rapid in vitro propagation. Explants of different genotypes, ages, and physiological states respond differently to culture, affecting the efficiency and cost of propagation.

SELECTION OF MOTHER TREES. It is necessary to select the mother tree with care. The explants used for breeding should be taken from mother trees possessing superior phenotypes, such as disease resistance, stress tolerance, high yield, and product quality.

PROLIFERATION OF EXPLANTS. Explants should proliferate readily. Even materials with strong ability to proliferate should be screened continuously during culture. At the same time, special media should be experimentally selected to suit different genotypes of the same species.

SIZE OF EXPLANTS. The surface size, volume, and cell number of an explant all can affect the results obtained by using tissue culture. For instance, intact plantlets could be regenerated only from *Manihot esculenta* shoot tips measuring over 2 mm in length, although tips less than 2 mm long produced only either callus or roots. In contrast, when eradication of viral infection is one of the culture objectives, small explants usually should be used. For example, when shoot tip explants of *Dianthus caryophyllus* were less than 2 mm, only roots were induced. If explants measuring 7.5 mm long were used, virus could not be removed. So, the optimum explants are 2–5 mm in length.

LOCATION OF EXPLANTS. Position of explants, i.e., developmental stage, is very important. For example, in shoot tip culture of *Rosa rugosa*, the success rate of terminal bud culture was higher than that using lateral buds.

SEASON AND TIME OF EXPLANT COLLECTION. The season during which the bud is collected is another factor to be considered. Both the dormant period and the sprouting time of in vitro cultured *Citrus* buds correspond to that occurring under field conditions (Altman and Goren, 1974), suggesting that buds about to sprout are suitable explants.

When embryos are used for culture in some species, their dormancy should first be broken. When *Pinus teada* seeds were cultured after stratification, their cotyledons could produce more buds. However, in some species fresh seeds do not have a dormant period (Sommer and Caldas, 1981).

ESTABLISHMENT OF ASEPTIC TISSUE CULTURES

Contamination is a common occurrence during culture. There are many causes of contamination, such as infected explants, insufficient sterilization of media and tools, and carelessness of technicians. In order to reduce losses and increase culture efficiency, contamination should be prevented at each step.

Cultures may be contaminated with bacteria or fungi. Characteristically, bacterial contamination appears as a glutinous spot 1–2 days after inoculation. Besides bacteria, the transport of materials, unclean medium, and the manipulator's carelessness can cause severe bacteria contamination. The tissue culturists should clean their hands frequently with 70% alcohol; tweezers and inoculating needles should be flamed in ethanol. Fungal contamination is characterized by mycelia of different colors growing from the initial point of contamination. Fungal contamination can be observed 3–10 days after inoculation. The factors that cause fungal contamination are mostly unclean work environment, failure of the air filter in the laminar flow hood, and oversized openings on culture vessels.

The greatest difficulty in perennial crop tissue culture is to establish aseptic materials, especially when using shoot tips as explants. As branches are exposed in the field for a long time, they harbor various microorganisms, many of which grow into the plant tissue. Cultures established from such tissues are thus easily contaminated.

Methods of Preventing Contamination of Materials

1. When shoot tips are used as explants, the collected branches can be precultured first in a clean room or under aseptic conditions. After the branches are washed clean with water, they are placed in sugar-free nutrient solution or running water to induce formation of new shoots. The newly developed shoots are then used as explants, thereby greatly decreasing contamination. Branches collected from the field can also be put in dark culture under sterile condition. When the etiolated shoots are cultured after sterilization, the contamination rate is greatly reduced.
2. Do not collect explants on a cloudy or rainy day. On a sunny day, explants collected in the afternoon have less contamination than those collected in the morning because growth of some bacteria and fungi harbored in the plant tissue are reduced by sunlight and resultant dessication.
3. At present, there is no satisfactory method for preventing

endogenous contamination of tissue. When microorganisms grow into the perennial crop tissue, both bud scale and phloem tissue should be excised. Only when the inner meristem is cultured can reliable results be achieved.

Agents Used For Explant Disinfection

Disinfectants that kill microorganisms but do not damage the plant materials should be selected. The commonly used disinfectants are bacteriocides, fungicides, and surfactants.

1. Ethanol has a strong ability to penetrate material and to kill microorganisms. A quick dipping (30 sec) of explants in 70–75% ethanol can be used as a first step for surface sterilization. However, explants cannot be thoroughly disinfected. Many explants of woody species should be sterilized with other disinfectants after they have been soaked in ethanol.
2. Good disinfection can be achieved when 0.1–0.2% mercuric chloride ($HgCl_2$) is used for 6–12 min. If this solution is used, it is important to note that residual mercuric chloride can only be removed after the treated materials are repeatedly washed, usually at least five times with aseptic water.
3. Explants can be soaked in 2–10% sodium hypochlorite (made by diluting commercial bleach) for 15–30 min. Explants must be rinsed three to four times. The disinfectant produces chloride by chemical reaction which kills microorganisms but does not damage plants.
4. Bleach solution containing 0.5–1.0% w/v calcium hypochlorite also is a good disinfectant. Explants are soaked in the disinfectant fluid for 20–30 min. This disinfectant does only minor damage to plant tissues and can readily be diluted.
5. Hydrogen peroxide (H_2O_2) solution, 6–12%, is also a good disinfectant, which does not damage plant tissue and can easily be removed from explant surfaces. It can be used to disinfect leaves.
6. Other published methods include soaking in bromine water (1–2%) for 5–10 min and in silver nitrate solution (1%) for 15–30 min. These, however, are not commonly used.
7. To increase the efficiency of disinfectants, surfactants should be added to the disinfecting solution. The concentration of the commonly used surfactants Tween 80 and Tween 20 ranges from several drops to 0.1%.

When these agents are used to disinfect materials, magnetic stirring, ultrasonic vibration, and other methods may be applied during soaking in order to achieve complete disinfection.

Examples of Explant Disinfection

ALTERNATE SOAKING IN MULTIDISINFECTANT SOLUTION. For some hard-to-disinfect but easy-to-contaminate materials, the following sterilizing procedures are desirable: Collect shoot tips, buds, or organs as explants. They should be well washed with running water and soap to remove surface dirt. Cut off useless parts with scissors and remove bud scales. Surface sterilization of explants can easily be accomplished with 5% sodium hypochlorite (Clorox or other commercial bleach containing 5% sodium hypochlorite). For softer tissues, if the surface of the tissue does not wet readily with the hypochlorite solution, a surfactant such as Tween 80 added to the solution at 0.1% may help. Alternatively, the material can be dipped in 80–95% ethanol and/or a 1:500 dilution of Roccal followed by an alcohol dip and then hypochlorite. The duration of such treatments must be determined empirically. Following surface sterilization, the tissue should be rinsed with sterile water, then with 0.1 N HCl and again with sterile water (Sommer and Caldas, 1981).

STERILIZATION PROCEDURES OF EXPLANTS (DE FOSSARD 1985). (1) Make mother plants responsive (e.g., by exposing them to either short or long days, by pruning to induce new shoot growth, or by using chemical treatments) and leave them in a clean environment in a clean potting mix, avoiding overhead watering. (2) Defoliate to one half petiole and brush in detergent. (3) Leave under running water for 1 h. (4) Use 1% w/v available chlorine in 0.01% concentrated detergent (20 min). (5) Decant chlorine and hold in sterile water. (6) Dissect to explant size. (7) Dip in a concentration of chlorine tolerated by the explant. (8) Place on a simple mineral-sucrose-gelrite medium.

Application of Antibiotics and Fungicide during Culture

There are reports of the inhibitory effect of antibiotics and fungicides to bacteria and fungi during culture. Two derivatives of penicillin G, namely ampicillin and carbenicillin, and three kinds of cephalospoxime, namely cephalothin, cefoxitin, and cefotaxime, were tested; it was found that they were relatively nontoxic to plant cells even up to 100 μg/ml, but they were also broad-spectrum antibiotics (Pollock et al., 1983). Sometimes when one kind of antibiotic is used in combination with another, the effects are synergistic. For instance, rifamycin (20 μg/ml) can be combined with trimethoprim to produce a broad-spectrum bactericidal and fungicidal cocktail. Benlate at 2–4 mg/l added to the medium for tissue culture of the rubber tree can markedly reduce the contamination frequency of fungi but was nontoxic to explants (C. Zhenghua, unpublished).

METHODS TO PREVENT BROWNING OF EXPLANTS

During the initial culture, explant browning is often a serious block to induction of dedifferentiation and bud regeneration, and it always makes the culture hard to maintain.

Factors Affecting Browning

As polyphenoloxidases in tissues are activated, cell metabolism is changed. After phenolic compounds are oxidized, they produce quinonic compounds which are brown. These diffuse gradually into the medium, inhibit activities of other enzymes, and thus damage the whole explant. It has been reported that two types of phenolic compounds are produced in cultures of chestnut, a concentrated type and another in hydrolyzed form (Chevre, 1983). Factors affecting explant browning are complex. As plant species, genotypes, locations of explants, and physiological conditions are different, the degree of browning may vary.

GENOTYPIC INFLUENCE. In rubber tree anther culture, we observed that only a few anthers of the cultivar Haiken 2 turned brown, and calli were readily induced. In some other strains, anthers browned easily, but callus induction was difficult. Such phenomena can often be found in plant tissue culture. One of the causes may be differences in the content of phenolic compounds and activity of polyphenoloxidases. In tree species that brown easily, selection of different genotypes should be considered, and only cultivars in which the browning rate is low should be used for culture.

PHYSIOLOGICAL STATE OF MATERIAL. As physiological states of material are different, their browning rates are also different. In *Castanea sativa*, cultures of juvenile materials contained less quinonic compounds than those from mature explants. The time during which buds are collected and their location are also important factors. Smaller quantities of quinones were formed in the bud cultures of *C. sativa* collected during late January than in cultures from bud explants collected in May and June. Many tannic compounds are synthesized during the growth of the first to fourth buds, and explants from these buds show very severe browning after inoculation (Chevre, 1983). Young explants (e.g., embryo) of oil palm seldom turn brown, although well-differentiated leaves readily turn brown after inoculation. In general, tissues undergoing cell division produce less quinonic compounds when cultured, but differentiated tissues form more quinones (Cui Yuanfang et al., unpublished).

EFFECTS OF MEDIUM COMPOSITION. Too high a concentration of inorganic salts causes phenolic oxidation of Palmae explants. For example, the inorganic salts in MS medium result in browning of oil palm explants, but when a modified MS medium with lower salt concentration was used, the browning rate was decreased, and calli and embryos could be cultured (see Chapter 31). Hormones can also cause browning if not properly used. BA or KIN can stimulate activity of polyphenoloxidases. This phenomenon has been observed in tissue culture of sugarcane (Chen Zhenghua, unpublished). Under optimal dedifferentiation conditions, massive proliferation of cells of some materials with good regenerating ability can inhibit phenolic oxidation. Browning also can be inhibited when buds are proliferating vigorously. Unsuitable incubation conditions, such as too high temperature or too strong light intensity, can increase polyphenoloxidase activity and thus accelerate the browning culture. This phenomenon has been observed in *Coffea arabica* cultures (Sondahl et al. 1981).

SUBCULTURE INTERVAL. Prolonged culture duration can also cause cultures to turn brown, leading to death.

Methods to Overcome Browning of Explants

SELECTION OF SUITABLE EXPLANTS AND OPTIMUM MEDIUM. Selection of suitable explants and establishment of optimum culture conditions are the major means of combating browning. Explants should have a strong regeneration ability, so that they will grow vigorously under dedifferentiating and redifferentiating conditions. The browning rate can thus be greatly reduced. The most important culture factors are appropriate concentrations of inorganic salts, sucrose, and hormones. Proper temperature and dark incubation can also reduce browning remarkably. If initial bud cultures are incubated in darkness or at a light intensity of about 150 lux for 1–6 weeks, phenolic oxidation can be inhibited (Hu and Wang, 1983).

FREQUENT SUBCULTURE. Frequent transfer of easily browning materials to fresh medium can reduce the damage of quinonic substances to cultures. In shoot tip culture of blackberry, the shoot tip explants were transferred to fresh medium 1–2 days after initial culturing (Broome and Zimmerman, 1978). In shoot tip culture of mountain laurel, the explants were transferred to fresh liquid medium 12 and 24 h after initial culturing; the cultures were then transferred every day for 1 week until browning was completely under control (Lloyd and McCown, 1980).

ADDITION OF ANTIOXIDANTS. Adding antioxidants to the culture medium and soaking or preculturing explants in antioxidants can prevent formation of quinonic compounds. Antioxidants include ascorbic acid, polyvinylpyrrolidone (PVP), and bovine serum albumin. Antioxidants are much more effective when added to liquid medium of static culture than when added to solid medium (Hu and Wang, 1983). In shoot tip culture of *Cattleya*, when 5 mM potassium cyanide, ascorbic acid, cysteine, and thiourea was added to liquid medium, polyphenoloxidase activity was effectively inhibited (Ichinhashi and Kako, 1977).

In shoot tip culture of *Fuchsia*, 0.01% PVP inhibited browning. When 0.7% PVP (both soluble and insoluble), 0.28 mM ascorbic acid, and 5% H_2O_2 was added together to 0.58% sucrose solution and agitated on a rotary shaker for 45 min, browning of *Tactona granis* explants could be remarkably reduced (Gupta, et al., 1980).

Active charcoal (0.1–0.5%) also effectively absorbs phenol oxidative compounds. Active charcoal has been used to prevent browning in tissue cultures of many tropical and subtropical trees (see Chapters 26 and 31).

CULTURE PROCEDURES

Procedure of In Vitro Propagation

Propagation through in vitro culture can be divided roughly into three stages: primary culture, shoot multiplication, and root regeneration. (These three stages are designated 1, 2, and 3 in Table 2.)

PRIMARY CULTURE. The primary culture stage comprises the initial development of the explant. The response of cultures at this stage varies with plant species and medium composition. Explants may develop into a single shoot or into multiple shoots. They also can develop into rooted plants or form calli and then differentiate shoots. When regenerating virus-free plants, the culture procedure is similar, but the explants used should be smaller (about 2 mm) than for other cultures (see Chapter 36).

Different plant species and medium additives used in primary culture are listed in Table 2. MS medium is frequently used as the basal medium for primary culture. Among the growth regulators used, the cytokinin BA is most effective, followed by KIN; ZEA and 2iP are used less frequently. In a few cases, no cytokinin or any other hormone is used in the first stage. The frequently used auxins are 2,4-D, NAA, and IAA. When

TABLE 2. Perennial Crops Propagated in Tissue Culture

Species	Media Composition (μM) and Procedure for Stage 1 (primary culture), Stage 2 (shoot multiplication), and Stage 3 (root formation)	Reference
Acacia sp.	1: MS (¾ macro, micro), 2,4-D 0.9, BA 4.4 2: MS (¾ macro, micro), BA 2.2, NAA 1.1 3: MS (½ macro, micro), BA 2.2, NAA 1.1, sucrose 58.4 mM, agar 1%, pH 5.8	Zhai, 1984b
Acacia auriculaeformis	1: ER, BA 13.2, IAA 8.6 2: same as above 3: ½ ER, ZEA 9.1, NAA 2.7	Yan and Chen, 1983
Acer saccharum	1: MS, 2,4-D 0.9–1.8, NAA 0.5–1.1, LH 400–500 mg/l, iron citrate 20 mg/l (for direct plant induction) 3: MS (½ macro), BA 0.4–0.7, NAA 1.1–1.6, LH 400–500 mg/l	Pang, 1981
Alchornea sp.	1: MS, NAA 1.1, BA 8.8 2: MS, BA 4.4 3: MS, NAA 1.1	Yang, 1982
Araucaria cunninghamia	1: B5 (for direct plant induction) 3: B5, IBA 5.9	Haines and de Fossard, 1977
Atriplex canescens	1: MS, KIN 0.05–0.5, IAA 0.06–0.6, GA 1.5 2: same as above 3: MS, IAA 11	Wochok and Sluis, 1980
Betula alleghaniensis	1: MS, ZEA 23, ADE 592 2: same as above 3: MS, IBA 4.9	Minocha, 1980
B. platyphylla	1: GD, BA 6.2, sucrose 292 mM 2: same as above 3: shoots moved into 1:1 peat/perlite mix for rooting	McCown and Amos, 1979
Camellia japonica	1: MS, BA 4.4, IAA 1.1, thiamine 3, calcium panthothenate 2.1, pyridoxine 5, ascorbic acid 5.7, ADE 148, sucrose 87.6 mM 3: IBA 2.5 mM (soak to induce rooting); culture in ½ MS	Samartin, 1984

Carica papaya	1: MS, BA 2, NAA 1.1 2: same as above 3: MS, NAA 5.4	Litz and Conover, 1979
Carica papaya	1: WH, NAA 1, 2ip 10 2: WH, NAA 0.1, BA 0.01	de Bruijne, 1984
Carica papaya	1: MS (hormone free), cut bud with cotyledon 2: MS, NAA 0.5, BA 2.2 3: ½ MS (hormone free)	
Castanea sativa	1: MS, thiamine 3, BA 4.4, IBA 0.5 2: MS, IAA 5.7, ADE 7.4, charcoal 0.5% 3: IBA 5 mM (soak the lower part of plantlets for 20 sec), MS (hormone free for root induction)	Chevre, 1983
Catalpa bungei	1: N6 or B5, BA 6.6–11, NAA 0.0025–0.05 2: N6 or B5, BA 0–4.4, NAA 0.0025–0.005 3: MS, N6 or B5, NAA 5 or soak in NAA 500–750 for 10 secs and transfer to hormone free media	Zhu et al., 1982
Cathaya argyrophylla	1: Monnier (macro), Ms (micro, organics, iron) KIN 2.3, 2,4-D 4.5–9 or NAA 27, glutamine 140, CH 200 mg/l (grown in dark) 2: medium same as above with 2,4-D 4.5 (light grown light for bud induction)	Chen and Guo, 1981
Celastrus Orbiculatus	1: MS, NAA 0.5–5, BA 8.8 2: same as above 3: MS, IBA 2.5	Wang et al., 1981
Cinnamomum albosericeum	1: Q culture medium, BA 8.8, IAA 5.7 3: ½ MS, IBA 1.0	Zhu et al., 1982
Citrus grandis	1: MS, NAA 1.1, BA 8.8 1: WH, GA 5.8–43.5, sucrose 146 mM	Yang, 1982
Citrus microcarpa	1: ½ MS, NAA 0.5, KIN 9.3 2: MS, IAA 2.9, KIN 9.3 3: MS (hormone free)	Huang, 1980
Citrus sinensis	1: Murashige-Tucker (MT), KIN 0.5, IAA 5.7	Kochba and Speigel-Roy, 1973

TABLE 2. Perennial Crops Propagated in Tissue Culture (*Continued*)

Species	Media Composition (μM) and Procedure for Stage 1 (primary culture), Stage 2 (shoot multiplication), and Stage 3 (root formation)	Reference
Citrus X poncirus	1: Knops (macro), MS (micro)	Kitto and Young, 1981
	2: same as above, BA, ADE 740	
	3: MT, NAA 5.4	
Coffea arabica	1: MS BA or ZEA 4.6–9.1, NAA 1.1	Kartha, 1981
	2: same as above	
	3: ½ MS, IBA 1.0	
Coffea arabica	1: MS, KIN 18.4, 2,4-D 4.5	Sondahl, 1981
	2: ½ MS (KNO_3 unchanged), NAA 0.27, KIN 2.3	
Coffea canephora	1: MS, KIN 0.2, 2,4-D 10	Sondahl, 1981
	2: MS, KIN 2.5, NAA 0.5	
Codiaeum variegatum	1: MS, BA 4.4–8.8, NAA 0.5-1.1	Tan, 1984
	2: MS, BA 8.8–13.2, NAA or IAA 0–1.1	
	3: MS, NAA 2.7	
Codiaeum variegatum	1: MS, BA 4.4–8.8, NAA 0.5–1.1	Tan, 1984
	2: MS, BA 8.8–13.2, NAA or IAA 0–1.1	
	3: MS, NAA 2.7	
Citrus grandis	1: MT, 2,4-D 9, BA 22, CH 1 g/l	Wang and Chan, 1978
	2: MT (2x), GA 29–43.5	
Cryptomoris japonica	1: MS	Hu and Wong, 1983
	2: MS, NAA 5.4	
	3: ⅓ MS, NAA 5.4	
Cunninghamia lancelolata	1: MS (¾ macro, micro), BA 2.2, NAA 1.1	Zhai, 1984b
	2: MS (½ macro, micro), IBA 7.4, NAA 2.5, sucrose 58.4 mM	
Cyphomandra betacea	1: MS, BA 4.4, NAA 2.7	Wu, 1984
	2: MS, BA 6.6, NAA 1.1	
	3: MS, NAA 1.1, sucrose 29.2 mM	
Daphne odora	1: M, KIN 0.5, NAA 0.5, CH 500 mg/l	Song et al., 1984
	2: MS, BA 2.2, NAA 2.7, CW 5%	
	3: MS, NAA 1.1	

Davidia involucrata	1: H or N6, 2,4-D 9, KIN 4.6	Bi et al., 1983
	2: N6, IBA 9.8–25, BA 4.4–8.8, H, N6, IBA 12.3, BA 4.4–8.8, IAA 2.9–5.7	
	3: H, IBA 10–25, BA 4.4–8.8	
Elaeis guineensis	1: WH, 2,4-D 4.5, KIN 2.3	Rabechault, 1970
	2: Heller, 2,4-D 4.5, KIN 2.3	
	3: WH or Heller, IAA 5.7	
Eucalyptus sp.	1: HB, BA 1.5, KIN 1.5, IBA 5	Hartney and Baker, 1980
	2: same as above	
	3: HB, IBA 5	
Eucalyptus botryodea	1: MS, IBA 1.25, BA 2.2	Wang et al., 1981
	2: same as above	
	3: WH, IBA 1, ADE 296	
E. citriodora	1: MS (liquid), KIN 1, BA 1.3, culture on roller drum	Gupta, 1981
	2: same as above but add agar	
	3: WH or MS	
Fortunella margarita	1: H, IBA 0.5, BA 5.3 for plantlet indution	Huang, 1980
	3: H, IBA 1–2.5	
Gingkgo biloba	1: modified WH, BA 2.2–4.4, NAA 2.7, NH_4Cl 10 mM, sucrose 117 mM, agar 0.5% Use of the following can also induce complete plantlets: N6, 2,4-D 13.5, BA 8.8, NAA 5.4, sucrose 88 mM, agar 0.5%	Luo, 1985
Gmelina arborea	1: MS, BA 4.4	Wang and Cao, 1983
	2: MS, BA 3.5, IAA 2.9–5.7	
	3: treat lower part of stem with IBA 612, culture on MS, IBA 2.5	
Grevillea rasmarinfolia	1: ½ MS, BA 2.2	Ben-Jaacov and Dax, 1981
	2: same as above	
	3: ½ MS (liquid), NAA 0.5 (on filter paper bridges)	
Hibiscus rosa-sinensis	1: MS, IAA 2.9, BA 8.8, KIN 9.3, sucrose 29.2 mM	Tian, 1984
	3: ½ MS, ZEA 0.9, NAA 2.2, sucrose 58.4 mM	
Lagerstroemia indica	1: MS (½ macro), BA 4.4, KIN 0.2, calcium panthothenate 0.2, biotin 0.4, CH 300, sucrose 44 mM	Huang, 1984b
	3: Plantlets should be at least 0.5–1.5 cm tall for root induction	

TABLE 2. Perennial Crops Propagated in Tissue Culture (*Continued*)

Species	Media Composition (μM) and Procedure for Stage 1 (primary culture), Stage 2 (shoot multiplication), and Stage 3 (root formation)	Reference
Leucaena leucocephala	1: MS, NH_4NO_3 changed to $(NH_4)_2SO_4$ 5.7 mM, BA 4.4, NAA 0.05, CH 300 mg/l, sucrose 146 mM	Huang, 1984a
	3: same culture medium as above, IAA 2.9–5.7, sucrose 58.4 mM	
Loropetalum chinense	1: MS, 2,4-D 4.5, NAA 5, BA 8.8, KIN 4.6	Song and Lian, 1984
	2: MS, NAA 5, IAA 2.9, BA 8.8, KIN 2.3 or MS, NAA 2.5, IAA 2.9, KIN 9.3	
	3: ½ MS, NAA 10	
Malus domestica	1: MS, BA 22	Lundergan, 1980
	2: MS, BA 13.2–22	
M. domestica	1: W, BA 22	Nemeth, 1981
	2: D, BA 0.9, ADE 300	
	3: QL, CDPPN 5 mg/l	
M. pumila	1: W, BA 22	Nemeth, 1981
	2: D, BA 0.9, ADE 300	
	3: QL, CDPPN 5 mg/l	
Malus sp.	1: MS, BA 2.2, NAA 0.05	Tan et al., 1981
	2: MS, BA 2.2, MS, BA 2.2, NAA 0.025	
	3: ⅓ MS, IAA 5.7–16.7	
M. sylvestris	1: ½ MS, BA 2.2	Werner and Boe, 1980
	2: same as above	
	3: ⅓ MS, IBA 9.8, agar 0.27%	
Malus sp.	1: MS, BA 4.4, IBA 4.9, GA 0.3	Jones and O'Ferrell, 1977
	2: same as above	
	3: same as above minus BA	
Malus sp.	1: LS, BA 4.4–8.8, IBA 2.5–4.9	James and Thurbon, 1981
	2: same as above	
	3: LS, IBA 15 + PG_3 1620 mg/l	
Manihot utilissima	1: MS, BA 0.4, NAA 1.1, GA 0.1 (entire explant forms plantlets)	Kartha, 1974
Manihot utilissima	1: MS, BA 0.4, ZEA 0.9, NAA 10, GA 0.1 (entire explant forms plantlets)	Nair, 1979

Maytenus hookari	1: MS, 2,4-D 27	Cheng et al., 1984
	2: MS, BA 8.8, 2,4-D 0.9, MS, BA 8.8, IAA 1.1	
	3: 1/8 MS, NAA 1, IAA 1.1	
Michelia alba	1. MS (2X vitamins), BA 8.8, ZEA 4.6, NAA 2.5, LH 1,000	Liu and Shu, 1983
Michelia macclurei	1. ½ MS, BA 2.2, sucrose 87.6 mM	Liu and Liang, 1985
	2: ¼ MS, BA 0.4–4.4, NAA 2.7 or IAA 5.7, sucrose 87.6 mM	
	3: SH, ZEA 2.8, NAA 7.5–10, IAA 11–14, CW 15–20%, sucrose 87.6 mM	
Momordica grosvenori	1: MS, BA 4.4, IBA 2.5	Gui et al., 1984
	3: MS, NAA 1.3–2.7	
Morus alba	1: MS, IAA 5.7–11, BA 4.4–8.8	Lin and Ji, 1985
	2: MS, IAA 2.9–5.7, BA 2.2–4.4	
	3: MS, BA 2.2, glutamic acid 13.6, lysine 13.7	
Murraya exotica	1: MS, KIN 4.6, NAA 2.7, ZEA 2.3, BA 2.2	Tian, 1985
	2: same as above	
	3: ½ MS, ZEA 0.9, BA 2.6	
Olea europaea	1: WH, NAA 22	Wang et al., 1981
	2: WH, NAA 1.1, BA 8.8, WH, NAA 2.7, BA 2.2	
Paulownia elongata x P. tomentosa	1: MS, BA 20, NAA 1.6, CH 500 mg/l, sucrose 146 mM	Huang, 1985c
	2: ½ MS, BA 8.8–13.2, NAA 0.5, sucrose 87.6 mM	
	3: ½ MS, NAA 4.3, sucrose 43.8 mM	
Paulownia fortunei	1: MS, IAA 5.7, BA 17.6	Yang, 1982
	2: same as above	
	3: ½ MS	
Pinus caribaea	1: MS or SH, BA 2.2–22	Divid, 1983
Prunus amygdalus	1: Knops (macro), MS (organics), BA 4.4	Tabachnick and Kester, 1977
	2: same as above	
	3: same as above	
P. armeniaca	1: MS, BA 8.8, NAA 0.5	Skirvin, 1979
	2: same as above	
	3: same as above	
P. armeniaca	1: Lloyd and McCown, 2ip	Suir, 1984 Skirvin, 1979
	2:½ MS (hormone free)	

TABLE 2. Perennial Crops Propagated in Tissue Culture (*Continued*)

Species	Media Composition (μM) and Procedure for Stage 1 (primary culture), Stage 2 (shoot multiplication), and Stage 3 (root formation)	Reference
Prunus avium	1: MS, BA 2.2–4.4, IAA 1.1, sucrose 87.6 mM	Xu et al. 1985
	3: ½ MS, IAA 5.7–16.7, sucrose 58.4 mM	
P. cerasifera	1: MS, BA 8.8, NAA 0.025, MS, GA 5.8, sucrose 87.6 mM	Cai, 1985
	3: ½ MS, NAA 2.5, IBA 4.9, charcoal 0.1%, sucrose 87.6 mM	
P. cistena	1: MS, BA 22	Loone, 1979
	2: same as above	
	3: ½ liquid MS, NAA 1.1	
P. persica	1: MS, NAA 1.1, BA 8.8	Skirvin, 1979
	2: MS, BA 4.4	
Punica granatum	1: N6, WH, 2,4-D 9, KIN 4.6, sucrose 87.6 mM	Wang et al., 1982
	2: B5, WH, BA 5.7, NAA 1.4–5, LH 500 mg/l, inositol 555	
	3: ½ MS, IBA 14.7, sucrose 58.4 mM	
Pyrus communis	1: MS, BA 4.4, CH	Lane, 1979
	2: MS, BA 4.4	
	3: ½ MS, NAA 11	
Rhododendron spp.	1: A (refer to Chapter 26), 2iP 9.8, IAA 5.7	Anderson, 1975, 1978
	2: same as above	
	3: ¼ A, AC 3 mg/l	
Rhododendron spp.	1: frax;1;2 A, 2ip 9.8, IAA 2.9, sucrose 87.6 mM	Economou and Read, 1984
	2: ½ A, BA 2.2, NAA 0.05, 2ip 4.9–74, IAA 2.9–23	
	3: ½ A, charcoal 0.6 g/1, IBA 2.5	
R. hybridum	1: MS with $(NH_4)_2SO_4$, ratio of NH_4:NO_3 changed to 1:1, KIN 2.3, NAA 0.5–1.25 or add low concentrations of ZEA	Que, 1985
Rosa chinensis	1: MS, NAA 0.5, BA 2.2	Yang, 1982
	2: MS, BA 4.4	
R. hybrida	1: MS, BA 6.6, IAA 1.7	Hasegawa, 1979, 1980
	2: same as above	
	3: ½ or ¼ MS, NAA or IAA 0.5	

R. hybrida	1: MS, BA 8.8, NAA 0.5	Skirvin and Chu, 1979
	2: same as above	
	3: ¼ MS	
Salix matsudana × *S. alba*	1: MS, BA 0.4, NAA 1	Bhojwani, 1980
	2: same as above	
	3: MS, NAA 1	
Santalum sp.	1: MS, BA 8.8, NAA 5	Barlass, 1980
	2: MS, BA 8.8	
	3: MS, IBA 4.9	
Santalum sp.	1: WH, 2,4-D 9.1, KIN 0.9–23	Rao, 1965
Sassafras randaiense	1: ½ LS, KIN 275, NAA 0.25	Wang and Hu, 1983
	2: ½ LS, BA 22	
	3: ½ LS, IBA 25	
Sedum telephium	1: MS, KIN 4.6, NAA 2.7, ZEA 2.3, BA 2.2	Tian, 1985
	2: same as above	
	3: ½ MS, ZEA 0.9, IBA 3.0, sucrose 58.4 mM	
Sequoia sempervirens	1: WH, NAA 0.5, KIN 2.3	Wang et al., 1981
	2: same as above	
	3: WH, BA 2.2	
Tectona grandis	1: MS (liquid), BA 0.4, KIN 0.5	Gupta, 1980
	2: MS, BA 4.4, KIN 2.3	
	3: WH, IAA 13.9, IBA 12.3, IPA	
Theobroma cacao	1: MS, CW 10%	Pence, 1979
	2: MS, NAA 6.4, CW 10%	
Vanilla planifolia	1: MS, thiamine 1.5, pyridoxine 2.4, nicotinic Kononowicz and acid 4, glycine 26.6, inositol 555, LH 1,000 mg/l, sucrose 87.6 mM	Janick, 1981
Vitis spp.	1: MS, 2, 4–D 4.5, BA 0.4, LH 500 mg/l	Kurl and Worley, 1977
	2: MS, NAA 10.7. BA 0.4	

BA is used in combination with auxin, explants usually can dedifferentiate, but in some cases, axillary buds sprout and proliferate.

During this stage, attention should be directed to the prevention of browning. The incubation temperature should be regulated according to the particular species of plant, usually ranging between 20° and 28°C. The photoperiod is a 12–16-hr day.

SHOOT MULTIPLICATION. The objective of the shoot multiplication stage is to produce the maximum number of useful buds and shoots. Shoot proliferation is usually employed because it provides genetic stability. Propagation coefficients can reach 100,000 clonal plants, or even 1 million plants per year from an initial explant.

In culture media for shoot multiplication, cytokinin is still indispensable (Table 2), and BA is most effective. The auxin used in most cases is NAA or IAA, but if the concentration is too high it will inhibit shoot proliferation and callus will be formed instead. A low concentration of GA can also accelerate shoot elongation.

The proliferation rate of shoots is the most important feature of in vitro propagation. In vitro culture techniques have practical value only when a fast proliferation rate is achieved in the Stage 2. The major factor influencing proliferation rate is the interaction of the physiological state of the plant material with the culture medium and its additives. Propagation rate may vary greatly among explants (Hu and Wang, 1983).

The physiological state of a given explant can be modified through various treatments (Hu and Wang, 1983). For example, in subculture of *Eucalyptus citriodora* explants, it was found that bud explants gradually died when cultures were incubated at 23–25°C. However, when cultures were preincubated under 15°C for 3 days prior to transfer to 25°C, the explants retained a good propagation rate (Gupta et al., 1981). The propagation efficiency is also significantly increased if subculture is performed once within a month (Hu and Wang, 1983). Incubating the plant material in liquid on a shaker appears to improve the physiological state of certain plant species and increase shoot proliferation. In the bud culture of apple rootstocks, a faster growth rate was observed when the in vitro produced shoots were cut and placed in liquid medium on an orbital shaker for 4 days (Snir and Erez, 1980). In bud proliferation of *Eucalyptus*, in addition to preincubation at 15°C, the bud explants were incubated in a shaking liquid medium for 2 weeks. This method increases bud proliferation rate. One single bud can produce 100,000 plants per year (Gupta et al., 1981).

ROOT REGENERATION. The purpose of the root regeneration stage is to regenerate adventitious roots from shoots obtained in the first and second stages and to form intact plants. In general, shoots measuring 10 mm long can be placed in rooting medium. Shoot growth is always inhibited by high cytokinin concentration in the previous stage. When cytokinin is removed in rooting medium, shoots undergo further development. Root formation can be induced readily in many herbaceous species, but it can be very recalcitrant in woody species, especially from mature trees. It is worth noting that callus may be formed at the basal part of the plantlet and block the direct contact between root and stem vascular tissue. Such plantlets cannot survive after transplantation. This problem may be overcome by root induction outside the test tube. Shoots may be placed directly in the greenhouse nursery or may be pretreated by predipping the cut shoot in 245–980 μM IBA for several seconds or longer. McCown and Amos (1979) reported that when the shoots of *Betula platyphylla* were placed in 1:1 peat/perlite in a warm (30–35°C) chamber with high humidity (over 80%), 100% rooting was obtained (Hu and Wang, 1983). A good rooting effect also can be achieved when shoots are planted in sand under intermittent mist spray.

Procedures of In Vitro Fertilization

In vitro fertilization techniques were developed by scientists at Dehli University, India, in the 1960s, using *Papaver somniferum* as material. Since then, many workers have studied the technique and have achieved success in Gramineae, Solanaceae, and Cruciferae. However, there have been only a few reports on woody species. The purpose of in vitro fertilization is to bring pollination and embryogenesis under artificial control so as to overcome blocks in distant hybridization and self-pollination. In vitro fertilization also can be used when seeds are hard to obtain under harsh environmental conditions. This technique is very useful in perennial crop breeding. For success, the pollen and ovules used should be at suitable developmental stages, and the optimal culture medium should be selected to ensure normal growth of the pollen tube and fertilization of the ovule. In vitro fertilization includes fertilization of pistil, ovule, and placenta.

PISTIL FERTILIZATION

1. Fertilization is performed under completely aseptic conditions. Collect flower buds 1 day prior to opening. Sterilize the bud, cut open the anthers on sterile filter paper, and collect pollen.
2. Remove the female flower bud 1 day prior to blossom and remove all flower organs except the pistil. For those species in which anthers

dehisce before flowering, emasculation and bagged collection should be completed beforehand in order to prevent prepollination of the pistil.

3. Pollinate the pistil aseptically and transfer the pistil to agar medium. Alternatively, preculture the pistil before flowering, then pollinate, and transfer to agar medium.
4. The pistil will be fertilized within 1–3 days. The ovary should then be transferred to a medium suitable for embryo development.

OVULE FERTILIZATION. On the basis of a method developed by Kameya in 1970, the practical procedures are as follows (Wang et al., 1981).

1. The pollen and unpollinated ovaries are collected under aspetic conditions as described above.
2. Preculture the pollen. For example, the pollen of *Brassica* plants can be precultured in agar medium ($CaCl_2$ 6.8 mM, H_3BO_3 0.2 μM, sucrose 58.4 mM, and agar 1%).
3. Dip the ovules in $CaCl_2$ (6.8 mM) solution for several seconds and then place in the medium in which the pollen has been inoculated.
4. Incubate the cultures at 20° ± 1°C for 24–48 hr, and then observe whether the pollen tubes have entered into the ovule.
5. Transfer the ovules containing pollen tubes into liquid medium. Nitsch medium is used as basal medium with the addition of 13.3 μM glycine, 8.1 μM nicotinic acid, 3 μM vitamin B_1, 5.9 μM vitamin B_6 and 58.4 mM sucrose.
6. After somatic embryos are formed, they are transferred into fresh Nitsch medium with 13.3 μM glycine, 8.1 μM nicotinic acid, 3 μM vitamin B_1, 5.9 μM vitamin B_6, and 0.8% agar.

FERTILIZATION OF OVULE CARRYING PLACENTAS

1. Cut down the whole placenta or part of it under aseptic conditions.
2. Inoculate the placenta into Nitsch medium with LH 500 mg/l.
3. Spread aseptic pollen on the placenta.
4. Transfer the placenta into embryo inducing medium.

Techniques to increase the fertilization rate are of great importance for in vitro fertilization. A suitable age of the explants (especially the ovules), a sufficient quantity of pollen to sprout, and the conditions favorable for the growth of pollen tubes and microspore development are major factors in increasing fertilization rate.

Procedure of Embryo Culture

Embryo culture of plants was initiated in the beginning of this century. Early investigators on perennial crops included Tukey (1933) and LaRue (1936). They found that in vitro embryo culture could be used to promote the germination of the seeds of some early maturing varieties.

SIGNIFICANCE. Embryos may be cultured in vitro at different developmental stages. There are several advantages of embryo culture. First, it can overcome nonviability, senescence or nongermination of seeds of some early maturing tree varieties. For instance, the embryos of early maturing peach varieties often do not germinate, making their hybrids difficult to obtain, but embryo culture in vitro can produce good hybrid seedlings. Second, it can be used to propagate rootstocks. Third, mature seeds can be obtained with in vitro embryo culture when they are difficult to achieve under harsh conditions. Fourth, embryo culture techniques can break through the dormancy of those cultivars (e.g., *Rosa rugosa* and apple) in which dormant periods are long and seedlings grow slowly (Yeung et al., 1981).

INCUBATION CONDITIONS. Incubation conditions vary with plant materials. In general, embryos at early developmental stages are difficult to culture, and success can be achieved more easily from cultures of mature embryos. If embryos have cotyledonary primordia and embryo buds have started to differentiate, they can be readily cultured. Different incubation conditions required for young and mature embryos are as follows:

1. Young embryo cultures require a higher salt level than those of mature embryos. For example, the whole salt concentration of MS medium is needed for young embryo cultures of rubber tree; for mature embryos, half-strength is adequate.
2. Young embryo culture demands a higher osmotic pressure. Therefore, sucrose should be present in higher concentration than for mature embryo culture.
3. As young embryo culture must accelerate organs to undergo further development, differentiation, and finally maturity, the culture medium should act as an endosperm. Organic components are thus indispensible. Mature embryos do not demand the organic compounds so strictly.
4. Early stage young embryos can be successfully cultured when ovules are cultured together with placentas and the culture medium is modified.

POSSIBLE RESULTS OF EMBRYO CULTURE

1. One embryo develops into a plant. This may or may not include some callus growth. This result is the same as that of the embryo developed in vivo.
2. The embryo first forms callus through dedifferentiation, then redifferentiates into a large number of somatic embryos or buds, and finally develops into many plants. In herbaceous plants, this technique has been used to obtain additional hybrid plants from distant or intervarietal hybridization. Somaclonal variation may also be obtained as a result of this technique.
3. One embryo produces multiple buds which can be used for rootstock propagation or for breeding to obtain more plants from hybrids of the same combination.

The sterilization methods and incubation conditions used for embryo culture are largely the same as those used for other explants.

CULTURE MEDIA

Two different measuring units are used for medium components in the literature. One is molarity in which the quantity of macroelements is given in millimolar units (mM) while those of microelements, organic additives, and hormones are in micromolar units (μM). Alternatively, concentrations may be expressed as milligrams per liter (mg/l).

MS and Related Media

MS MEDIUM. MS medium is most frequently used in tissue culture of perennial crops. It contains high concentrations of inorganic salts (such as potassium, ammonium, and nitrate) and microelements. MS medium has been most successfully used for the in vitro culture of woody species since 1983 by Chinese workers.

Minor modifications of MS medium have also had good effects in perennial crop tissue culture. The following media, which are very similar to MS medium, are often found in tissue culture literature.

LS MEDIUM. (Linsmaier and Skoog, 1965) The macroelements, microelements, and iron-salt of LS medium are similar to those of MS; of organic substances, only 1.2 μM thiamine chloride, 555 μM inositol, and 87.6 mM sucrose of MS medium are retained: glycine, nicotinic acid, and pyridoxal chloride are not included. Chalupa (1974) used LS medium plus 10 μM NAA, 4.4 μM BA, and 574 μM arginine to culture poplar explants.

When calli were formed, they were transferred into NAA-free LS medium with 0.17–3.0 μM BA. Shoots were regenerated in 10–14 days. LS medium with 4.6 μM KIN and 5.4 μM NAA was also used to culture callus and suspended cells of *Picea abies*. The dry weight of the cells increased 10–14 times in 2 weeks. Small somatic embryos were formed after the cells were transferred to fresh medium (Chalupa and Durzan, 1983).

BL MEDIUM. (Brown and Lawrence, 1968) Its components are similar to those MS. It has been used to culture *Pseudotsuga douglasii* explants.

BM MEDIUM. The composition of BM medium is largely the same as that of MS, with only thiamine chloride removed. The sucrose concentration is 87.6 mM. Using BM medium, plants have been regenerated from suspended cells of *Citrus* (Button and Botha, 1975).

ER MEDIUM. (Eriksson, 1965) It is frequently used in tissue culture of legume crops. In perennial crops, it has been used in tissue culture of *Acacia auriculaeformis* (Table 2).

Media with High Content of Potassium Nitrate

B5 MEDIUM. (Gamborg et al., 1968) B5 medium contains a low concentration of the ammonium form of nitrogen and a high concentration of thiamine chloride. It has been used in tissue culture of *Araucaria Cunninghamii* and *Vitis*.

N6 MEDIUM. (Zhu et al., 1975) N6 medium is extensively used in anther culture of cereal crops in China. It has also been used in tissue culture of some woody species, e.g., anther culture of *Citrus* (see Chapter 24) and tissue culture of *Catalpa bungei*. Modified N6 medium has been used for tissue culture of some coniferous trees (Liisa et al., 1982).

SH MEDIUM. (Schenk and Hildebrandt, 1972) SH medium contains a high concentration of mineral salts, among which ammonium and phosphate are provided by $NH_4H_2PO_4$.

Media with Moderate Content of Inorganic Salts

H MEDIUM. (Bourgin and Nitsch, 1967) H medium contains half the strength of macroelements of MS medium, with only potassium dihydrogen phosphate and calcium chloride slightly lower. There are fewer microelements in H medium than in MS, but concentrations are higher. However, H medium contains more types of vitamins.

NITSCH (1969) MEDIUM. The composition of Nitsch medium is basically the same as that of H medium, except that the concentration of biotin is 10 times higher.

MILLER (1963) MEDIUM AND BLAYDES (1966) MEDIUM. Miller and Blaydes media are identical.

Media with Low Content of Inorganic Salts

In most cases, the following media are used for rooting.

WHITE (WH) MEDIUM. The inorganic salt composition of White (1943) medium is inconsistent in many published papers. The formulation listed in this paper is a standard salt solution of White medium recently reevaluated (Evans et al., 1983). In perennial crop tissue culture, WH medium has been used for *Elaeis guineensis* embryo culture, *Citrus grandis* endosperm culture, and *Eucalyptus* tree root initiation.

WS MEDIUM. (Wolter and Skoog, 1966) WS medium is suitable for root initiation. It has been used for root induction from calli of *Populus davidiana, Cryptomeria fortunei*, and *Prunus pseudocerasus*. Complete plants have been regenerated from *P. davidiana* calli in WS medium.

HE MEDIUM. (Heller, 1953) The potassium and nitrate of HE medium are provided separately through different compounds.

MODIFIED NITSCH (1951) MEDIUM. Modified Nitsch medium is often used for anther culture of *Nicotiana tabacum*, but seldom for perennial crops. After modification, it has been used to culture *Hevea brasiliensis* embryos.

HB MEDIUM*. (Holley and Baker, 1963) Macroelement levels of HB medium are a little more than one-half those in Knop's solution.† One-half milliliter of Berthelot's solution‡ is used in 1-l medium as microelements. It has been used to eliminate virus from some flower plants (e.g., *Dianthus caryophyllus*). In perennial crops, it has been used to culture *Eucalyptus* shoot tips for rapid shoot propagation.

---

*Components of HB medium are 125 mg/l KNO_3, 500 mg/l $Ca(NO_3)_2 \cdot 4H_2O$, 200 mg/l $MgSO_4 \cdot 7H_2O$, 125 mg/l KH_2PO_4, 25 mg/l $FeSO_4$, 1 mg/l thiamine HCl, 8 mg/l ADE, 40 g glucose, and 0.5 ml Berthelot's solution.

†Knop's solution contains 800 mg/l $Ca(NO_3)_2 \cdot 4H_2O$, 200 mg/l $H_2PO_4 \cdot 7H_2O$, 200 mg/l KNO_3, 200 mg/l KH_2PO_4, and a trace of $FeSO_4 \cdot 7H_2O$.

‡Berthelot's solution contains 2000 mg/l $MnSO_4 \cdot 7H_2O$, 100 mg/l $ZnSO_4 \cdot 7H_2O$, 50 mg/l H_3BO_3, 500 mg/l KI, 50 mg/l $CoCl_2 \cdot 6H_2O$, 50 mg/l $NiCl_2 \cdot 6H_2O$, 50 mg/l $CuSO_4 \cdot 5H_2O$, 100 mg/l $BeSO_4 \cdot 4H_2O$, 1 mg/l Conc $\cdot$ H_2SO_4 and 1000 ml H_2O.

GERM PLASM PRESERVATION BY TISSUE CULTURE

The potential advantages of preserving perennial crop germ plasm via tissue culture are as follows: (1) a large number of asexual offspring can be obtained within a short period from stored cultures with good multiplication ability, e.g., one meristem of grape can produce several million plants in 1 year, and of apple can give hundreds of thousands of plants; (2) as the tissue cultures are virus- and pathogen-free, they can be used for international germ plasm exchange, thus preventing the obstructions imposed by quarantine systems on the movement of living plants across national boundaries; and (3) relatively little space is needed to preserve large quantities of clonal germ plasm. For example, a space of only 2 m^2 is needed to store shoot tip cultures of 800 grape cultivars, with 6 replicates of each cultivar (Morel, 1975). The limitations of this method are the following: (1) a long time is necessary for a cultured plantlet to develop into a fruit-bearing plant; (2) storage of large quantities of plant materials requires expensive facilities; and (3) there is a risk of the materials' being lost as a result of accidental hazards (such as power failure or fire). However, at present this technique can at least be used as a supplementary method for preserving germ plasm of woody species, especially for species which cannot produce fertile seeds, as well as some rare and precious species which are in danger of extinction.

Two approaches are usually used for plant germ plasm preservation through tissue culture.

Growth Inhibition

The basis of the growth inhibition approach is to control medium composition and incubation temperature. Good effects can be achieved by adding growth inhibitors (e.g., abscisic acid) or some components with cell osmotic function (such as glycitol, sorbitol, and cyocel) to the medium. For instance, potato culture can be effectively preserved in medium with 4.7 μM abscisic acid + 175 mM sucrose at 10°C, subcultured at yearly intervals, or in medium with 300 mM glucose + 14.6 mM sucrose transferred every 6 months. The workers in the International Institute of Tropical Agriculture (IITA) successfully maintained *Manihot utilissima* cultures on medium containing high cytokinin and giberellin, and low sugar at 20–22°C for over 2 years. Grape shoot tips retained regenerating ability when they were stored at 9°C and subcultured once a year (Morel, 1975). Apple shoot tips did not lose their growing ability when they were maintained at 1–4°C for 1 year. An ordinary 0.28 m^3 refrigerator could store 2000 culture tubes. It would require several hectares of land to accommodate the same number of trees (Lundergan and Janick, 1979). In general, cultures of crops growing in temperate zones are preserved at 0–6°C, and tropical woody species are maintained at 15–20°C.

Reducing air pressure and oxygen content is another method of germ plasm preservation which inhibits growth of cultures (Bridgen and Staby, 1984).

Freeze Preservation

In the freeze preservation method the cells or tissues are frozen and maintained at the temperature of liquid nitrogen (LN), −196°C. At this temperature the cells stay in a completely inactive state. Therefore, one can not only avoid the occurrence of genetic variation but also maintain the morphogenic potential of plant material. Up to now, little work has been done on germ plasm storage of perennial crops by this method. Freeze preservation involves the following steps.

PREPARATION AND PRECULTURE OF PLANT MATERIAL. When using embryos as explant material, it is better to select early-phase embryos, because their cells are undergoing rapid mitosis. If using buds, winter buds are better as their cells contain less water. Callus or suspension cells should be precultured. The objectives of preculture are (1) to increase the proportion of dividing cells, because the contents in dividing cells or protoplasts are condensed and they can tolerate low temperatures; (2) to reduce the content of freezable water in cells through preculture. Tissue with a low water content regenerates more readily than that with high water level. The vigorous cell division period for callus usually occurs 9–12 days after incubation; that of suspension cultures is 5–7 days.

CRYOPROTECTANTS. Cryoprotectants are useful for (1) preventing cell damage due to dehydration during the period from freezing to thawing; (2) increasing the survival rate and regenerating ability of the cells; (3) reducing the water content of the culture solution so as to increase its viscosity; and (4) slowing ice crystallization during freezing. The commonly used cryoprotectors include dimethyl sulfoxide (DMSO), lactoprotein hydrolysate, proline, polyethyleneglycol, glycerol, and sucrose.

A combination of several cryoprotectants is better than a single protective agent. Jian Lingcheng has used a mixed solution containing 2.5% DMSO, 10% polyethyleneglycol (with a molecular weight of 6000), 14.6 mM sucrose, and 20.5 mM $CaCl_2$ to preserve calli of rice and sugarcane. The survival rate can reach 90% or even 100%. DMSO can prevent the formation of ice crystals inside the cells. Polyethyleneglycol slows the growth of ice crystals outside the cells. Sucrose or glucose can protect cell membranes. $CaCl_2$ can stabilize the cell membrane system. Combinations of these agents can thus promote high survival rates of the callus after freezing storage.

FREEZING. Plant material is usually frozen by one of two methods: rapid cooling or slow cooling. Rapid cooling does not damage organelles and membranes of frozen cells: the extracellular solution ices up during rapid cooling, freezable water inside the cells diffuses to the crystal surfaces of the outer solution, and the water content of the cells decreases so that ice crystals do not form inside them. This process is also called protective dehydration of cells. In rapid cooling of *Manihot utilissima* meristem, shoot tips were precultured in culture medium for 4–6 days and then transferred into 4-ml sterile ampules containing 0.5 ml of 5% DMSO liquid medium. The ampule bottles were sealed in an ethanol flame and quickly placed into the LN tank. When the frozen shoot tips were thawed after 1 hr and then cultured, 30–40% of the tips could grow. When 10% glycerol and 5% sucrose was used in the culture medium for rapid cooling of shoot tips in LN, 13% of the frozen tips developed into plantlets when cultured, and 8% of the tips formed callus and roots. However, this method is applicable only to some plants. For others, a slow cooling method is usually required.

In slow cooling methods, the plant material should first be conditioned under low temperature. The material is cooled gradually or stepwise to an optimum intermediate temperature, then rapidly cooled by plunging into LN. This initial slow cooling reduces the amount of intracellular freezable water by dehydrating the cells. It thus prevents the lethal cell damage caused by intracellular ice formation. This method has been used for cryopreservation of strawberry meristems. The meristems were first cultured in medium containing 5% DMSO for 2 days, then cooled in 5% DMSO solution gradually to −40°C at the rate of 0.8°C/min, and finally plunged into LN. After 8 weeks, the plant regeneration rate was 95% with extensive shoot proliferation. A slow cooling method for freeze preservation of apple buds has been demonstrated. Annual shoots were collected from mature trees in late December to early January, with shoots cut into 15-cm segments and put into plastic bags. The packed shoot segments were stored at −3°C for 14 days, −5°C for 3 days, −10°C for 1 day, then −40°C, and finally placed into LN. After 23 months of storage, the frozen material was slowly thawed to 0°C and placed in a culture container with water at 20°C. After this treatment, 77% of the shoot segments developed buds (Kartha, 1981).

THAWING. The frozen material should be rapidly thawed by plunging it into water at 37–40°C under aseptic conditions. Slow thawing damages organelles and cell membranes and causes death due to frequent recrystallization of intracellular water during this thawing process. However, winter buds of perennial crops must be thawed slowly at 0°C, letting water return into the cells. It may be that in winter buds intracellular water has evaporated during the slow cooling period in winter. The slow thawing can prevent osmotic shock and protect cell membranes from damage (Jian, 1985).

EVALUATION OF SURVIVAL RATE, DEGREE OF INJURY, AND REGENERATING ABILITY. The thawed plant material is recultured by routine methods. The viability and injury of freeze preserved tissue cultures are commonly tested by using the fluorescein diacetate (FDA) staining technique developed by Widholm (1978). The stained material is observed by using a fluorescent microscope. The viable cells can be stained and show fluorescence. Another method is to use 2,3,5-triphenyltetrazolum chloride (TTC) staining technique. The mitochondria of viable cells reduce TTC to a red compound, formazam, which does not dissolve in water but is soluble in ethanol. TTC testing can thus be done spectrophotometrically. The degree of cell injury can be estimated indirectly by calculation of a standard curve. However, from a practical point of view, the most important criterion of viability for freeze preserved meristems is the ability to regenerate plants. The final frequency of plant regeneration should be calculated and compared with that of controls.

HISTOLOGICAL AND CYTOLOGICAL OBSERVATIONS

In vitro experiments frequently require the collection of histological and cytological data, such as physiological state, microspore development, callus formation, and chromosome number. The commonly used techniques for fast preparation of slides are discussed below.

Histological Observation

Aceto carmine or PICCH (propiono-iron-carmine-chloral hydrate) staining techniques can be used to observe the developmental stages and pathways of microspores, the physiological state of cells, early developmental stages of the embryo, and the process of small callus formation. PICCH staining techniques were originally used for chromosome observation. However, the method is also very suitable for observation of microsporogenesis and somatic embryogenesis of perennial crops. This staining technique has been employed in the study of microsporogenesis and somatic embryogenesis of rubber tree (Chen, 1983). The PICCH staining technique is described below.

PLANT MATERIAL. Plant material comprises inducing cells, small embryos, anthers, small calli, etc.

FIXATION. The material is fixed in a mixture of 95% alcohol, chloroform, and propionic acid (6:3:1). A $Fe(OH)_3$ saturated solution of propionic acid can be used as a mordant. The material is fixed for 12–14 hr at room temperature. After fixation, the material is washed with 95% alcohol several times and stained with PICCH solution. If it has been treated by mordant during fixing, iron is not needed in the PICCH solution. If it has not, several drops of $Fe(OH)_3$ saturated propionic acid solution should be added to the PICCH solution. If the stained material is not to be observed immediately, it should be soaked with 95%, then 85–75% alcohol and preserved in 70% alcohol.

PREPARATION OF PICCH STAINING SOLUTION

1. Propionic carmine solution is first prepared. Carmine, 0.5 g, is added to 45% propionic acid solution, 100 ml, and then boiled for 3–4 hr with a reflux condenser. The boiled solution is cooled and then filtered.
2. Chloral hydrate, 2 g, is added into 0.5% propionic carmine solution, 5 ml, fully dissolved by stirring; and then mixed with several drops of $Fe(OH)_3$ saturated propionic acid solution. The mixture can then be used for staining.

STAINING AND SLIDE PREPARATION. One anther or a piece of callus is placed on a glass slide, squashed with a needle or tweezers, and treated with 1–2 drops of PICCH solution. If cultured cells are used, the slide should be smeared with another glass to create a thin layer of cells, and then 1–2 drops of PICCH solution is added. It usually takes 20–30 min to achieve good staining of the material. Finally, a cover glass is put on for microscopic observation.

PREPARATION OF PERMANENT SLIDES. The stained slide is soaked in a mixture of 1:1 glacial acetic acid/*n*-butanol until the cover glass can be removed, transferred to *n*-butanol for several seconds, then taken out and sealed with acacia.

Chromosomal Observation

PLANT MATERIAL. Plant material includes calli, root tips, or young leaves.

TIME OF SAMPLE COLLECTION. Different woody species have different mitotic peaks. Differences in the peak time also appear in different varieties of the same species or different locations and growing conditions of the same variety. The time of peak mitosis should be determined by testing samples at different times. Samples can be collected for observation every 30 min or 1 hr. In rubber trees, the optimum time for collecting young leaves from field-growing trees is 9:00–10:00 A.M.; the root tips and young leaves of plantlets under in vitro culture are better collected at 11:30–12:00 A.M. Samples should be taken when the plant material is growing vigorously. In vitro cultures should be incubated under conditions which encourage vigorous mitotic division.

SAMPLING TECHNIQUES

Callus Collection. Calli sometimes grow to very large sizes, but only a few regions contain cells that are actively dividing. The optimal sampling location should be carefully chosen in order to obtain maximum number of mitotic figures. Suitable locations can be selected under a dissecting microscope. Good growing locations are generally smooth, bright, and slightly protuberant.

Leaf Collection. In broadleaf trees, young leaves with rich mitotic figures are usually less than 1 cm in length. Small leaves of this size can be taken for observation from shoot tips at a suitable time of day.

Root Tip Sampling. New roots about 1 cm long are collected at an appropriate time of day. The root tips, 0.5 cm in length, are cut carefully as material for observation.

PRETREATMENT OF THE MATERIAL

P-dichlorobenzene (PDB) Treatment. Add 0.7 mg PDB crystals ($C_6H_4Cl_2$) to 50 ml distilled water to prepare a PDB solution. The prepared solution is placed in a thermostat at 60°C for 4 hr; then it is taken out, cooled, and put into a brown bottle. It should be warmed up to 20°C before use. Plant material with large chromosomes is dipped in the solution for 4–5 hr; material with small chromosomes is soaked for 1–2 hr. For pines and cypresses, the duration of treatment is 12–24 hr. The PDB treatment can shorten chromosomes, but the chromosome will not double even if the treatment is prolonged.

Treatment With 8-hydroxyquinoline. Soak the plant material in 3 mM 8-hydroxyquinoline for 2–4 hr.

Cold Treatment. Place the plant material into a refrigerator for 12–24 hr. For lacquer tree, 24 hr of cold storage gives good results.

Staining and Slide Preparation. If iron haematoxylin is used for staining, the cells should first be treated with iron, as a mordant, for 30 min, and then soaked in haematoxylin solution for 2–4 hr. They are then decolorized with 45% glacial acetic acid, and then slides are made for observation. For some plant cells, Feulgen staining method can be used.

In the tissue culture of rubber trees, after Feulgen staining, the cells were stained with PICCH (Chen Zhizheng and Chen Zhenghua, unpublished). The technique is as follows:

1. The cultured cells are treated with saturated PDB solution for 4 hr: 1 hr at room temperature and 3 hr in the refrigerator (4°C).
2. The cells are fixed with mixed solution of chloroform–glacial acetic acid–ethanol (1:3:6) for 24 hr.
3. After fixation, the materials are soaked through 95%, 75%, and 50% alcohol, and then rinsed clean with water.
4. The materials are immersed in 1 M HCl solution under room temperature for 5–10 min, then transferred into 1 M fresh HCl solution for 20 min, and finally washed with water and centrifuged.
5. The materials are stained with Schiff's reagent and placed in the dark overnight. They are then fully washed with bleach solution, rinsed with water, and stained with PICCH. Smears are made for observation.

Preparation of Schiff reagent solution: Basic fuchsin, 1 g, is added to 200 ml boiling distilled water for full dissolution. When the temperature is reduced to 50°C, 1–2 g of $Na_2S_2O_5$ or $NaHSO_3$ is added to the solution. Flasks are sealed and placed in darkness for 12–24 hr.

Preparation of washing solution: In 100 ml distilled water, place 93 mM $NaHSO_3$ or 5 ml $Na_2S_2O_5$ and 5 ml 1 M HCl.

REFERENCES

Altman, A. and Goren, R. 1974. Growth and dormancy cycles in *Citrus* bud cultures and their hormonal control. *Physiol. Plant* **30**:240–245.

Anderson, W. C. 1975. Propagation of *Rhododendron* by tissue culture: 1. Development of culture medium for multiplication of shoots. *Proc. Int. Plant Propag. Soc.* **25**:129–135.

———. 1978. Rooting of tissue cultured rhododendrons. *Proc. Int. Plant Propag. Soc.* **28**:135–139.

Barlass, M., Grant, W. J. R., and Skene, K. G. M. 1980. Shoot regeneration in vitro from native Australian fruitbearing trees—Quandong and bush. *Aust. J. Bot.* **28**:405–409.

Beevers, L. 1976. Senescence. In: *Plant Biochemistry*. (J. Bonner and J. E. Varner, eds.), pp. 771–794. Academic Press, Orlando, Fla.

Ben-Jaacov, J. and Dax, E. 1981. In vitro propagation of *Grevillea rosmarinifolia*. *Hortic. Sci*. **16**:309–310.

Bhojwani, S. S. 1980. Micropropagation method for a hybrid willow (*Salix matsudana* × *alba* NZ-1002). *N. Z. J. Bot*. **18**:209–214.

Bi, S., He, L., Kong, F., Xu, Z., and Su, C. 1983. Culture of dovetree (*Davidia involucrata*). *Plant Physiol. Commun*. **4**:43–44 (in Chinese).

Bonga, J. M. 1982. Vegetative propagation in relation to juvenility maturity and rejuvenation. In: *Tissue Culture in Forestry*. (J. M. Bonga and D. J. Durzan, eds.), p. 387–412.

Broome, O. C. and Zimmerman, R. H. 1978. In vitro propagation of blackberry. *Hortic. Sci*. **13**:151–153.

De Bruijne, E., De Langhe, E., and Van Rijk, R. 1974. Action of hormones and embryoid formation in callus culture of *Carica papaya*. *Meded Fac. Landbouwwet. Rijksuniv. Genet*. **39**:637–645.

Button, J. and Botha, C. E. J. 1975. Embryogenic maceration of *Citrus* callus and the regeneration of plant from single cell. *J. Exp. Bot*. **26**:723–729.

——— and Kochba, J. 1977. Tissue culture in the *Citrus* industry. In: *Applied and Fundamental Aspects of Plant Cell, Tissue, and Organ Culture* (J. Reinert and Y. P. S. Bajaj, eds.), pp. 70–92. Springer-Verlag, Berlin.

Cai, J. 1985. Tissue culture of *Prunus cerasifera*. *Plant Physiol. Commun*. **1**:33 (in Chinese).

Chalupa, A. and Durzan, D. J. 1973. Growth of Norway spruce (*Picea abies* (L) Karst) tissue and cell cultures. *Commun. Inst. For. Cech*. **8**:111–125.

Chen, W. and Guo, L. 1981. Embryo culture, callus induction and bud formation on callus of *Cathaya argyrophylla*. *Acta Bot. Sin*. **23**(3):249–250 (in Chinese).

Cheng, Z., Wang, J., and Meng, G. 1984. Culture of stem segments of *Maytenus hookeri*. *Plant Physiol. Commun*. **2**:38 (in Chinese).

Chevre, A. M. 1983. In vitro vegetative multiplication of chestnut. *J. Hortic. Sci*. **58**:23–29.

De Fossard, R. A. 1985. Principles of plant tissue culture. In: *Tissue Culture as a Plant Production System for Horticultural Crops*. (R. H. Zimmerman, R. J. Griesbach, F. A. Hammerschlag, and R. H. Lawson, eds.), pp. 1–14. Martinus Nijhoff, Dordrecht.

Economou, A. S. and Read, P. E. 1984. In vitro shoot proliferation of Minnesota deciduous azaleas. *Hortic. Sci*. **19**:60–61.

Eichholtz, D. A., Robitaille, H. A., and Hasegawa, P. M. 1979. Adventive embryony in apple. *Hortic. Sci*. **14**:699–700.

Evans, D. A., Sharp, W. R., Ammirato, P. V., and Yamada, Y. 1983. Formulation of culture media used most often for plant cell culture. In: *Handbook of Plant Cell Culture*, Vol. 1, pp. 971. Macmillan, New York.

Gui, Y. L., Gu, S. R., and Xu, T. Y. 1984. Organogenesis in tissue culture of *Monordica grosvenori* leaf explants. In: *Cell and Tissue Culture of Plants* (Institute of Botany, Academia Sinica, ed.), pp. 68–73 (in Chinese).

Gupta, P. K., Nadgir, A. L., Mascarenhas, A. F., and Iagannathan, V. 1980. Tissue culture of forest trees: Clonal multiplication of *Tectona grandis* L. (teak) by tissue culture. *Plant Sci. Lett*. **17**:259–268.

———, Mascarenhas, A. F., and Jagannathan, V. 1981. Tissue culture of forest trees—clonal propagation of mature trees of *Eucalyptus citriodora* hook by tissue culture. *Plant Sci. Lett*. **20**:195–201.

Haines, R. J. and De Fossard, R. A. 1977. Propagation of hoop pine (*Araucaria Cunninghamii* Ait.) by organ culture. *Acta Hortic*. **78**:297–302.

Hasegawa, P. M. 1979. In vitro propagation of rose. *Hortic. Sci*. **14**:610–612.

———. 1980. Factors affecting shoot and root initiation from culture rose shoot tips. *J. Am. Soc. Hortic. Sci*. **105**:216–220.

Hu, C. Y. and Wang, P. J. 1983. Meristem, shoot tip, and bud cultures. In: *Handbook of Plant Cell Culture*. Vol. 1 (D. A. Evans, W. R. Sharp, P. V. Ammirato, and Y. Yamada, eds.), pp. 177–227. Macmillan, New York.

Huang, C. 1980. A preliminary experiment of test tube plantlets obtained from 4 flower tree species by tissue culture. *Plant Physiol. Commun*. **6**:37–39 (in Chinese).

——— and Li, J. 1983. Tissue culture of lateral bud of *Fortunella margarita. Plant Physiol. Commun*. **3**:47 (in Chinese).

———. Huang, C. 1984a. Tissue culture of axillary bud from crape myrtle. *Plant Physiol. Commun*. **3**:44 (in Chinese).

———. 1984b. Tissue culture of axillary bud from *Leucaena leucocephala. Plant Physiol. Commun*. **3**:45 (in Chinese).

———. 1985. In vitro culture and plantlet regeneration from immature embryo of *Paulowinia elongata* × *P. tomentosa. Plant Physiol. Commun*. **1**:39 (in Chinese).

Ichihashi, S. and Kako, S. 1977. Studies on clonal propagation of *Cattleya* through tissue culture method: II. Browning of *Cattleya. J. Jap. Soc. Hortic. Sci*. **46**:325–330.

James, D. J. and Thurbon, I. J. 1979. Rapid in vitro rooting of the apple rootstock, *M. 9. J. Hortic. Sci*. **54**:309–311.

——— and Thurbon, I. J. 1981. Shoot and root initiation in vitro in the apple rootstock M.9. and the promotive effects of glucinol. *J. Hortic. Sci*. **56**:15–20.

Jian, L. 1985. Freeze preservation of germplasms. In: *Methods for Plant Tissue Culture*. pp. 221–232. Botany Society of China, Beijing (in Chinese).

Jones. O. P. and Hopgood, M. E. 1979. The successful propagation in vitro of two rootstocks of *Prunus*: The plum rootstock pixy (*P. insititia*) and the cherry rootstocks F12/1 (*P. avium*). *J. Hortic. Sci*. **54**:63–66.

——— and O'Ferrell, D. 1977. Propagation in vitro of M26 apple rootstocks. *J. Hortic. Sci*. **52**:235–238.

Kartha, K. K., Gamborg, O. L., Constabel, F., and Shyluk, J. P. 1974. Regeneration of cassava plants from shoot apical meristems. *Plant Sci. Lett*. **2**:107–113.

——— and Gamborg, O. L. 1975. Elimination of cassava mosaic disease by meristem culture. *Phytopathol*. **65**:826–828.

———. 1981. Meristem culture and cryopreservation—methods and application. In: *Plant Tissue Culture, Methods, and Applications in Agriculture* (T. A. Thorpe, ed.), pp. 181–211. Academic Press, New York.

———, Mroginski, L. A., Pahl, K., and Leung, N. L. 1981. Germplasm preservation of coffee (*Coffea arabica* L) by in vitro culture of shoot apical meristems. *Plant Sci. Lett*. **22**:301–307.

Kitto, S. L. and Young, M. J. 1981. In vitro propagation of *Carrizo citrange. Hortic. Sci*. **16**:305–306.

Kochba, J. and Spiegel-Roy, P. 1973. Effect of culture media on embryoid formation from ovular callus of "Shamouti" orange (*Citrus sinensis*). *Z. Pflanzenzuecht*. **69**:156–162.

Kononowicz, H. and Janick, J. 1984. In vitro propagation of *Vanilla planitolia*. *Hortic. Sci*. **19**:58–59.

Krul, W. R. and Worley, J. F. 1977. Formation of adventitious embryos in callus cultures of 'Seyval,' a French hybrid grape. *J. Am. Soc. Hortic. Sci*. **102**:360–363.

Lane, W. D. 1979a. In vitro propagation of *Spirea bumalda* and *Prunus cistena* from shoot apices. *Can. J. Plant Sci*. **59**:1025–1029.

———. 1979b. Regeneration of pear plants from shoot meristem tips. *Plant Sci. Lett*. **16**:337–342.

Li, M. 1980. Application of *p*-dichlorobenzene in pretreatment for plant chromosomes. *Hereditas* **2**(6):30–32 (in Chinese).

Lin, S. and Ji, D. 1985. Plantlet regeneration from anther culture of *Morus alba*. *Plant Physiol. Commun*. **1**:33–34 (in Chinese).

Litz, R. E. and Conover, R. A. 1978. In vitro propagation of papaya. *Hortic. Sci*. **13**:241–242.

Liu, M. and Shu, J. 1983. Tissue culture of *Sedum ergthro-stictum, Michelia alaba* and *Cedrus*. *Plant Physiol. Commun*. **6**:38–39 (in Chinese).

Liu, M. and Liang, S. 1985. Tissue culture of stem segments of *Michelia macclurei*. *Plant Physiol. Commun*. **1**:37 (in Chinese).

Lloyd, G. and McCown, B. 1980. Commercially feasible micropropagation of mountain laurel, *Kalmia latifolia*, by use of shoot tip culture. *Proc. Int. Plant Propag. Soc*. **30**:421–427.

Lobber, P. and Muller, W. H. 1976. Volatile growth inhibitor produced by *Salvia leucophylla*: Effects of seedling root tip ultrastructure. *Am. J. Bot*. **63**:196–200.

Lundergan, C. 1980. Regulation of apple shoot proliferation and growth in vitro. *Hortic. Res*. **20**:19–24.

——— and Janick, J. 1979. Low temperature storage of in vitro apple shoots. *Hortic. Sci*. **14**:514.

Luo, Z. 1985. Tissue culture of stem segments of *Ginkgo biloba*. *Plant Physiol. Commun*. **1**:35–36 (in Chinese).

Ma, F. and Liu, Y. 1985. Tissue culture of winter buds of mature *Morus alba*. *Plant Physiol. Commun*. **1**:34 (in Chinese).

McCown, B. and Amos, R. 1979. Initial trials with commercial micropropagation of birch selections. *Proc. Int. Plant Propag. Soc*. **29**:387–393.

Minocha, S. C. 1980. Cell and tissue culture in the propagation of forest trees. In: *Plant Cell Cultures: Results and Perspectives*. (F. Sala, B. Parisi, R. Cella, and O. Ciferri, eds.), pp. 295–300. Elsevier/North-Holland Biomedical, Amsterdam.

Morel, G. M. 1975. Meristem culture techniques for the long-term storage of cultured plants. In: *Crop Genetics Resources for Today and Tomorrow* (O. H. Frankel and J. G. Hawkes, eds.), pp. 327–332. Cambridge University Press, Cambridge.

Nair, N. G., Kartha, K. K., and Gamborg, O. L. 1979. Effect of growth regulators on plant regeneration from shoot apical meristems of cassava and on the culture of internodes in vitro. *Z. Pflanzenphysiol*. **95**:51–56.

Nemeth, G. 1981. Adventitious root induction by substituted 2-chloro-3-phenyl-propionitriles in apple rootstocks cultured in vitro. *Sci. Hort.* **14**:253–259.

Owens, L. D. 1979. Kanamycin promotes morphogenesis of plant tissue. *Plant Sci. Lett.* **16**:225–230.

Pence, V. C., Hasegawa, P. M., and Janick, J. 1979. Asexual embryogenesis in *Theobroma cacao* L. *J. Am. Soc. Hortic. Sci.* **104**:145–148.

Pang, D. 1981. Preliminary report of tissue culture of *Acer saccharum. Plant Physiol. Commun.* **1**:56 (in Chinese).

Pollock, K., Barfield, D. G., and Shields, R. 1983. The toxicity of antibiotics to plant cell culture. *Plant Cell Rep.* **2**:36–39.

Que, G. 1981. Shoot tip culture of *Sassafras tsumu. Plant Physiol. Commun.* **4**:44 (in Chinese).

———. 1985. In vitro propagation of young shoots of *Rhododendron hybridum. Plant Physiol. Commun.* **4**:39 (in Chinese).

Rabechault, H. 1970. Colonies cellulaires et formes embryoids obtenues in vitro a partir de culturbs de'embryons de palmier a huile (*Elaeis guineensis* Jacq. var. dura Becc.) *C.R. Acad. Sci. Paris* **270**:3067–3070.

Rao, P.S. 1965. In vitro induction of embryonal proliferation in *Santalum album* L. *Phytomorphology* **15**:175–179.

Samartin, A. 1984. In vitro propagation of *Camellia japonica* seedlings. *Hortic. Sci.* **18**(2):225–226.

Shu, Y. 1984. Tissue culture of *Serissa serissoides. Plant Physiol. Commun.* **3**:45–46 (in Chinese).

Shi, Z. and Li, Y. 1982. Plantlet formation from tissue culture of *Sapium sebiferum. Plant Physiol. Commun.* **2**:38–39 (in Chinese).

Simola, L. K. 1982. Ultrastructure of callus cultures from *Betula pendula* and *Picea abies. Proc. 5th Int. Congr. Plant Tissue Cell Cult.*, pp. 173–174.

Skirvin, R. M. and Chu, M. C. 1979. In vitro propagation of 'Forever Yours' rose. *Hortic. Sci.* **14**:608–610.

Skirvin, R. M. and Rukan, H. 1979. The culture of peach, sweet and sour cherry, and apricot shoot tips. *Ill. State Hortic. Soc.* **113**:30–38.

Snir, I. and Erez, A. 1980. In vitro propagation of Malling Merton apple rootstocks. *Hortic. Sci.* **15**:597–598.

———. 1984. In vitro propagation of Canino Apricot. *Hortic. Sci.* **19**(2):229–230.

Sommer, H. E. and L. S. Caldas. 1981. In vitro methods applied to forest trees. In: *Plant Tissue Culture Methods and Applications in Agriculture*. (T. A. Thorpe, ed.), pp. 349–358. Academic Press, New York.

Sondahl, M. R., Monaco, L. C., and W. R. Sharp. 1981. In vitro methods applied to coffee. In: *Plant Tissue Culture Methods and Applications in Agriculture*. (T. A. Thorpe, ed.), pp. 325–347. Academic Press, New York.

Song, C. and Lian, W. 1984. Shoot tip culture of *Daphne odora* in vitro. *Plant Physiol. Commun.* **5**:39 (in Chinese).

Song, P., Peng, C., and Zhang, T. 1982. Callus formation and organogenesis in tissue culture of *Loropetalum chinense. Plant Physiol. Commun.* **4**:33 (in Chinese).

Tabachnik, L. and Kester, D. E. 1977. Shoot culture for almond-peach hybrid clones in vitro. *Hortic. Sci.* **12**:545–547.

Tan, L., Xu, G., and Deng, S. 1981. Tissue culture of apple dwarf rootstocks. *Plant Physiol. Commun.* **5**:35–36 (in Chinese).

Tan, W. 1984. Culture of axillary bud of *Codiaeum variegatum*. *Plant Physiol. Commun*. **2**:36 (in Chinese).

Thorpe, T. A. and Biondi, S. 1984. Conifers. In: *Handbook of Plant Cell Culture*. Vol. 2 (W. R. Sharp, D. A. Evans, P. V. Ammirato, and Y. Yamada, eds.), pp. 435–470. Macmillan, New York.

Tian, H. 1984. Tissue culture of *Hibiscus rosa-sinensis* stems. *Plant Physiol. Commun*. **2**:35 (in Chinese).

———. 1985. Tissue culture of *Sedum telephium* and *Murray exotica*. *Plant Physiol. Commun*. **1**:38–39 (in Chinese).

Wang, B. and Cao, Y. 1983. Tissue culture of shoot apex from Malay bushbeech (*Gmelina arborea*). *Plant Physiol. Commun*. **2**:38 (in Chinese).

Wang, H. 1982. Plantlet regeneration from pomegranate leaf. *Plant Physiol. Commun*. **6**:36 (in Chinese).

Wang, J., Jin, B., Gao, X., and Jia, C. 1981. Embryo culture. In: *Tissue Culture of Vegetables*. (L. Shuxuan and C. Shouchun, eds.), pp. 76–103. Shanghai Science and Technology Press, Shanghai (in Chinese).

Wang, K., Zhang, P., Ni, D., and Bao, C. 1981. Callus formation and organ regeneration in tissue culture of several woody plants. *Acta Bot. Sin*. **23**(2):97–103 (in Chinese).

Wang, T. and Chan, C. 1978. Triploid *Citrus* plantlets from endosperm culture. *Proc. Symp. Plant Tissue Cult.*, Science Press, Beijing, pp. 463–468.

Welb, D. T. and Santiago, O. D. 1983. Embryo culture in *P. caribaea*. *Plant Sci. Lett*. **32**:17–21.

Werner, E. M. and Boe, A. A. 1980. In vitro propagation of Malling 7 apple rootstocks. *Hortic. Sci*. **15**:509–510.

Widholm, J. M. 1971. The use of fluorescein diacetate and phenolsafranin for determining viability of cultured plant cells. *Stain Technol*. **47**:189–194.

Wochok, Z. S. and Sluis, C. J. 1980. Gibberellic acid promotes *Atriplex* shoot multiplication and elongation. *Plant Sci. Lett*. **17**:363–369.

Wu, Y. 1984. Induction of callus and regeneration of plantlets in embryogenic bud of *Cyphomandra betacea*. *Plant Physiol. Commun*. **4**:33–34 (in Chinese).

Xu, G., Zhou, J., and Deng, S. 1985. In vitro culture of *Prunus avium*. *Plant Physiol. Commun*. **4**:36 (in Chinese).

Yan, M. and Chen, P. 1983. Tissue culture and plantlet regeneration of *Acacia acuriculaeformis*. *Plant Physiol. Commun*. **1**:29 (in Chinese).

Yang, N. 1982. Tissue culture and organogenesis of some woody plants in vitro. *Plant Physiol. Commun*. **4**:23–27 (in Chinese).

Yeung, E. C., Thorpe, T. A., and Jensen, C. J., 1981. In vitro fertilization of embryo culture. In: *Plant Tissue Culture Methods and Applications in Agriculture*. (T. A. Thorpe, ed.), pp. 253–271. Academic Press, New York.

Zhai, Y., Zhou, Z., and Li, T. 1982. Tissue culture of selected trees of *Cunninghamia lanceolata*. *Plant Physiol. Commun*. **4**:31–32 (in Chinese).

———. 1984. Tissue culture and plantlet regeneration of *Acacia*. *Plant Physiol. Commun*. **4**:32 (in Chinese).

Zhu, J., Jiang, J., Tang, K., Song, Y., and Wang, Y. 1982. Tissue culture of *Cinnamomum albosericeum*. *Plant Physiol. Commun*. **2**:40–41 (in Chinese).

Zhu, L., Jin, Y., and Luan, Y. 1982. Culture of axillary bud of *Catalpa bungei. Plant Physiol. Commun*. **2**:38 (in Chinese).

Zhu, Z., Wang, J., Sun, J., Xu, Z., Zhu, Z., Yin, G. C., and Bi, F. Y. 1975. Establishment of a better medium for anther culture through comparison of nitrogen resources. *Sci. Sin*. **2**:484–490 (in Chinese).

CHAPTER 3
Haploid Induction in Perennial Crops

Zhenghua Chen

ACHIEVEMENTS

Over the last decade, pollen plantlets have been obtained from more than 20 species, 11 genera, and 9 families of perennial crops (Table 1). The first successful anther culture of a perennial crop was carried out with the medicinal plant *Lycium halimifolum* Mill (Zenkteler, 1972). In China, *Populus nigra* L. was the first tree species used for the production of pollen plantlets (Wang et al., 1975). Pollen plantlets have subsequently been obtained from more than 10 other species and interspecific hybrids. The callus induction frequency of poplar pollen plantlets is currently around 20% up to a maximum of 50%. The callus differentiation frequency is around 20%, though the most successful is above 60% (Zhu et al., 1980). Transplanted trees now are 7–8 years old. Somatic chromosome variation in pollen plantlets has been systematically studied (Lu et al., 1979; Wu et al., 1981). Rapidly growing plants have been screened to determine whether plants are derived from pollen cells (Chen et al., 1981). Haploid plantlets were obtained from unpollinated poplar ovaries and survived after transplanting in 1984. This result opens another pathway for obtaining haploid plantlets from perennial crops (see Chapter 8).

In 1977 pollen plantlets were obtained from *Hevea brasiliensis* (Chen et al., 1978). To date, pollen plantlets have been obtained from 10 rubber

TABLE 1. Successful Induction of Haploids from Perennial Crops

Plant	References
Pollen-derived Plantlets	
Populus (more than 20 species)	Wang et al., 1975; Group of Tree Breeding, North Eastern Forestry Academy, China, 1977; Zhu et al., 1980; Wu et al., 1981; Lu et al., 1978.
Hevea brasiliensis Muell.-Arg.	Chen et al., 1978
Aesculus hippocastanum L.	Redojevic, 1978
Citrus microcarpa Bge.	Chen et al., 1980
Vitis vinifera L.	Zou and Li, 1981
Malus prunifolia (Willd.) Borkh.	Wu, 1981
Malus pumila Mill	Fei and Xue, 1981
Litchi chinensis Sonn.	Fu and Tony, 1983
Euphoria longan Lour.	Yang and Wei, 1984
Poncirus trifoliata (L.) Raf.	Hidaka et al., 1979
Lycium halimifolum Mill	Zenkteler, 1972
barbarum L.	Fan et al., 1982; Gu, 1981
chinense Mill	
Camellia sinensis (L.) D. Kuntze	Chapter 34
Embryo Sac–derived Plantlets	
Poplus	Chapter 8
Hevea brasiliensis Muell.-Arg.	Chapter 32

clones. The first transplanted pollen plant is now 6 years old. Cytological investigations have been systematically performed on pollen callus, embryos, test tube plantlets, and transplanted pollen plants (Cen et al., 1981; Chen et al., 1982a). The first pollen-derived tree and its clone have blossomed. An F_1 hybrid seedling derived from a pollen plant crossed with another clone has been obtained. The pollen-derived trees have been tapped for latex (Xu, 1984). In 1984, haploid plantlets were again obtained through culture of unpollinated ovules of *Hevea* trees (see Chapter 32).

It was reported in 1978 that pollen plantlets were obtained from *Aesculus hippocastanum* L. (Radojevic, 1978). From 1978 to 1986, induction of pollen plants was successful in the following fruit trees: *Citrus microcarpa* Bge., *Vitis vinifera* L., *Malus prunifolia* (wild.) Borkh., *M. pumila* Mill., *Litchi chinensis* Sonn., and *Euphorbia longana* Lour. Among them pollen plantlets of two species from *Citrus, Vitis*, and two species of *Malus* survived after transplanting. A pollen plantlet of *Malus prunifolia* reached the age of 4 years, and budding and top-grafting clones were obtained (Chen et al., 1980; Zou and Li, 1981; Wu, 1981; Fei and Xue, 1981).

In the case of medicinal perennial crops, pollen plantlets have been induced successfully in *Poncirus trifoliata* (L) Raf. (Hidaka, 1979). They have also been obtained from *Lycium barbarum* L. and *L. chinense* Mill. and survived after transplanting in China (Gu, 1981; Fan et al., 1982).

Pollen plantlets have been obtained from tea *Camellia sinensis* (L) D. Kuntze. Many plantlets survived after transplanting (Chapter 34). This technique offers an alternative method for tea plant breeding.

CRITICAL VARIABLES

Progress has been made in the last decade in anther culture of perennial crops. Anther culture techniques, androgenesis, and morphogenesis of pollen plantlets and variation in somatic chromosome number have all been extensively studied. In comparison with anther culture of herbaceous plants, haploid induction of perennial crops has the following characteristics:

Increased Induction Period of Haploid Plantlets and Distinct Phases of Morphogenesis

Usually, 110–200 days is required from anther inoculation to induction of complete haploid plantlets from perennial crops; herbaceous plants such as *Nicotiana tabacum, Oryza sativa*, and *Brassica napus* require only 36–88 days (Table 2).

This longer period for pollen plant induction in perennial crops may be caused by distinct stages of embryogenesis, embryo differentiation and plantlet formation, lack of synchronicity of differentiation of embryos and shoots.

DISTINCT STAGES OF EMBRYOGENESIS. The duration of androgenesis for tree species is longer than for herbaceous plants. In the case of rubber tree microspores, division of the multicellular mass reached its climax 20–30 days after inoculation; in *Brassica napus*, it occurred 5–7 days after inoculation (Chen et al., 1978; Chen and Chen, 1983). A similar trend was found in *Malus prunifolia* and *Aesculus hippocastanum* as well (Wu, personal communication; Radojevic, 1978). The development of pollen through embryos preceded young and mature embryo stages. Intact plantlets could be formed only when the mature embryo was transferred onto a shoot induction medium (Table 2). Morphological investigations of pollen and embryo sac embryos of rubber tree showed that the development of young embryos began with distinct polarity: they could differentiate into organs but had no organ primordia. A sign that the young embryo had converted to a mature embryo was the gradual completion of organ differentiation. Anther culture experiments and anatom-

TABLE 2. Number of Days in Various Morphogenetic Stages for Pollen-Derived Plantlets from Perennial Crops

Plants	Callus	Embryo		Shoot	Plantlet	Rooting	Total
		young	mature				
Woody Plants							
Hevea brasiliensis Muell.-Arg	50	20–30	40–60		30–40		140–180
Citrus microcarpa Bge.	—	70	20		20–30		110–120
Litchi chinensis Sonn.	60	30	30–40		40		160–170
Vitis vinifera L.	30		20–30		60		110
Malus prunifolia (Willd.) Borkh	30	—		90–120	30	20	180–200
Herbaceous Plants							
Nicotiana tabacum	—	21			15–20		36–45
Oryza sativa L.	21–28	—			45–60		66–88
Brassica napus L.	—	20–30			20–30		40–60

ical observations on rubber trees indicated that organogenesis within an embryo was nonsynchronous. Embryos could be seen by the naked eye within 10–20 days after transfer of anther-derived callus onto the differentiation medium. However, organ primordia did not form at that time. Primordia of the main root and cotyledons could be visualized clearly only 30–50 days after transfer onto the differentiation medium, and terminal bud primordia were formed after 60–90 days (Chen et al., 1982b). In herbaceous plants, young embryos develop to mature embryos very quickly and have very similar nutrient demands; however, for anther culture of *Citrus*, litchi, rubber tree, etc., the culture conditions should be altered as the embryos mature. The media should be changed several times, to induce the pollen plantlets successfully and to enhance the plantlet induction frequency (Chen et al., 1980; Fu and Tong, 1983).

EMBRYO DIFFERENTIATION AND PLANTLET FORMATION. The development of embryos in vitro was similar to that of zygotic embryos in vivo: both pass through the globular-shaped stage, heart-shaped stage, torpedo stage, and cotyledon stage. The formation of pollen plantlets was similar to seed germination. These two stages progressed very rapidly in herbaceous plants and sometimes could not be distinguished. In perennial crops, complete plantlets could be formed only if the mature embryos were transferred onto appropriate shooting medium. Embryo germination generally involved root formation, then elongation of hypocotyl, greening of cotyledon, and emergence of the young stem with terminal buds between the two cotyledons. True leaves were differentiated continuously from the terminal bud to form the complete plantlet. However, in some plants shoots would first be formed from somatic cell-derived embryos followed by root induction (see Chapter 29).

The other mechanism for forming pollen plantlets involves differentiation from pollen callus (without going through embryo stages). This route also takes longer in perennial crops than it does in herbaceous plants. For instance, more than 3 months was required for bud differentiation of *Malus prunifolia*, before rootless plantlets could be transferred onto rooting medium to form entire plantlets (Table 2).

DIFFERENTIATION IN EMBRYOS AND SHOOTS IS NOT SYNCHRONOUS. Perennial crop morphogenesis may take one of the following paths:

1. Embryos develop directly from pollen grains.
2. Callus forms from pollen grains and then gives rise to many embryos from which plantlets develop.
3. Numerous secondary embryos develop from hypocotyl or cotyledon primodia of pollen embryos.
4. A bud or a cluster of buds form directly from a callus, without an embryo stage.
5. Embryonic cell masses originate from pollen and are continuously subcultured to produce plantlets.

Pollen and embryo sac embryos of rubber tree and pollen embryos of litchi, longan, and grape follow paths (2) and (3) (Chen et al., 1978; 1984; Fu and Tong, 1983; Yang and Wei, 1984; Zou and Li, 1980). Paths (1), (2), and (4) are followed in *Lycium barbarum* (Fan et al., 1982). The fourth path was most common in poplar and *Malus prunifolia* (Tree Breeding Group, Northeastern Forestry Academy, China, 1977; Wu et al., 1981). Embryonic cell mass had been obtained by means of anther culture. Embryo and plantlets could then be formed (Fu and Tong, 1983).

In paths (2), (3), and (4) all embryos and buds did not develop synchronously. The occurrence and development of secondary embryos as well as the quality of the embryos were usually affected by different factors, such as explant source and environmental conditions.

In order to achieve higher induction frequencies for embryos, buds, or plantlets, it was necessary to avoid transferring the immature embryo onto shooting medium too early and to provide suitable media at different developmental stages.

Media

BASAL MEDIUM. MS medium or modified MS medium has been used as the basal dedifferentiation medium in anther culture of most tree species (such as *Populus, Malus prunifolia, Lycium barbarum*, and *Aesculus hippocastanum*). MS medium has a high total amount of inorganic salts and also a high concentration of NH_4 ions. In rubber tree anther culture, the induction frequency was sharply decreased if the NH_4NO_3 in MS medium was cut in half. However, the induction frequency of pollen embryos was remarkably increased if the KNO_3 in MS medium was reduced by 50% and the other components were kept at their original levels (Chen et al., 1982a, b). Modified Medium B_5 and Medium N_6 have been used as dedifferentiation media for a few species (e.g., grape and citrus), possibly because of a need for higher concentrations of nitrate nitrogen. For all tree species in which pollen culture was successful, MS medium or modified MS medium was used as the basal medium for further differentiation of pollen embryos or for callus differentiation.

HORMONES. Simultaneous auxin and cytokinin supplementation in dedifferentiation medium for perennial crop anther culture is critical. The absence of either is unfavorable to dedifferentiation of the microspore. In herbaceous plants, in contrast, the demand for hormones is irregular. There are different protocols for use of cytokinins in differentiation media: KIN is used in combination with 2,4-D in *Malus prunifolia*, rubber tree, and *Aesculus hippocastanum*. BA is used in combination with 2,4-D but in the absence of KIN in *Vitis vinifera* (Zou Changjie et al., 1980). KIN or BA is used together with 2,4-D in *Citrus* and *Populus* (see Chapters 7 and 24).

ORGANIC SUPPLEMENTS. Most researchers have paid great attention to the application of organic supplements. Addition of 8 mg/l insulin and 20.5 μM biotin to the differentiation medium favored bud differentiation in *Malus prunifolia* (see Chapter 16). The presence of L-glutamine and 400 mg/l bee royal jelly promoted bud differentiation and plantlet formation in litchi. It was necessary to add 5–10% coconut water to the dedifferentiation medium of rubber tree (Chapter 32). Anther culture of *Aesculus hippocastanum* was successful only when MS medium was supplemented with 5.93 μM thiamine hydrochloride, 40.6 μM nicotinic acid, 21 μM Ca-pantothenate, 555 μM *myo*-inositol, and 200 mg/l casein hydrolysate, while the inorganic components in MS medium were maintained (Radojevic, 1978).

Mixing of Somatic Cell Tissue from Anther Walls with Pollen Calli or Embryos

The important tasks of anther culture of perennial crops are inhibiting the growth of somatic cells and obtaining haploid embryos and callus. The following methods have been used to improve the development of pollen embryos and the differentiation of pollen callus.

OPTIMIZATION OF HORMONES IN THE MEDIUM. In *Poncirus trifoliate*, a medium containing 0.93–9.3 μM KIN and 1.1 μM IAA was used to obtain pollen embryos. No propagation of somatic cells was observed in this case. In anther culture of *Lycium barbarum*, pollen embryos were obtained when 0.93–4.6 μM KIN and 0.54 μM NAA or 5.7 μM IAA was used (see Table 3).

ORIGIN OF PLANTLETS. Sometimes anther wall derived callus becomes mixed with pollen callus or pollen embryo. Microscopic examination is able to distinguish pollen (microspore) embryogenesis from proliferation of somatic cells. Various methods have been developed to promote microspore embryogenesis and to inhibit growth of somatic cells. Higher concentrations of sucrose (204 mM), 5–10% coconut water, and appropriate concentrations of hormones (4.6 μM KIN, 5.4 μM NAA, 4.5 μM 2,4-D) have proved to be effective. These factors depress excessive callus growth and promote pollen embryogenesis. The ratio of multicellular masses of pollen to somatic anther wall callus is then very high so that the embryo induction frequency is also high. Propagation of somatic cells cannot be suppressed entirely because microspore development is then also inhibited.

IDENTIFICATION OF CALLUS ORIGIN. Histological observations indicated that somatic cell callus generally appeared earlier than pollen callus when they were together. In poplar, for example, the callus appearing around 20 days after anther inoculation was mostly of somatic cell origin, since the pollen multicellular mass was unable to break through the anther wall at this time. Only the callus occurring 30–40 days after inoculation was derived from pollen (Zhang et al., 1979). Pollen derived embryos of *Aesculus hippocastanum* were at the granule-shaped and cotyledon stages 56 days after culture and could only be seen through the magnifying glass (Radojevic, 1978).

Sometimes only the anther wall somatic cells propagated and no development of the microspore was seen. In such cases, the plantlet formed

TABLE 3. Anther Culture Procedure and Medium for Some Tree Species

Tree Species	Process	Basal Medium	Hormones (μM)	Sucrose (mM)	Supplement (mg/l)	Culture result	No. of Days of Culture
Lycium barbarum	I	MS	2.3–4.6 KIN; 0.9–9.0 2,4-D	88–146		Callus	30
			0.93–4.6 KIN, 0.54 NAA or 0.57 IAA	88–146		Embryo	30
	II	MS	4.4 BA, 0.54 NAA	88		Differentiated bud	20–30
	III	MS	0.57 IAA	88		Rooting	
Vitis vinifera	I	Modified B_5*	6.6–8.8 BA, 2.3 2, 4-D	88	500 LH	Callus	30
	II	Modified B_5 or	2.2–17.6 BA, 1.1 NAA	58	500 LH	Embryo	20
	III	MS	0.44 BA, 0.54 NAA			Plantlet	60
		MS (½ Macro-elements)					
Poncirus trifoliata	I	MS	0.93–9.3 KIN, 1.1 IAA	146		Embryo	28
	III	MS		58		Plantlet	
Aesculus hippocastanum	I	Modified MS	4.6 KIN, 4.5 2,4-D	58	See text	Embryo	56
	III	Modified MS		58	See text	Plantlet	

*Modified B_5: 1.1mM $(NH_4)_2SO_4$, 1μM $CuSO_4 \cdot 5H_2O$, 1μM $CoCl_2 \cdot 6H_2O$. The other components are the same as in B_5 medium.

was of somatic cell origin. This event was probably due to inappropriate selection of anthers for inoculation, suppression of microspore development, or unfavorable medium. If this occurs, changes in media and development stage of the anther should be investigated to correct the situation.

Pollen Plants Were Often Mixoploids

Research on anther culture of numerous tree species indicates that chromosome number varies in the development of pollen embryos to plantlets. Study of root-tip cells from pollen plantlets of *Vitis vinifera* revealed aneuploid, diploid, and tetraploid cells, although the majority of mitotic metaphases were haploids, with chromosome number 19 (Zou Changjie et al., 1981). Most of the pollen plantlets of *Poncirus trifoliata* were also mixoploids containing cells with chromosome numbers X + 1, X + 2, and 2X in addition to the haploid cells of X = 9 (Hidaka et al., 1979). Cytological observations of root tips of rubber tree embryos and plantlets showed that the embryos and most cells of the pollen plantlets were haploid, but that mixoploids (with the chromosome number X = 9, 2X = 18, 3X = 27, 4X = 36, 5X = 45) also existed (Cen et al., 1981). These experiments demonstrated that mixoploidy in somatic cells occurs at early developmental stages of embryos and pollen plantlets.

The somatic cell chromosome number of transplanted pollen trees changed continuously toward diploid. Haploid and diploid somatic cells of pollen plantlets of poplar in the year of transplantation were 58% and 20.1%, respectively. The remaining somatic cells were aneuploids. After 1 year, the proportion of diploid somatic cells had increased to 50%, haploid somatic cells had decreased to 30%, tetraploids to 8%, and the remaining were aneuploids (Lu, 1979). In leaves of rubber tree pollen plantlets, most cells just after transplantation had chromosome numbers of 18–27. However, more than 60% of the cells had chromosome numbers of 28–36 when the plantlets reached a height of 160 cm.

The rate of diploidization varied in various pollen plantlets in poplar. Thus more than 1 year after transformation, some plantlets were mixoploids in which diploidy predominated, some were mixoploids in which haploidy predominated, and very few plantlets were aneuploids. However, chromosome number of most of the cells in control plants was 38 (Lu et al., 1979; Chen et al., 1981; Wu Kexian et al., 1981).

Chromosome variation in vitro has been studied by many scientists. It has been suggested that chromosome numbers in haploid cultures change more often than in diploid cultures (Sacristan, 1971). The mechanism of chromosome number variation in pollen embryos, callus, and plantlets in anther culture has also been investigated. Variation in chromosome number may be related to endomitosis, nuclear fusion, or multipolar mitosis (J.A. McComb, 1978; Hu et al., 1978). Cytological observations of pollen plantlets of rubber tree showed that the chromosome numbers of the pollen embryos and plantlets were mostly multiples of 9. Rubber tree is an amphidiploid with a basic chromosome number of 9 (9 = X) with 18 chromosomes in its gametes (n = 18). This phenomenon is probably related to multipolar mitosis. We observed the abnormal division of a cell with 18 chromosomes. There were two poles on one side of the equatorial plate, one with 18 chromosomes, and the other with 9, and there was another pole with 9 chromosomes on the other side of the equatorial plate. This configuration could lead to a cell with 27 chromosomes and a cell with 9 chromosomes (Chapter 32). Many plant hormones in the medium might cause abnormalities in cell division (Sunderland, 1973). In perennial crops, abnormal cell divisions occurred more often because of the longer induction period to form pollen plantlets. In addition, this effect was prolonged to a certain extent after transplantation because of the long growth period of perennial crops. This may result in pollen plantlets from mixoploids and may explain why mixoploids are so common among perennial crops. In herbaceous plants, abnormal somatic cells were seldom seen and were usually eliminated during meiosis: only around 10% of the chromosomal variation was sustained through the sexual process (Hu Han et al., 1982). Chromosomal variation could be observed during the development of individual pollen plantlets. Cells with the chromosome number (n) of the gametes gradually decreased along with a gradual increase of cells with the chromosome number ($2n$) of spores. In pollen plantlets artificial chromosome doubling could be used to accelerate the spontaneous rate of diploidization and reduce mixoploidy (Wu et al., 1980).

Importance of Haploids and Homozygous Diploids

It is usually impossible to obtain pure lines of a perennial crop without many generations of inbreeding. Production of haploids and homozygous diploids is thus a more important time saving technique for perennial crops than for herbaceous plants. The establishment of perennial crop haploids would promote the development of tree genetics. As a result of the shortage of homozygous material, genetic research on perennial crops has proceeded more slowly than for annual plants, so that there are many gaps in our knowledge of the economic characteristics of some important tree species. It would be more convenient to use haploids rather than diploids as material for genetic research on partial chromosome homology. In addition, haploids are also important for tree species improvement. Crossing experiments using homozygous diploids could clarify whether the genes which control certain characters are recessive or dominant, whether they were monogenic or polygenic, etc. These methods could be used in combination with sexual hybridization to improve the possibility that progeny would be of the desired type. In addition, if desirable gametoclonal variants could be selected from pollen-derived trees, vegetative methods could be used for rapid propagation. Chromosomal variants of pollen-derived trees are easier to maintain than those of herbaceous plants because the favorable characteristics could be maintained by vegetative propagation for a longer time.

Characteristics of Pollen Plantlets

Studies of pollen-derived plantlets of poplar, rubber tree, and *Malus prunifolia* have found a great deal of variation in leaf shape, growth vigor, girth of tree, etc., of pollen plants originating from the same donor. In 1-year-old pollen-derived poplars, for example, the tallest reached 80 cm, and the shortest was only 20 cm. In the case of 1-year-old pollen-derived rubber trees, the tallest was 131 cm and the shortest only 19.5 cm.

This variation resulted from many factors, including carry-over effects of plant hormones which could appear in the pollen-derived plants (H_1). Wu et al. (1981) observed that variation of leaf shape in young pollen plants gradually disappeared as the plant grew, though some were retained in the process of vegetative propagation. Some of the variations among pollen-derived plantlets may be related to ploidy levels (Lu et al., 1980), and others may be related to growing conditions, such as transplantation time, soil conditions, and acclimatization. It is quite difficult to find a tree appropriate for control, because of the remarkable variation among seedlings of certain tree species. The morphological investigation of pollen-derved plants (H_1) alone is insufficient to show the variation among individuals. It is necessary to propagate donor and plants asexually so that the morphological and economic characteristics of both can be compared. The variation caused by environmental factors can thus be minimized to show the genetic variation among pollen plants more clearly. This method was used in the study of pollen plants of poplar and rubber trees, and desirable clones of pollen-derived plants surpassing the donor plant were selected. The appropriate number of clones of each pollen-derived plant depends on the cultivation conditions of clones, the plant size, the characteristics to be identified, etc. In general, rubber tree ideally should have 20 clones, with at least 5 clones in order to save space and labor.

PERSPECTIVES

Research on perennial crop anther culture only began recently. The time and procedures necessary to go from establishment of cultures to plantlet regeneration are longer and more complex than those of herbaceous plants. However, improvements in culture techniques during the last 10 years have enhanced the induction frequency of haploids and made anther culture successful in many tree species. Future achievements will be even greater.

As to the recalcitrant tree species, it is helpful to follow the procedures used for the successful anther culture of perennial crops. Observation of androgenesis and morphogenesis in the process of anther culture is of significant importance not only for the establishment of new anther culture techniques for tree species but also for the enhancement of haploid induction frequency of tree species in which anther culture is already successful. Ploidy of pollen-derived plants should be studied to learn the mechanisms of chromosomal variation and to discover consistent methods for obtaining various chromosomal variants and homozygous diploids.

It is necessary to speed the incorporation of pollen-derived plants into breeding practice. Poplar pollen trees are now 7–8 years old, and rubber trees have grown for 6 years. Both these species have produced blooming pollen plants, and pollination experiments are being carried out. Pollen plants of both *Malus prunifoliata* and *Vitis vinifera* have been growing for more than 3 years. Anther culture must be combined with conventional breeding to shorten the time needed for conventional breeding alone and to enhance selection efficiency. If large numbers of pure lines with different genotypes are obtained via anther culture, hybrids should be produced and tested as soon as possible to select the most desirable combinations for early release. If interspecific or intergeneric hybridization were combined with anther culture techniques, major new tree species might be obtained quickly.

Stable homozygous mutants with good characteristics could also be obtained after physical and chemical mutagenesis, screening, and chromosome doubling of haploid cell lines with differentiation ability. Chromosome engineering of perennial crops could be developed by using laser microbeam irradiation and micromanipulation to transfer or remove specific chromosomes.

REFERENCES

Cen, M., Chen, Z., Qian, C., Wang, C., He, Y., and Xiao, Y. 1981. Investigation of ploidy in the process of anther culture of *Hevea brasiliensis*. *Acta Genet. Sin*. **8**:169–174 (in Chinese).

Chen, H., Dong, Y., Guo, Y., and Dong, H. 1981. A study on character heredity and its selective application of H_1 pollen-derived plantlet of poplar. *Yangshu Keji Tongxun* **5**:7–15 (in Chinese).

Chen, Z., Wang, M., and Liao, H. 1980. The induction of *Citrus* pollen plants in artificial media. *Acta Genet. Sin.* **7**:189–191 (in Chinese).

———, Chen, F., Chien, C., Wang, C., Change, S., Xu, X., Ou, H., He, Y., and Lu, T. 1978. Induction of pollen plants of *Hevea brasiliensis* Muell. Arg. *Acta Genet. Sin.* **5**:99–107 (in Chinese).

———, Qian, C., Cen, M., Wang, C., Suo, C., Xiao, Y., and Xu, X. 1981. Relationship between somatic cells and microspores in anther culture of *Hevea brasiliensis* Muell. Arg. *Xibei Zhiwu Yanjiu* **1**:31–37 (in Chinese).

———, Qian, C., Cen, M., Xu, X., and Xaio, Y. 1982a. Recent advances in anther culture of *Hevea brasiliensis* Muell. Arg. *Theor. Appl. Genet.* **62**:103–108.

———, Qian, C., Xu, X., and Deng, Z. 1982b. Anther culture of rubber tree and sugar cane. *Proc. 5th Int. Congr. Plant Tissue Cell Cult.*, (Fujiwara, A., ed.) Japanese Association for Plant Tissue Culture, 533–534. Tokyo, July 11–17, 1982.

Chen, Z. Z. and Chen, Z. 1983. High frequency induction of pollen-derived embryoids from anther cultures of rape (*Brassica napus* L). *Kexue Tongbao* **28**:1690–1694.

Fan, Y., Zang, S., and Zhao, J. 1982. Induction of haploid plants in *Lycium chinense* Mill. and *L. barbarum* from anther culture. *Hereditas* **4**:25–26 (in Chinese).

Fei, K. and Xue, G. 1981. Induction of haploid plantlet by anther culture in vitro in apple cv. "Delicious." *Sci. Agric. Sin.* **4**:41–44. (in Chinese).

Fu, L. and Tong, D. 1983. Induction pollen plants of litchi tree (*Litchi chinensis* Sonn). *Acta Genet. Sin.* **10**(5):369–374 (in Chinese).

Group of Tree Breeding, Northeastern Forestry Academy, China. 1977. Induction of haploid poplar plantlets from pollen. *Acta Genet. Sin.* **4**(1):49–54 (in Chinese).

Gu, S. 1981. The obtaining of pollen-derived plantlet of *Lycium barbarum* L. *Acta Bot. Sin.* **23**:246–248 (in Chinese).

Hidaka, T., Yamada, Y., and Shichijo, T. 1979. In vitro differentiation of haploid plants by anther culture in *Poncirus trifoliata* L. Raf. *Jap. J. Breed.* **29**:248–254.

Hu, H., Hsi, T., and Chia, S. 1978. Chromosome variation of somatic cells of pollen calli and plants in wheat (*Triticum aestivum* L.) *Acta Genet. Sin.* **5**:23–30 (in Chinese).

———, Xi, Z., and Wang, X. 1982. Production of aneuploid and heteroploid of pollen-derived plants. *Proc. 5th Int. Congr. Plant Tissue Cell Cult.*, (Fujiwara, A. ed.) Japanese Association for Plant Tissue and Cell Culture, 467–473. Tokyo.

Lu, Z., Chang, X., Liu, Y., and Zhang, P. 1979. Somachromosomal observations of pollen-derived plantlet of poplar. *Acta North Forest. Coll.* **2**:14–22 (in Chinese).

McComb, J. A. 1978. *Variation in Ploidy Levels of Plant Tissue Culture*. pp. 167–180. Science Press, Beijing.

Radojevic, L. 1978. In vitro induction of androgenetic plantlets *Aesculus hippocastanum. Protoplasma* **96**:369–374.

Sacristan, M. D. 1971. Karyoptic changes in callus cultures from haploid and diploid plant of *Crepis capillaris* L. Wallr. *Chromosoma* **33**:173–183.

Sunderland, N. 1973. Nuclear cytology. In: *Plant Tissue and Cell Culture*. (H. E. Street, ed.), pp. 161–190. Blackwell, Oxford.

Wang, C., Chu, Z., and Sun, C. 1975. The induction of *Populus* pollen plants. *Acta Bot. Sin*. **17**:56–59 (in Chinese).

Wu, J. 1981. Haploid plantlet obtained by anther culture of *Malus pumila* mill. *Acta Northeast Agric. Coll*. **2**:105–108 (in Chinese).

Wu, K., Xu, M., Ma, H., Li, W., and Zhu, W. 1981. Preliminary analysis of genetic expression of pollen-derived plantlet of *Populus simoniinigra. Linye Keji* **2**:5–6 (in Chinese).

Xu, X., Zhang, S., Ou, X., Pang, R., and Chen, Z. 1984. Study of pollen plants and their clones in Hevea. *Int. Symp. Genet. Manipul. Crops*, p. 120 (abstract). Science Press, Beijing.

Yang, Y. and Wei, W. 1984. Induction of pollen plantlets in longan. *Acta Genet. Sin*. **11**:288–293. (in Chinese).

Zenkteler, M. 1972. Development of embryos and seedlings from pollen grains in *Lycium halimifolium* Mill. in the in vitro culture. *Biol. Plant* **14**:420–422.

Zhang, X., Liu, Y., and Lu, Z. 1979. Cytological observation of microspore in anther culture of poplar. *Acta Northeast Forest. Coll*. **1**:5–13 (in Chinese).

Zhu, Y., Wang, R., and Liang, Y. 1980. Induction of pollen-derived plantlet of poplar. *Linye Kexue* **16**:190–197 (in Chinese).

Zou, C. and Li, P. 1981. Induction of pollen-derived plantlet of *Vitis vinifera. Acta Bot. Sin*. **23**:79–81 (in Chinese).

CHAPTER 4

Cell Suspension Culture and Mutant Screening in Perennial Crops

Zhenghua Chen, Yuhua Song, and *Wenbin Li**

INTRODUCTION

The need for saplings of forest trees, fruit trees, and other economically important woody species has been increasing. Sufficient quantities are required for introductions of newly developed superior varieties if they are to replace old varieties promptly. Cell suspension culture is an important method for rapid propagation of improved varieties. Artificial seed engineering may be used with cells produced by suspension culture in bioreactors. The induction of embryogenesis from cell suspension cultures is a foundation for development of artificial seed technology. Up to now, however, there have been few studies on regeneration of embryos or plantlets in woody plants. Embryos and plantlets have been regenerated from liquid suspension cultured cells in only about a dozen genera and families.

Because suspended cells are cultured under artificially controlled conditions and can be uniformly exposed to physical and chemical

*English translation by Kaiwen Yuan.

mutagens, the mutagenesis procedure is similar to that used for microorganisms. Established mutation induction and screening procedures have high efficiency and good reproducibility.

Extensive studies on screening mutants useful to agriculture have been conducted in herbaceous plants and are now being extended to perennial crops. Cell suspension culture techniques have also provided useful materials for genetic engineering.

TECHNIQUE OF SUSPENSION CULTURE

Induction and Subculture of Callus

SELECTION OF SUITABLE EXPLANTS. Callus with embryogenetic potential can be induced from culture of explants. Suitable explants vary, depending on the plant. Calli with regenerating ability have been obtained by using hypocotyl explants from bacteria-free plants of *Liquidambar Styraciflua* (Sommer and Brown, 1980); in *Carica papaya*, calli from which plantlets could be regenerated were induced by using ovules and flower peduncles as explants (Litz and Conover, 1980, 1982). The first task of suspension culture is thus to evaluate suitable explants.

Techniques for surface sterilization of explants are described in Chapter 2.

EXPLANT INOCULATION AND PRIMARY CULTURE. The sterilized explants are placed in agar medium. The media used successfully in woody species will be described later in this chapter. The inoculated explants are incubated at a temperature of 25°C, under low light or dark conditions.

SUBCULTURE OF EXPLANTS. Callus can be induced from cultured explants in about 3 weeks. The callus is then dissected and subcultured. The subculture medium is generally the same as the primary induction medium. Two factors that should be considered are callus selection and medium composition. Two types of callus are usually induced from stem explants of rubber tree. One is a loose, yellow callus which is suitable for cell suspension culture; the other callus is hard and green and should be discarded (C. Zhenghua, unpublished). The medium composition should be regulated to improve callus quality. Besides looseness or hardness, the dedifferentiation ability of callus is an important aspect of callus quality. Components and concentrations of hormones in the medium are important factors. A slight increase of 2,4-D will make the cell texture loose, and supplementation with certain natural organic substances (coconut water, yeast extracts, lactoprotein hydrolysate) usually improves the ability of cells to differentiate.

Isolation of Single Cells

MECHANICAL ISOLATION OF SINGLE CELLS. Calli are put into 50-ml flasks after they are weighed. One gram of callus should have 4 to 8 ml of liquid medium added. The flasks are placed on a rotary shaker (100–110 rpm) for suspension. Enzymes also can be used for isolation. For instance, cells have been isolated with 1% macerozyme from *Citrus* plants (Button and Botha, 1975). As enzymatic isolation often damages cells, many investigators prefer not to employ this method.

FILTRATION OF SUSPENSION AND COLLECTION OF SINGLE CELLS. The cultures are in their best physiological state after the calli in the medium have been shaken continuously for 7–10 days at around 100–150 rpm. The cells can then be filtered with a nickel sieve. Usually a sieve with 200–300 mesh pore size is used for large cells, and 400 mesh size for small cells. The filtered cell suspensions are centrifuged at 500 rpm for 5–10 minutes. Single cells are then precipitated and collected. Aseptic technique must be strictly followed during the whole process. The centrifuge tubes should be sterilized beforehand and centrifuged only after they are covered with sterile caps. Precisely graduated 0.01-ml centrifuge tubes should be used as cell collectors.

Density of Cultured Cells

APPROPRIATE DENSITY. In general, the density of cultured cells is 10^5–10^9/ml. In practice, cell size differs greatly. For example, suspended cells of *Idesia polycarpa* multiply rapidly at a density of 10^6 (Chen Zhenghua, unpublished). One milliliter of packed culture containing approximately 2×10^6 cells could produce over 5000 uniform embryos on 2,4-D-free medium (Lutz et al., 1985).

CALCULATION OF INITIAL CELL DENSITY. Hemacytometers can be used for plant cell counting. Under the microscope, the cells in the four angular and the central squares are counted five times. The mean number of cells can be calculated according to the following equation (Ye, 1985):

$$\text{Cell number/ml} = \text{cells in the 5 squares} \times 50{,}000$$

DETERMINATION OF LIVING CELL PERCENTAGE. The percentage of living cells determines the quality of the suspension culture. A good cell suspension contains at least 60% living cells. The proportion of living cells can be determined by using the following procedure: the culture medium is used as solvent for making a 0.1% phenol safranal solution. Five-milligram per liter stock solution of fluorescein diacetate is prepared by using acetone and stored in a refrigerator. It should be diluted to a 0.01% concentration before use. The cells and the stock solution are mixed on a slide, stained with a drop of 0.1% phenol safranal solution, and covered for observation under the microscope. The dead cells will be stained by the phenol safranal solution, but the living cells will not be stained. After 15–30 min, the smear can also be observed under phase-contrast and fluorescent microscopes. Living cells will be stained; dead cells will not fluoresce. Dead cells can also be identified with 0.5% Evans blue staining (Ye, 1985).

NUMBER OF CELLS SHOWN BY VOLUME. The number of cells can be measured by packed cell volume, i.e., total cell volume/milliliter medium. If 0.2 ml *Idesia polycarpa* cells is added to 1 ml liquid medium, the cell density is at about 10^6 (calculated by hemacytometer). With this method, we can determine the cell density without use of the hemacytometer. It also can be used to measure large-sized cells which are difficult to count with a hemacytometer. Precisely graduated 0.01-ml centrifuge tubes should be used.

Culture Methods

The following methods are commonly used in cultures of herbaceous and woody plants.

STATIC CULTURE WITH SHALLOW LIQUID LAYER. Suspensions with a certain density of isolated cells are put into petri dishes or specially made flat flasks and placed in an incubation room for static culture. This method gives the cultures the advantage of a large air space, so ventilation is good and toxic metabolic products can diffuse easily. It is convenient to transfer the cultures or to add fresh medium, and making observations and photographs is easy. The disadvantage of this method is that the density of suspended cells can easily be too high or too low. Sometimes concentrated cells become linked together, potentially affecting cell division. In addition, with this method, it is difficult to conduct detailed observations.

SOLID PLATE CULTURE. Cells are first suspended in liquid medium (at twice the required cell density), then mixed with melted agar medium containing the same components as liquid medium. After cooling, the suspension cells are embedded in the solidifed agar in the petri dish. This method has the advantage that the isolated cells are evenly spread in the medium and will not aggregate. This facilitates the observation of cell development. Unlike static liquid culture, this technique does not provide much aeration and does not facilitate diffusion of accumulated toxic products. Therefore, the developmental speed of isolated cells is usually slow.

SUSPENDED DROP (MICRODROP) CULTURE. Cell suspensions are placed in drops on a petri dish. Each drop should be about 0.1 ml. The petri dish is then turned over so that the small drops are suspended from the dish. This method is usually used when the density of cultured cells is low. With this method, several media can be put onto one petri dish to conduct comparisons under controlled conditions. When this method is used, the humidity must be maintained in the culture dish to prevent the drops from drying as a result of evaporation.

FEEDER LAYER CULTURE. Two culture layers are used with this method. The lower layer is a plate made up of solid medium; the upper layer is liquid medium.

The first two of these four methods are fundamental; the second two have been developed from the first two. The first method is the one most frequently used. However, the best results can be achieved if the first two methods are used in combination: small cell masses are formed and are then transferred to solid plate culture.

Cell Growth

DETERMINATION OF MITOTIC INDEX. The cellular mitotic index (MI) refers to the percentage of cells undergoing mitosis (including the prophase of cell division). MI is very low in callus, usually below 1%. Repeated subculture can increase the MI of suspended cells within a short period, sometimes to as much as 10% (Evans et al., 1981). If a 4-day subculture period is used, visible and countable mitotic figures can be obtained at any time. The highest MI often appears 24 hr after subculture. When the culture duration is prolonged, the MI of cultured cells gradually decreases and may become as low as that of callus. Methods for testing MI are described by Evans et al. (1981).

Mitotic metaphases in 1000 cells are counted repeatedly three to five times according to the following equation:

$$\text{Mitotic index (\%)} = \frac{\text{total number of divided cells}}{\text{total observed cells}} \times 100$$

DETERMINATION OF CELL GROWTH RATE. Cell growth rate is usually depicted by a growth curve. Cultured cells are pipetted into sterile calibrated tubes for centrifugation. Packed cell volume can be calculated by this technique. Cells and medium then are returned to the dish for culture. This can be done repeatedly. If five measurements are needed, five dishes should be counted. The initial cell densities in the dishes and the medium used should be constant. Only one dish should be used for each measurement to prevent errors caused by loss of cells during pipetting.

DETERMINATION OF PLATING RATE. The plating rate is the percentage of inoculated cells that have developed into cell masses:

$$\text{Plating rate (\%)} = \frac{\text{number of cell masses in each plate}}{\text{number of cells inoculated in each plate}} \times 100$$

The total number of cells inoculated in each plate equals the number of cells contained in each milliliter of culture solution times the volume of culture solution in the plate (Cytological Lab, Shanghai Institute of Plant Physiology, 1978).

Selection of Cultured Cells

In a population of cells, not all the cells are viable. Some cells are dead; others have stopped mitosis and appear to accumulate many starch grains. Cells of different density can be separated by discontinuous density gradient centrifugation, and the most viable cell layer can be selected. The media used for density gradient centrifugation are sucrose, serum albumin, ficoll (usually ficoll-400), and percoll, which has been most frequently used in recent years. The merit of ficoll and percoll is that they cannot enter cells and are not metabolized. Each layer of cells isolated by percoll is rinsed with nutrient solution two to three times. When the most viable layer of cells is cultured, the induction frequency of callus and embryos can be increased remarkably. For example, among the four layers of *Nicotiana tabacum* pollen grains separated by percoll, multicellular masses were formed only in the second layer (Huang et al., unpublished). A high-frequency (about 85–90%) embryogenesis system from single carrot cells fractionated by percoll has also been established (Komamine and Nomura, 1986). Six different concentrations of ficoll-400 ranging from 3 to 20% were used, and free pollen grains of rice were separated by discontinuous density gradient centrifugation. Then each layer was cultured individually. A high induction frequency for pollen callus was achieved only in the layer separated with 12% ficoll-400 (Zuo et al., 1982). In *Idesia polycarpa*, among the four layers of cells separated by density gradient centrifugation, high induction callus frequencies were obtained in the second and third layers (Chen Zhenghua et al., unpublished). The procedures of discontinuous density gradient centrifugation are as follows:

TABLE 1. Preparation of Ficoll Solution

Concentration of ficoll solution (%)	5	10	15	20	25
Grams of ficoll	2.5	5	7.5	10	12.5
Milliliters of 58.7 mM sucrose solution	50	50	50	50	50

PREPARATION OF FICOLL SOLUTION

1. Prepare 250 ml of 58.7 mM sucrose solution.
2. Prepare different concentrations of ficoll sucrose solution.
3. Filter sterilize the solutions.
4. Put the discontinuous density gradient solutions into 10-ml centrifuge tubes. Use an aseptic 1-ml syringe to inject 1 ml solution into each centrifuge tube. The solution should be injected layer by layer and from high density to low density. The manipulation should be done with great care. Do not break the boundary between layers. When all 5 ml of the gradient solutions is in the tube, the clear boundaries of four layers should be visible.

ISOLATION OF CELLS WITH DIFFERENT DENSITIES

1. Centifuge cell suspensions at 500 rpm for 5 min to precipitate cells.
2. Add 0.5 ml of precipitated cells to the upper layer (the layer with lowest density) of the discontinuous gradient solution.
3. Centrifuge for 10 min at low speed in bench centrifuge. Four to five layers of cells will be isolated.

CELL CULTURE

1. After centrifugation, cells are collected respectively from each layer in turn. Each collection is put into a small centrifuge tube with an appropriate amount of culture medium and then centrifuged to wash away the ficoll.
2. The supernatant fluid is removed after centrifugation, and cells are resuspended in a suitable volume of medium.

OBSERVATION OF EXPERIMENTAL RESULTS. Cell proliferation is observed under an inverted microscope. Fresh medium is added to the petri dishes after 4–7 days of culture. Small calli may be visible in 15 days.

CAUTIONS

1. Starting with the second filtration of the ficoll density gradient solution, all manipulations should be performed aseptically.

2. When filling the gradient solution, the syringe should be emptied slowly; the syringe needle should be next to the tube wall.
3. The gradient solution layers should not be stored for long periods of time or disturbed.
4. When absorbing cells from gradient layers, the needle should remain on the surface of the layer. Do not insert too far into the layer to prevent mixing with cells from other layers.

Synchronization of Cell Division

Synchronization of cell division is very important for in vitro culture. It is needed for studying the cell cycle, for obtaining large volumes of cells in the same metabolic state, and for synchronizing embryo morphogenesis.

An obvious peak of the mitotic index indicates that cell divisions are synchronous.

The following methods have been used for the synchronization of cell division:

DNA INHIBITORS. DNA inhibitors include hydroxyuridine, 5-Budr, and excess thymine. Cells are blocked in the G_1 phase after treatment with the inhibitor and enter the S phase synchronously after removal of the inhibitor. This method is most commonly used.

GROWTH REGULATORS. Tobacco cells in S phase were transferred to fresh medium with 2,4-D and KIN for 24–72 hr to induce synchronization of cell division. KIN is the key growth regulator that causes synchronized cell division in this treatment. The mitotic index of *Acer pseudoplatanus* L. reached a peak after cells were transferred to medium with KIN or BA for 40–72 hr.

CELL DENSITY. Low initial cell density followed by normal cell density can also induce synchronization of cell division. In *Acer pseudoplatanus* L., cells with low density (2×10^4 cells/ml) were cultured for 24–28 days, followed by subcultures at normal cell density. Cells divided synchronously for five to six generations (Cytological Lab, Shanghai Institute of Plant Physiology, 1978).

MORPHOGENESIS OF EMBRYOS

Modes of Morphogenesis

There are two routes of morphogenesis in in vitro culture. Embryos may form directly from embryogenic cells, or differentiated cells may first redifferentiate, then be determined as embryogenic cells through callus proliferation, and finally form embryos. Nucellar cells first develop into embryogenic cell mass or callus through proliferation and form embryos again. They can be considered embryogenic cell clones determined before embryogenesis. Evans et al. (1981) considered that cortical cells of plantlets developed from embryogenic cells have an important function in vegetative proliferation. The differentiated cells should be switched on again by cytokinin and auxin under artificial culture. They will then undergo mitosis, develop into callus, and finally form embryogenic cells. A long culture period will be necessary to induce embryos by this means. For example, embryoids take 70 days to arise from cultured calli of *Coffea arabica* leaf explants (Sondahl et. al., 1979). Bapat and Rao noted that five passages of culture was required to induce embryos from *Santalum album* (Zhou, 1982).

When liquid medium is used for cell suspension culture, cells can divide and freely develop into cell masses. These cells first become embryogenic cell complexes, and then the surface cells of the complex develop into embryos. As the embryogenic cell complex originally arises from a single cell through repeated division, the developed embryos can still be considered to come from a single cell. This developmental path is commonly observed, although direct development of embryoids from single cells is rarely seen (Zhou, 1982).

Conditions for Embryo Formation

1. Physiological isolation of a cell from its peripheral tissue is a precondition for embryo morphogenesis (Steward et al., 1958). If a border cell layer is formed as soon as the embryo begins to develop, the embryo can then maintain its relative independence.
2. Hormone levels should be appropriate. In many perennial crops, cytokinin used in combination with auxin can regenerate embryos, e.g., in *Citrus* (Button, 1975), and *Coffea arabica* (Sondahl, 1979). In some species, such as red currant, embryos can be induced with cytokinin alone (Zhou, 1982). In *Palmae* plants, embryos can be regenerated only when 2,4-D is added to the medium initially and reduced or removed in the second passage (see Chapter 31).
3. Interaction among cells is also an important condition. In plate culture, the plating density is an essential factor. An appropriate density can stimulate formation of embryoids.
4. NH_4^+ is an important nutrient for embryo formation. All media used thus far for successful embryogenesis contain NH_4^+ (Zhou, 1981, 1982).

5. Embryos can be induced only from some tissues in a species. Induction is easy in some species or varieties, but difficult in others.

Synchronization of Embryogenesis

The development of the embryos formed via cell suspension culture in vitro is usually not synchronized. In the same culture, single initial cells, multicellular proembryos, spherical embryos, torpedo-shaped embryos, and even mature embryos with cotyledons can be observed. In order to produce artificial seeds it will be necessary to develop separation procedures that permit large-scale collection of synchronized embryos at the same stage of development. Filter sieves and beads can be used to separate and collect uniformly developed embryos (Warren and Fowler, 1977). The separated embryos at different developmental phases should be cultured separately for particular stages of development. The embryos to be propagated should have superior genotypes that are needed in production.

The embryos used for producing artificial seeds should remain dormant so that they can be stored in artificial seed coats (Ammirato, 1983). Therefore, mechanisms for inducing embryo dormancy also require investigation.

Cold treatment and mitotic inhibitors can inhibit germination of embryos (Evans et al., 1981). However, only when synchronously developed mature embryos are so treated will they have the potential for synchronous germination later. During sowing, somatic embryos should first be induced to germinate under artificially controlled optimum environmental conditions. The uniformity of embryo germination under artificial conditions is directly correlated to synchronization of embryo development.

Bioreactors for Large-Scale Embryogenic Cell Suspensions

Using bioreactors, culture conditions can be optimized for growth of the plant cells, and very high yields can be expected. Batch cultures of 1–5 l can easily produce hundreds of thousands of plants, and the embryos are already morphologically distinct, therefore, easy to separate by size (Styer, 1985; Rowe, 1985). However, this system has some problems. Very few species or genotypes can produce cell suspension cultures with high and predictable frequencies of regeneration, and genetic variation seems to increase with the amount of time cells and tissues remain in culture (Rowe, 1985).

Somatic Embryo Delivery Systems

Fluid drilling, encapsulation, and seed tapes have all been suggested as potential delivery systems.

FLUID DRILLING. Fluid drilling involves suspending the seeds in a carrier gel, supplemented with additives, that is then pumped into the soil. Drilling devices that separate the precision-drill pregerminated seeds with high efficiency have been developed. As a delivery system, fluid drilling should work well if the embryos have sufficient vigor to become autotrophic (Gray, 1981; Styer, 1985; Rowe, 1985).

Encapsulation involves the formation of a gel matrix around individual embryos. This enables nutrients and agricultural chemicals to be delivered simultaneously. The benefit of this delivery system in the field or greenhouse is that conventional seeding machines can be used (Redenbaugh et al., 1984; Rowe, 1985).

EXAMPLES OF SUCCESSFUL CELL SUSPENSION CULTURE

Carica papaya

Ovules have been used as explants, in a callus induction medium of MS (with half concentration of macroelements) or WH medium, with 176 mM sucrose, 20% v/v filter sterilized coconut water, and 2.76 mM glutamine, at pH 5.7. Somatic embryogenesis occurred in cell suspensions of *C. stipulata* in these media with 2.2 μM BA, 0.54 μM NAA, and 10 mg/l activated charcoal. The suspended cells were incubated on 100-rpm shakers to regenerate embryos (Litz and Conover, 1980, 1982).

Liquidambar styraciflua

Hypocotyls from aseptic seedlings have been used as explants. The callus induction medium was modified Blaydes medium containing 58.7 mM sucrose, 0.7% agar, and 0.55 mM inositol, 0.44–7.0 μM BA, and 0.54–5.4 μM NAA. It was prepared in a liquid form. Cell suspensions were isolated from callus and incubated on 90-rpm shakers until embryos formed. The embryos were then transferred into modified Blaydes solid medium without hormones to initiate buds and roots (Sommer et al., 1980).

Santalum album

Young shoots of sandalwood from 20-year-old trees were used as explants. For callus induction, the explants were placed first into

MS + 9.0 μM 2,4-D + 2.3 μM KIN, at pH 6.0. Suspended cells isolated from callus were incubated in liquid MS + 0.29–5.8 μM GA to regenerate embryos. When endosperm was used as explant, callus was still induced in MS medium with 4.5–9.0 μM 2,4-D, 2.2–8.8 μM BA, and 5.4 μM NAA. The pH was 6.0. Embryos were regenerated from suspended cells in MS medium + 1.3 μM BA, 5.7 μM IAA, 1 μM GA, 1.4 μM KIN. Finally, the embryos were rooted in solid White medium without hormones, and complete plants were formed. These experiments showed that the combination of high concentrations of minerals and GA plays an important role in induction of embryogenesis. In sandalwood, differentiation of embryos occurred on a medium supplemented with 1.45–5.8 μM GA (Lashmi Sita et al., 1979, 1980a, b).

Citrus sinensis

Calli were induced first in BM medium, and freely suspended cells were produced by macerozyme treatment. For this procedure, BM medium was supplemented with 1% macerozyme plus 0.6 M of sucrose, mannitol, and sorbitol. The suspended cells were incubated on a 120-rpm shaker for 6 hr at 25°C, the macerozyme-containing medium was removed, and the cells were cultured in BM medium. The differentiation medium was BM + CM 10% agar medium. Intact plantlets were regenerated in BM medium containing 2.9 μM GA (Button, 1975). Spiegel-Roy and Vardi (1984) transferred MT medium containing 0.055 M galactose to calli. To increase the size of embryos, they were transferred to MS suspension culture medium containing 0.055 M galactose, 30 μM GA, and 500 mg/l malt extract. After 5 weeks the embryos were transferred to MT medium with 0.06 M sucrose, 30 μM GA, and 50 mg/l malt extract. After 5 weeks the cotyledonary embryos were subcultured onto the MT medium with 0.15 M sucrose, 0.1 mM ADE, 3 mM GA, and 0.7% agar for development of shoots and roots.

Pinus

Shoot tips of seedlings have been used as explants. Cell suspension was cultured in LS medium plus 5.4 μM NAA and 4.6 μM KIN on a 1-rpm shaker. Embryos were obtained by using this medium.

EXAMPLES OF MUTANTS ISOLATED FROM CELL CULTURES

Much work has been performed on selection of mutants from cultured cells. Most of this research has concentrated on selecting mutants resistant to amino acids or their analogues, nucleic acid base analogues, antibiotics, nitrogen metabolites, and auxotrophs. A *p*-fluorphenylalanine-resistant cell line has been selected in *Acer pseudoplatanus* L. (Gathercole and Street, 1976), and an aminopterin-resistant cell line has been selected in *Datura innoxia* (Mastrangelo and Smith, 1977). In *Rosa damascena*, a mutant resistant to ultraviolet light has been isolated (Murphy et al., 1979).

Stable salt-tolerant callus lines and embryos regenerated in *Citrus sinensis* and *C. aurantium* have been selected. Embryos regenerated from salt-tolerant callus cells survived and grew better in saline medium than did embryos regenerated from the original salt-sensitive callus (Kochba et al., 1982). Mutants that are useful to agriculture mainly include those resistant to pathogenic toxins, pathogens, and herbicides and tolerant to salt and cold environments.

Mutants resistant to fungal and bacterial toxins may be selected in cell culture, using disease-inducing toxins. Fungal and bacterial toxins may be glycosides, polysaccharides, terpenes, phenols, amino acids, or a combination of several compounds. Pathogenic toxins may be host-specific or non-host-specific. Host-specific toxins have a high biological activity only in the host, e.g., leaf blotch of apple and lemon; non-host-specific toxins can harm both the host and other taxonomically similar species (Zhou, 1983).

Preparation of Materials

Suspended cells capable of plant regeneration and having a stable chromosome number should be selected as materials. If the aim is to select recessive mutants, it is better to use haploid cell lines. The growing index of materials should be maintained, so suspended cells should be subcultured every 3–4 days. The cells should be filtered to remove large cell masses (Flick, 1983).

Mutagenesis

Ultraviolet light, gamma-ray, ethylmethane sulphonate (EMS), ethyleneimine (EI), N-methyl-N^1-nitro-N-nitrosoguanidine (NTG), and sodium azide can be used for mutagenic treatment. However, mutants can also be produced under in vitro culture without any mutagenic treatment. Flick (1983) has shown that of mutants isolated in vitro, 54 were isolated by using mutagens and 70 were obtained without any mutagen. Half the lethal dose of mutagens is usually used. After the treatment, the mutagenic agent should be washed away (by rinsing three times at least).

Buffering Growth of Cells

After mutagenic treatment, the cells should be grown for several generations before in vitro selection. Better results can be achieved from cells grown for 3–4 days to permit expression of the mutants.

Selection and Preparation of Toxin

The selected pathogenic bacteria are first cultured, then crude toxin extracts are prepared by using filtration and sterilization. They are then diluted into different concentrations and the selective medium containing different concentrations of toxin is prepared.

Preparation of Selective Media

Selective media can be prepared by using different concentrations of one toxin or a mixture of several toxins (Zhou, 1983). To select an appropriate concentration for the treatment, different concentrations of toxin can first be injected into the plant, then an estimation can be made from the response to the infection (Wang, 1985). The mutagenized cells can be cultured in liquid medium first, then transferred to solid plate medium or can be cultured directly on solid plate medium (Flick, 1983).

Selection System for Screening Disease-Resistant Mutants

Disease-resistant mutants can be screened by continuously increasing toxin concentration in the selective medium. The cultured cells may also be transferred to other toxin containing media to obtain mutants resistant to several diseases (Zhou, 1983).

Plant Regeneration and Field Evaluation

The selected cells are proliferated and plants regenerated in the selective medium. The plants are then planted in the field, inoculated with toxin, and evaluated by conventional methods. Disease-resistant mother trees are asexually propagated, and further evaluation of disease resistance is conducted in the field. Finally, new varieties with disease resistance and other superior economic characteristics are produced.

REFERENCES

Ammirato, P. V. 1983. Embryogenesis. In: *Handbook of Plant Cell Culture*. Vol.

1 (D. A. Evans, W. R. Sharp, P. V. Ammirato, and Y. Yamada, eds.), pp. 82–123. Macmillan, New York.

Button, J. and Botha, C. E. J. 1975. Embryogenic maceration of *Citrus* callus and the regeneration of plant from single cell. *J. Exp. Bot.* **26**:723–729.

Cytological Lab of Shanghai Institute of Plant Physiology, Academia Sinica. 1978. Research techniques on cytology and biochemistry. In: *Plant Tissue and Cell Culture*. pp. 48–57. Shanghai Science and Technology Press, Shanghai.

———. 1978. Growth of cultured cells. In: *Plant Tissue and Cell Culture*. pp. 73–107. Shanghai Science and Technology Press, Shanghai.

Evans, D. A., Sharp, W. R., and Flick, C. E. 1981. Growth and behavior of cell culture: Embryogenesis and organogenesis. In: *Plant Tissue Culture Methods and Applications in Agriculture*. (Thorpe, T. A., ed.), pp. 45–113. Academic Press, New York.

Flick, C. E. 1983. Isolation of mutants from cell culture. In: *Handbook of Plant Cell Culture*. Vol. 1 (D. A. Evans, W. R. Sharp, P. V. Ammirato, and Y. Yamada, eds.), pp. 393–441. Macmillan, New York.

Gathercole, R. W. E. and Street, H. E. 1976. Isolation, stability, and biochemistry of a *P*-fluorophenylalanine-resistant cell line of *Acer pseudoplatanus* L. *New Phytol.* **77**:29–41.

Gray, D. 1981. Fluid drilling of vegetable seeds. *Hortic. Rev.* **3**:1–27.

Kochba, J., Ben-Hayyim, G., Spiegel-Roy, P., Saad, S., and Neumann, H. 1982. Selection of stable salt tolerant callus cell line and embryos in *Citrus sinensis* and *C. aurantium*. *Z. Pflanzenphysiol.* **106**:111–118.

Komamine, A. and Nomura, K. 1986. Mechanism of somatic embryogenesis in a high frequency embryogenesis system from single carrot cells. *VI Int. Congr. Plant Tissue Cell Cult.* (D. A. Somers, B. G. Gengenbach, D. D. Biesboer, W. P. Hackett, C. E. Green, eds.) University of Minnesota, Minneapolis, 158 (abstract).

Lakshmi Sita, G., Raghava Ram, N. V., and Vaidyanathan, C. S. 1979. Differentiation of embryos and plantlets from shoot callus of sandalwood. *Plant Sci. Lett.* **15**:265–270.

———, Shobha, J., and Vaidyanathan, C. S. 1980a. Regeneration of whole plants by embryogenesis from cell suspension culture of sandalwood. *Curr. Sci.* **49**:196–198.

———, Raghava Ram, N. V., and Vaidyanathan, C. S. 1980b. Triploid plants from endosperm culture of sandalwood by experimental embryogenesis. *Plant Sci. Lett.* **20**:63–69.

Litz, R. E. and Conover, R. A. 1980. Somatic embryogenesis in cell culture *Carica stipulata*. *Hortic. Sci.* **15**:733–735.

——— and Conover, R. A. 1982. In vitro somatic embryogenesis and plant regeneration from *Carica papaya* L. ovular callus. *Plant Sci. Lett.* **26**:153–158.

Lutz, J. D., Wong, J. R., Rowe, J., Tricoli, D. M., and Lawrence, R. J. 1985. Somatic embryogensis for mass cloning of crop plants, pp. 105–116. In: *Tissue Culture in Forestry and Agriculture*. (R. R. Henke, K. W. Hughes, M. J. Constantin, and Hollaender, A., eds.), Plenum Press, New York.

Mastrangelo, I. A. and Smith, H. H. 1977. Selection and differentiation of aminopterin resistant cell of *Datura innoxia*. *Plant Sci. Lett.* **10**:171–179.

Murphy, T. M., Hamilton, C. M., and Street, H. E. 1979. A strain of *Rosa damascena* cell resistant to ultraviolet light. *Plant Physiol.* **64**:936–941.

Redenbaugh, K., Nicho, J., Kossler, M. E., and Paasch, B. 1984. Encapsulation of somatic embryos for artificial seed production. *In Vitro* **20**:256.

Rowe, W. J. 1985. New technologies in plant tissue culture. In: *Tissue Culture as a Plant Production System for Horticultural Crops*. (R. J. Zimmerman, R. J. Griesbach, F. A. Hammerschlag, and R. J. Lawson, eds.) pp. 35–52. Martinus Nijhoff, Dordrecht, Netherlands.

Sommer, H. E. and Brown, C. L. 1980. Embryogenesis in tissue culture of sweet-gum *Liquidambar styraciflua* L. *Forest. Sci.* **26**:257–260.

Sondahl, M. R., Spahlinger, D. A., and Sharp, W. R. 1979. A histological study of high frequency and low frequency induction of somatic embryos in culture leaf explants of *Coffea arabica* L. *Z. Pflanzenphysiol.* **94**:101–108.

Spiegel-Roy, P. and Vardi, A. 1984. Citrus. *Handbook of Plant Cell Culture*. Vol. 3. (D. A. Evans, W. R. Sharp, P. V. Ammirato, and Y. Yamada, eds.), pp. 355–372. Macmillan, New York.

Steward, F. C., Mapes, M. O., and Mears, K. 1958. Growth and organized developed of cultured cells: II. Organization in cultures grown from freely suspended cells. *Am. J. Bot.* **45**:705–708.

Styer, D. J. 1985. Bioreactor technology for plant propagation. *Basic Life Sci.* **32**:117–130.

Wang, J. J. 1985. Isolation of cell mutants: I. Isolation of disease-resistant mutants. In: *Methods of Plant Tissue Culture*. pp 140–143. Botanical Society of China, Beijing.

Warren, G. S. and Fowler, M. W. 1977. A physical method for the separation of various stages in the embryogenesis of carrot cell cultures. *Plant Sci. Lett.* **9**:71–76.

Ye, H. C. 1985. Plant cell suspension culture. In: *Methods of Plant Tissue Culture*. (Botanical Society of China, ed.), pp. 170–176. Botanical Society of China, Beijing.

Zhou, J. P. 1983. Isolation of disease-resistant mutants at cell level in higher plants. *Hereditas* **5**(6):46–48 (in Chinese).

Zhou, J. Y. 1981. Embryoids produced from somatic plant cells in vitro: I. In vitro embryogenesis of somatic plant cells. *Acta Phytophysiol. Sin.* **7**(4):389–397 (in Chinese).

———. 1982. Embryoids produced from somatic plant cells in tissue cultures: II. Factors affecting the initiation and development of plant embryoids. *Acta Phytophysiol. Sin.* **8**(1):91–99 (in Chinese).

Zuo, Q., Chen, Y., Li, S., and Wang, M. 1982. Application of density gradient centrifugation in isolated pollen culture of rice. *Annual Report of the Institute of Genetics, Academia Sinica*, Beijing University Press, p. 31 (in Chinese).

CHAPTER 5

Protoplast Culture and Fusion in Perennial Crops

Wenbin Li and *Zhenghua Chen**

Protoplasts of higher plants have been comprehensively studied since they were first isolated by Cocking in 1960. To date, complete plants have been regenerated from protoplasts of about 60 species (Binding et al., 1984). Somatic hybrid plants have also been obtained by use of protoplast fusion (Gamborg et al., 1981; Schieder and Vasil, 1980; Vasil et al., 1979). The combination of protoplast culture and genetic engineering is opening vast opportunities for genetics and plant breeding. At present, increased attention is being directed to protoplasts of woody plants, though there are only a few successful examples (Krikorian, et al. 1988).

POTENTIAL OF PROTOPLAST RESEARCH IN PERENNIAL CROP BREEDING

In comparison with herbaceous plants, perennial crops are woody with highly lignified stem tissue. Breeding new varieties of forest or fruit

*English translation by Xiaxian Zhou.

trees requires a longer time. In *Citrus sinensis*, for example, about 10–12 years elapses from sowing to first bloom. Ten additional years is required before the yield and quality of fruit can be fully evaluated. An even longer period is required for backcrossing hybrid progeny to stabilize parental traits. If protoplast fusion is used, hybrid plants may be obtained within a single year, and protoplasts from the hybrid plant can be fused again with protoplasts of either parent. Somatic backcrosses can be achieved more rapidly than sexual hybridization. Chromosome elimination has been documented in a wide variety of somatic hybrids (see Evans, 1983).

Protoplast fusion has been used successfuly to overcome hybrid incompatibility in herbaceous plants. It may be extremely important for breeding perennial crops, in which cross-incompatibility is common and synchronization of flowering period of the parents is very difficult to obtain. Moreover, the sexual organs of some perennial crops degenerate, making sexual hybridization impossible. Protoplast fusion could also be used for transferring specific desirable genes from wild species. For instance, the commercially important species *Pinus resinosa* (cultivated widely in North America) has a limited range of intraspecific genotypes and is incompatible with other species. If interspecific crosses could be achieved through protoplast fusion, useful traits, such as disease resistance, could be transferred into *Pinus resinosa*. Protoplast fusion might also be used for sexual hybridization of triploids, which are usually sterile.

Somaclonal variation has been detected in plants regenerated from protoplasts (Shepard et al. 1980). Somaclonal variation may involve changes in chromosome number, chromosome structure, and nucleotide sequence as well as in important agricultural characteristics, such as disease resistance. Without doubt the variation frequency is higher in protoplast-derived plants than in plants derived from bud culture because cultured protoplasts are exposed for a longer time to the culture environment. The production of plants infected with microorganisms may be preventable because of the completely single-cell origin of plantlets regenerated from protoplasts. Protoplast fusion clearly has the potential to improve existing varieties directly.

Protoplast fusion has an important role in genetic engineering of perennial crops. The basic techniques of genetic engineering isolate a piece of exogenous DNA and combine it with the recipient's genome via an engineered vector. Expression of exogenous DNA in the recipient plants is the goal of gene engineering. Protoplasts are desirable recipients for engineered vectors because they lack cell walls. The character determined by the exogeneous DNA can be expressed when plantlets are regenerated from the protoplasts.

STATUS OF PROTOPLAST RESEARCH IN PERENNIAL CROPS

Studies of protoplast isolation, culture, and fusion in perennial crops have covered 30 genera, including forest trees (conifers and broadleaf trees), fruit trees, medicinal trees, and other economically important plants. The majority of the studies have concentrated on protoplast isolation (Table 1). Protoplasts of some tree species have been cultured (Table 2), and various results have been obtained. Regenerated plantlets were obtained by culturing protoplasts from nucellar calli of two varieties of *Citrus sinensis* L. (Nucellar Shamouti and Shamouti Landau), three varieties of *C. aurantium* L. and one variety of *C. reticulata* Blanco. (Murcott, Dancy, and Ponkan), one variety of *C. paradisi* Macf. (Duncan), and one variety of *C. lemon* L. Burm (Villafranca) (Vardi et al., 1975, 1982, 1983). Variants were obtained from the protoplast-regenerated plantlets. Regenerated plants have also been successfully obtained from suspension cultures of *Santalum album* protoplasts by Rao et al. (1984). Regeneration using fused protoplasts of perennial crops has seldom been reported (Table 3).

KEY POINTS OF PROTOPLAST CULTURE

The leaves of perennial crops, either seedlings or adult trees, usually do not survive removal of their lower epidermis. The kinds and concentration of enzymes which are used to separate cells and degrade cell walls are thus different from those used for herbaceous plants. Attention

TABLE 1. Protoplast Isolation in Perennial Crops

Species	Reference
Acer pseudoplatanus	Rona and Grignon, 1972
Pseudotsuga menziesii	Winton et al., 1975
Pinus Taeda	Winton et al., 1975
Pinus echinata	Winton et al., 1975
Tsuga heterophylla	Winton et al., 1975
Populus x euramericana	Saito, 1980a
Paulownia taiwaniana	Saito, 1980a
Leucana leucocephala	Venketeswaran and Ganhi, 1980
Sapium sebifera	Venketeswaran and Ganhi, 1980
Copaifera multijuga	Venketeswaran and Ganhi, 1980
Betula pendula	Steinhauer et al., 1980
Populus tremuloides	Verma and Wann, 1983
Ulmus sp.	Dorion et al., 1983
Larix decidua	Ahuja, 1984b
Quercus petraea	Ahuja, 1984b
Hevea	Othman, 1980

TABLE 2. Protoplast Culture in Perennial Crops

Species	Culture Result	Reference
Pseudotsuga menziesii	Cell mass	Kirby and Cheng, 1979 Kirby, 1982
Pinus Pinaster	Callus	David and David, 1979
Hevea	Cell division	Li Wenbin et al., 1983
Lycium chinensis	Callus	Sun Yongru et al., 1982
Citrus sp.	Plantlet regen.	Vardi and Spiegel-Roy, 1982
Alnus glutinosa	Callus	Huhtinen et al.. 1982
Alnus indica	Cell mass	Huhtinen et al., 1982
Coffea sp.	Cell mass	Orozco and Schieder, 1983
Pinus contorta	Cell mass	Hakman and von Arnold, 1983
Betula platyphylla	Cell division	Smith and MacCown, 1983
Rhododendron sp.	Cell division	Smith and MacCown, 1983
Populus tremula	Cell mass	Ahuja, 1983
Populus tremuloides	Cell mass	Ahuja, 1983
Fagus sylvatica	Cell division	Ahuja, 1983
Malus pumila	Callus	Niizeki et al., 1983
Santalum album	Plantlet regen.	Rao and Bapat, 1984
Actinidia chinensis	Callus	Cai Qigui et al., 1985 (personal communication)

TABLE 3. Protoplast Fusion in Perennial Crops

Species	Reference
Paulownia taiwaniana + *Populus euramericana*	Saito, 1980b
Citrus sp. + *Citrus* sp.	Vardi et al., 1983
Populus tremula + *Populus tremuloides*	Ahuja, 1984a
Populus tremula + *Fagus sylvatica*	Ahuja, 1984a

should be directed to selection of appropriate plant materials and to type and concentration of enzymes used if large amounts of protoplasts with high viability are to be obtained.

Selection of Donor Materials for Protoplast Isolation

Donor materials are generally divided into two types. One type includes cultivated or wild plants and plants grown in greenhouse or in pots. The other type is derived aseptically, including explant-derived callus, suspension-cultured cells, plantlets regenerated from tissue culture, as well as aseptic plantlets germinated from seeds in test tubes.

Materials should be chosen as follows:

1. Select materials with the potential for morphological differentiation, from which plant regeneration is readily induced by protoplast culture. This is most important.
2. Select materials which yield large quantities of protoplasts after enzyme treatment.
3. The protoplasts should be stable after enzyme treatment and during experimental manipulation.
4. The protoplasts obtained should be highly viable, vigorous, and able to divide continuously.

Natural and artifically cultured materials are evaluated as follows:

Young leaves of naturally growing plants are desirable materials for protoplast isolation. They are abundant and easily collected. Large numbers of protoplasts with a high percentage of viable protoplasts can be obtained from young leaves under appropriate conditions. It is difficult to isolate perennial crop protoplasts if the leaf is already fully expanded; instead, only young leaves should be used for protoplast isolation. Leaves of *Ulmus campestris* were used for protoplast isolation by Dorion et al. (1983). They isolated the protoplasts of the first (younger), second, and third (older) leaves from the top. It was found that the protoplast yield of the first leaf was higher than that of the second leaf, and the protoplast yield of the second leaf was again higher than that of the third. The viability of protoplasts from the young leaf was much greater than for the old ones (Fig. 1). Ahuja (1984a) found that protoplasts with high yield and activity could be isolated only from young leaves. However, the age of the maternal plant was irrelevant. Numerous viable protoplasts could be obtained from the young leaves of mature (more than 70 years old) beech (*Fagus sylvatica*) trees. The division of regenerated cells was observed in culture.

Ahuja (1984a) proposed a protocol for preparing the original protoplast materials. Using quaking aspen (*Populus tremuloides*) as an example, in late winter and early spring branches 40–50 cm long bearing dormant buds were taken from adult trees and stored at 4°C for several months. Prior to protoplast isolation, the branches were dipped in tap water after the two ends of each branch were cut off. For 8–10 days, the branches were kept at 22–24°C, light intensity of 3400–4000 lux, and humidity of 60–70%. The first leaf to emerge would yield protoplasts without the use of high enzyme concentrations, because of its thin cutin layer. With this protocol, young leaves for protoplast isolation could be obtained throughout the year. This protocol could also be applied to other tree species (Ahuja, 1984a).

Although large numbers of protoplasts can be obtained from young leaves, their potential to redifferentiate after dedifferentiation in culture

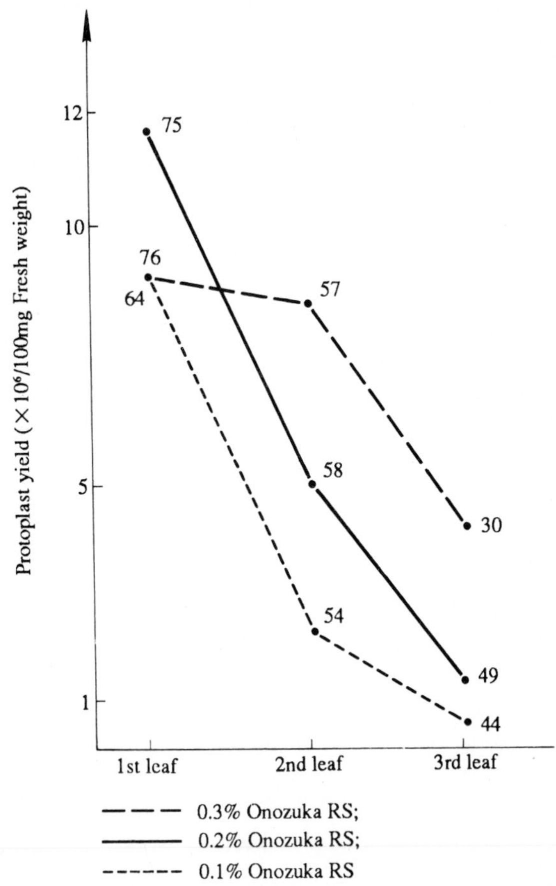

Figure 1. Protoplast yield as affected by position (and age) of leaf.

has not been proved for perennial crops. To date, no plantlet has been regenerated from the leaf-derived protoplast of a perennial crop. This is a problem for further investigation.

The other naturally growing materials—the stems, roots, etc.—are not suitable for protoplast isolation.

Cotyledons of aseptic seedlings are also good materials for protoplast isolation. This material need not be infiltrated by bacteriocides because it does not require surface sterilization treatment. High yields have been obtained in isolating protoplasts from cotyledons. Isolation of protoplasts from cotyledon, hypocotyl, young root, old root, young leaf, and callus derived from young stems was compared in *Populus simonii* by Zhang Yiwen et al. (1981). With appropriate enzyme concentrations (1.5%

TABLE 4. Protoplast Yield of Materials from Different Parts of *Populus simonii*

	Protoplast Yield (%)				
Material	**2 hr**	**4 hr**	**6 hr**	**8 hr**	**10 hr**
Cotyledon	88	99	—	—	—
Hypocotyl	0	0	—	—	—
Young root	Very few	40	43	—	—
Old root	0	0	0	—	—
Young leaf	0	Very few	18	—	—
Callus	—	68	84	92	51

cellulase, 0.75% pectinase), cotyledons gave large quantities of protoplasts in a short time, more than can be derived from young leaves and callus (Table 4).

David et al. (1979) isolated many protoplasts from cotyledons of *Pinus pinaster*. Callus was formed after culture. Protoplasts isolated from cotyledons may meet the requirements of activity, yield, and stability. However, for young leaves, totipotency (organ redifferentiation) has not yet been demonstrated. In addition, the preparation of aseptic seedlings is quite complicated. Cotyledons are small and many seeds must be used to obtain large numbers of protoplasts.

Explant-derived callus and suspension cultures are commonly used as materials in protoplast research on perennial crops. In vitro methods have the following advantages: they are not affected by environmental factors; results are reproducible; protoplast yield, activity, and stability are good; and protoplast differentiation potential can be evaluated by using other experiments in tissue culture, such as explant culture, organ culture, and pollen culture, as well as cell culture. If the explant-derived callus of a given tree species can be induced to organogenesis, this suggests that the callus-derived protoplast may have a higher differentiation potential. Vardi et al. (1975) used the early experience in regenerating plantlets from *Citrus* nucellus culture as the basis for the first regeneration of plantlets from protoplasts in perennial crops. They used nucellus-derived callus as source material. Suspension cultures were successfully used as materials for protoplast isolation in *Santalum album* by Rao et al. (1984).

Experience in tissue culture can, of course, only be used for reference; plant species which are successful in tissue culture will not necessarily be successful in protoplast culture too. For example, wheat among herbaceous plants and poplar among perennial crops have been very successful in anther culture, but their protoplast culture is still difficult.

It is possible to obtain a higher yield of protoplasts by using calli or suspension cultures than by using young leaves or cotyledons as materials. Othman et al. (1980) compared young leaves and suspension cultures in *Hevea* and found a much higher yield of protoplasts from suspension culture than from young leaves.

Both callus and suspension cultures usually must be subcultured for several generations to be suitable for protoplast isolation. Othman et al. (1980) found that suspensions which had undergone more generations of subculture gave higher protoplast yields than those experiencing only a few subcultures. Experiments with *Lycium chinensis* (Sun Yongru et al., 1982) in which young leaf- and stem-derived callus was subcultured every 7 days demonstrated that the callus was suitable for protoplast isolation only after eight generations of subculture. The potential for morphological differentiation would probably decline after so many subcultures. This is a significant problem in isolating protoplasts from callus and suspension cultures. However, it can be solved by careful selection of materials for callus induction. Young embryo tissues, for instance, are a good choice, as they are outstandingly capable of division and differentiation.

In conclusion, young leaves, cotyledons, calli, and suspension cultures have been used as materials for protoplast isolation. Each of these materials has advantages and disadvantages. The latter two materials are widely used at present. In addition to the materials discussed, pollen and pollen mother cells have also been employed for protoplast isolation.They are seldom used since the necessary protocol is quite complicated and collecting the source materials in large quantities is difficult.

The Use of Enzymes

The correct selection of enzymes, enzyme concentrations, and enzyme combinations affects protoplast yield and quality of protoplasts, such as activity and stability.

Cellulase, hemicellulase, pectinase, macerozyme, and driselase are enzymes commonly used for cell isolation and cell wall degradation. Enzymes made in Japan are widely used in China.

Cell walls could be removed and protoplasts released from herbaceous plants when two to three of these enzymes were used. Because perennial crop cell walls are different from those of herbaceous plants, the kinds and concentrations of enzymes used for cell wall degradation and protoplast isolation must also be different than those used for herbaceous plants. Different source materials require different enzyme compositions. Figure 1 shows that in a single enzyme solution, young leaves yielded far more protoplasts than adult materials.

Table 5 lists enzyme solutions used in protoplast isolation from perennial crops. Cellulase was used in concentrations of 0.5–3%, as for herbaceous plants. Pectinase was used more often for perennial crops than

TABLE 5. Enzymes Used in Protoplast Isolation of Some Perennial Crops

Material	Cellulase	Macerozyme	Pectinase	Hemicellulase	Driselase	Mannitol (M)	pH	Reference
Young leaf (*Santalum album*)	2% R-10	0.5% R-10	0.5% Sigma	0.5% Sigma	—	0.55	5.6–5.8	Lakshmi Sita et al., 1983
Callus (*Santalum album*)	1% R-10	—	0.5% Sigma	—	—	0.55	5.6–5.8	Lakshmi Sita et al., 1983
Young leaf (*Ulmus campestris*)	0.1–0.2% RS	—	0.01–0.03% Y-23	—	0.05%	0.6	5.6	Dorion et al., 1983
Young leaf (*Populus tremuloides*)	0.5% R-10	0.1% R-10	—	—	—	0.7	5.6	Ahuja, 1983
Young leaf (*Coffea*)	3% R-10 or 2230	—	0.5% Y-23	—	—	0.6	5.8	Orozco and Schieder, 1983
Young leaf (*Fagus sylvatica*)	0.1% R-10	0.5% R-10	0.1% Serva	—	—	0.7	5.6	Ahuja, 1984b
Callus (*Lycium chinensis*)	0.5% R-10	0.2% R-10	—	—	0.2%	0.6	5.6	Sun Yongru et al., 1982
Anther-derived callus (*Malus pumila*)	2% R-10	—	0.1% Y-23	—	—	0.7	5.8	Niizeki et al., 1983
Cotyledon (*Pinus Pinaster*)	0.8–1% R-10	—	0.04–0.5% Sigma	0.04–0.5% Rhozyme	0 or 0.5%	0.6–0.7 Glucose		David and David, 1979

for herbaceous plants. Macerozyme, which is largely used for herbaceous plants, was employed for only a few perennial crop species.

Solutions for protoplast isolation—in addition to the components concerned with cell isolation and cell wall degradation, inorganic salts, or chemicals such as $CaCl_2$, KH_2PO_4, and dextran sulfate potassium salt—were necessary to strengthen stability and activity and to reduce protoplast breakage.

In some experiments, enzyme treatments were in two steps. Isolated cells were first obtained from cultures treated with an enzyme solution in which pectinase predominated. Protoplasts were obtained from these cells after treatment with enzyme solution in which cellulase predominated. Sometimes the cultures were treated with many steps. Tables 6 and 7 show the solutions used to isolate protoplasts from *Populus alba* L. seedlings by Verma and Wann (1983). Figure 2 shows the protocol for protoplast isolation. The protocol also consisted of two steps in enzyme treatment. However, the first step resulted in an undesirable isolation of single cells, while high yield of protoplast resulted from the second treatment with enzyme solution. By this method, protoplasts were obtained at the rate of 5.4×10^6/g of fresh tissue.

David and David (1979) found in protoplast isolation from *Pinus pinaster* cotyledons that the composition and concentration of the enzyme solution affected not only yield of protoplasts but spontaneous protoplast fusion (Table 8).

As mentioned previously, kinds, concentration, and combinations of enzymes required for protoplast isolation from perennial crops are more complicated than those for herbaceous plants. This increased complexity extends to the conditions of enzyme treatment.

The isotonic agent in the enzyme solution and its concentration significantly affect the results of protoplast isolation. Mannitol was the most commonly used isotonic agent; sorbitol and glucose could also be used. The osmotic concentration varied, depending on the particular plant spe-

TABLE 6. Solution Used for Dissolving Enzymes in Protoplast Isolation of *Populus alba* L. Seedlings

Component	Concentration
$NaH_2PO_4 \cdot H_2O$	0.7 mM
$CaCl_2 \cdot 2H_2O$	6.8 mM
$Na_3C_6H_5O_7 \cdot 2H_2O$	5.8 mM
Ampicillin	0.4 mg/ml
Bovine serum albumin	1.0 mg/ml
Mannitol	0.7 M
MES buffer	0.9 mM

TABLE 7. Enzymes Used To Isolate Protoplasts from *Populus alba* L. Seedlings

Enzyme	Concentration (%)	
	Enzyme Solution I	Enzyme Solution II
Macerozyme R-10	0.2	0.2
Macerozyme	0.2	0.2
Pectinase (Sigma)	0.2	0.2
Rhozyme HP-150	0.1	1.0
Cellulase R-10	—	1.0

cies, the material, even the physiological state of the material (see Table 5). In general, the osmotic concentration in the enzyme solution, if a little bit higher than that in medium, increases protoplast stability.

Protoplast Culture

One of the principal goals in protoplast research is to regenerate plantlets from protoplast culture. The selection of the medium culture methods and the control of culture conditions are important.

MEDIUM. There are limited experimental data on protoplast culture of perennial crops. Regenerated plants have been obtained only from protoplast cultures of *Citrus* and *Santalum*. It is quite difficult to discuss the role of medium composition in protoplast culture of perennial crops.

Studies of plant tissue culture have been carried out for several decades. Experience involving tissue culture media is useful for protoplast culture. The media commonly used in protoplast culture must often be modified according to the material and condition. Dorion et al. (1983) found that the survival rate of *Ulmus* protoplasts increased when half-strength inorganic salt MS medium was used with 13.6 mM $CaCl_2 \cdot 2H_2O$ added to the basal medium. For culturing protoplasts from *Coffea* leaves, Orozco and Schieder (1983) used A-43 as a basal medium, containing high concentrations of potassium. They modified the experiment to completely remove KCl or to replace it with 15 mM NH_4Cl.

Table 9 shows media used for protoplast culture of some perennial crops and notes modifications to the media. Table 10 shows inorganic salt components in the basal media.

For herbaceous plants, the concentration of inorganic salts in protoplast culture media should usually be higher than in media for tissue culture or cell culture, and an increase in the concentrations of Ca^{2+} and Mg^{2+} and a decrease in NO_3^- concentration strengthen protoplast stability and promote cell division. These guidelines are probably applicable to

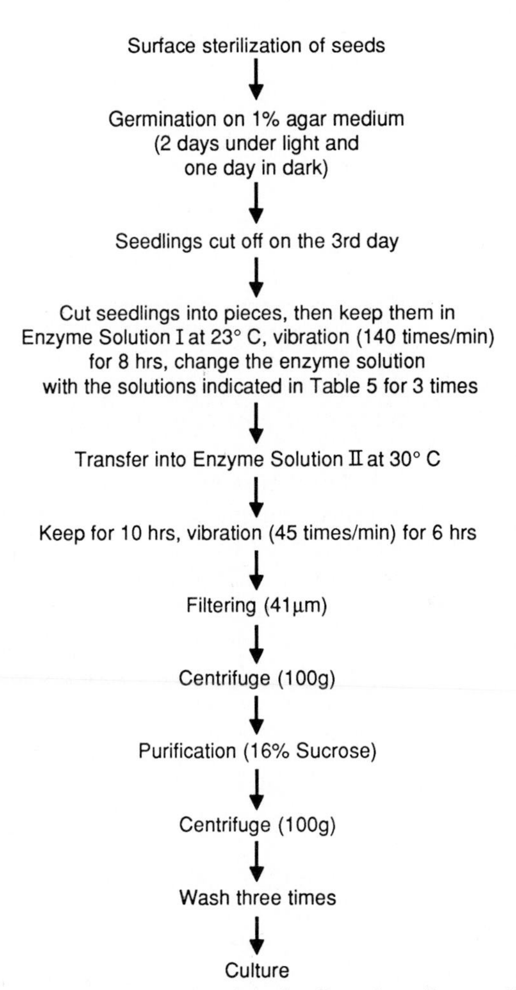

Figure 2. Protoplast isolation procedure for *Populus alba* seedlings.

TABLE 8. Effect of Enzyme Solutions on Spontaneous Protoplast Fusion in *Pinus Pinastrer*

Enzyme Concentration	Nuclear Types (%) (1000 Protoplasts Calculated)		
	Mononuclear (%)	Binuclear (%)	Multinuclear (%)
0.5 Cellulase R-10 0.5 Driselase 0.5 Pectinase (Sigma) 0.5 Hemicellulase	83	10	7
1 Cellulase R-10 0.5 Pectinase (Sigma) 0.5 Hemicellulase	93	5	2
0.08 Cellulase R-10 0.04 Pectinase (Sigma) 0.04 Hemicellulase	98	2	0

perennial crops as well. Table 10 shows that the concentrations of inorganic salts in the medium are comparatively low and that of Ca^{2+} high, except in MS medium.

Organic substances and plant hormones are also important. Most organic supplements are vitamins, though lactalbumin hydrolysate, yeast extract, and coconut water are added to many media. Appropriate concentrations and combinations of plant growth regulators favor the continuous division of cells and their organogenesis.

The osmotic concentration of the medium and its pH value affect culture efficiency. Table 11 shows the effect of osmotic concentration on protoplast survival rate in *Ulmus* (Dorion, 1983).

CULTURE CONDITIONS. The effects of protoplast density, illumination, temperature, and other environmental factors on protoplast culture are discussed below.

Protoplast Density in Inoculation. It is known from herbaceous protoplast culture that the protoplast density in culture affects the division of regenerated cells. Both high and low protoplast densities are unfavorable to cell division. Densities of 10^4–10^6 protoplasts/ml are often used. This reflects the interrelationships between cells. Where no specific selection system is incorporated, mechanical methods such as micromanipulation are used to select hybrid cells after protoplast fusion, and the cell density should be appropriately reduced to a certain degree. At the same time, the undesirable effect of low cell density may be complemented by the modification of culture medium and culture procedure. For example, Kao's medium could be supplemented with high levels of vitamins, amino acids, and organic acids to give good results in low-

TABLE 9. Media Used in Perennial Crop Protoplast Culture

Material and Tree Species	Basal Medium	Plant Hormone (μM)	Isotonic Agent Concentration	Organic Supplements (μM)	Result of Culture	Reference
Suspension cell (*Pinus contorta*)	Kao et al. (1974) or K_3 (Nagy and Maliga, 1976) $CaHPO_4$ removed, + 2 mM $CaCl_2 \cdot 2H_2O$	2.3 2,4-D, 27 NAA, 1.8 BA	0.4 M Glucose	30 Thiamine, 6.0 pyridoxin, 3.6 vitamin F	Cell masses	Hokman and Arnold, 1983
Young leaf (*Alnus glutinosa* and *A. incanus*)	N_6 macro-elements, MS micro-elements	2.3 KIN, 9 2,4-D	0.5 M Mannitol	7.2 Vitamin F, 0.4 biotin, 28.4 vitamin C, 1.1 folic acid, 6.0 pyridoxin, 15 thiamine	Callus and cell masses	Huhtinen et al., 1983
Nucellar callus (*Citrus* sp.)	BM (with same inorganic salts as in MS)		0.3 M Mannitol, 0.3 M sucrose	30 Thiamine, 59 pyridoxin, 18 vitamin F, 26.6 glycine, 500.0 malt extract	Regenerated plantlet	Vardi et al., 1975
Cotyledon (*Pinus pinaster*)	MS + 4mM $CaCl_2 \cdot 2H_2O$	16.4 NAA, 4.4 BA, or 33 NAA, 8.8 BA	0.6 M Glucose or 0.7 M glucose	Kao's sugars (1979) (sucrose excluded), vitamins, and organic acids, 20 mg/l coconut water, 125 mg/l CH, 10 μM glutamine	Callus	David and David, 1979
Anther-derived callus (*Malus pumila*)	8P	0.9 2,4-D, 2.3 ZEA, 5.4 NAA	0.7 M Mannitol	8P	Callus	Niizeki et al., 1983
Leaf (*Coffea* sp.)	A-43 15 mM NH_4Cl added, and/or KCl removed	4.5 2,4-D, 2.2 BA	0.7 M Mannitol, 0.6 M sucrose	18 Vitamin F, 3.0 pyridoxin, 1.5 thiamine, 1.1 folic acid, 2.0 biotin	Cell masses	Orozco and Schieder 1983
Callus (*Santalum album*	MS	9.0 2,4-D, 0.9 BA	0.55 M Mannitol	0.3 Thiamine, 3.0 pyridoxin, 1.8 vitamin F, 26.6 glycine	Cell masses	Lakshmi et al., 1982
Callus (*Lycium chinensis*)	D_2	8.1 NAA, 3.5 BA	0.4 M Glucose, 0.05 M sucrose	12 Thiamine, 24 pyridoxin 14.4 vitamin F, 18.6 glycine, 0.2 biotin, 0.9 folic acid, 5% coconut water	Callus	Sun Yongru et al., 1982

TABLE 10. Basal Media for Perennial Crop Protoplast Culture

	Basal Medium					
Component	MS	D_2	A-43	K_3	8P	N_6
In mM						
NH_4NO_3	20.6	3.4	10	3.1	7.5	
KNO_3	18.8	14.7	19.8	24.7	18.8	28.0
$CaCl_2 \cdot 2H_2O$	3.0	6.1	8.8	6.1	4.0	1.1
KH_2PO_4	1.25	0.6	1.5		1.25	3.0
$Na_2HPO_4 \cdot H_2O$			1.1			
$CaHPO_4 \cdot 2H_2O$			0.3			
$MgSO_4 \cdot 7H_2O$	1.5	3.2	1.6	1.0	1.2	0.75
$(NH_4)_2SO_4$	1.0					
KCl			20.1		4.0	
$FeSO_4 \cdot 7H_2O$	0.1	0.1	0.05	0.1		0.1
Na_2EDTA	0.1	0.1	0.05	0.1		0.1
Sequestrene 330Fe	0.1					
In μM						
$MnSO_4 \cdot H_2O$	100	22 ($\cdot 4H_2O$)	220 ($\cdot 4H_2O$)	60	20 ($\cdot 4H_2O$)	
$ZnSO_4 \cdot 7H_2O$	30	6 ($\cdot 4H_2O$)	86 ($\cdot 4H_2O$)	7	7	5
H_3BO_3	100	32.2	322	48	48	26
KI	5	1.5		4.5	4.5	5
$Na_2MoO_4 \cdot 2H_2O$	1	0.4	2	1	1	
$CaCl_2 \cdot 6H_2O$	0.11	0.07		0.11	0.11	
$CuSO_4 \cdot 5H_2O$	0.1	0.06		0.1	0.1	

TABLE 11. Effect of Osmotic Concentration on Protoplast Survival Rate

Concentration of Mannitol (M)	0.4	0.5	0.6	0.7
No. of viable protoplasts ($\times 10^5$/100 mg fresh weight)	19	32	32	28
Survival rate	49	64	60	59

density cultures. In protoplast culture of perennial crops, the culture efficiency was affected by protoplast density during inoculation. Table 12 shows the effect of protoplast density of *Citrus* on the formation of cell masses. The best results were achieved at densities of 10^5 and 8×10^4 protoplasts/ml (Vardi et al., 1975).

Illumination and Culture Temperature. Illumination is a vital factor in the life of green plants; temperature affects metabolic rates and process. These factors also affect protoplast culture. Culture room temperatures of 24–28°C are generally suitable except for temperature-sensitive species. Early in a given protoplast culture experiment, relatively weak continuous illumination is better; after the formation of cell masses—especially during the induction of organogenesis—illumination should be strengthened. Table 13 gives illumination and temperatures for the in-

TABLE 12. Effect of *Citrus* Protoplast Density on Formation of Cell Masses

	Protoplast Density in Inoculum (No. of Protoplasts/ml of Medium)						
	4×10^5	2×10^5	10^5	8×10^4	6×10^4	4×10^4	2×10^4
Experiment 1	—	—	5655 ± 345*	5064 ± 231	3382 ± 227	2100 ± 182	864 ± 115
Experiment 2	17955 ± 643	11927 ± 400	7355 ± 318	—	—	3609 ± 223	0

*No. of cell masses that appeared 4 weeks after culture contained in 1 ml of medium ($\bar{X}$ ± SD).

TABLE 13. Illumination and Temperature for Protoplast Culture of Perennial Crops

Species	Illumination	Temperature(°C)
Citrus L.	Continuous scattering light	26 ± 1
Lycium chinensis	Continuous weak illumination, less than 300 lux	25
Populus alba L.	Continuous scattering light	25
Coffea L.	Darkness	25

duction of division of regenerated cells in protoplast culture of some perennial crops.

Protoplast Fusion

Protoplast isolation and culture are only the first steps in protoplast research. Protoplast fusion and various genetic manipulations of protoplasts really open a broad prospect for the study of genetics and breeding of perennial crops.

To date, experiments have been carried out on interspecific protoplast fusion in very few perennial crop species (see Table 3) and no experiments on culture of hybrid cells have been reported.

In herbaceous plants, protoplast fusion has been used to form somatic hybrids between incompatible species. There are many methods for the induction of protoplast fusion. The polyethyleneglycol (PEG) method, as modified by Kao and Michayluk (1974), is often used. This method has been adopted in protoplast fusion experiments carried out in perennial crops. Kao and Michayluk demonstrated that high concentrations of calcium ion and high pH values may increase the fusion frequency, which was used to develop a simple and effective method for fusion (Li Hsiang-hui et al., 1978).

1. Dilute the two kinds of cleaned, isolated protoplasts in washing liquid (Table 14) to a density of 1×10^6/ml protoplasts, then mix equal

TABLE 14. Solutions Used for Protoplast Fusion

Component	Washing Liquid	PEG Solution	Diluting Solution
$CaCl_2 \cdot 2H_2O$	3.5 mM	10.5 mM	50 mM
KH_2PO_4	0.7 mM	0.7 mM	—
NaOH	—	—	50 mM
PEG 1500	—	0.33 M	—
$NH_2CH_2CO_2H$	—	—	50 mM
Mannitol	0.5 M	0.1 M	0.4 M
pH	5.6	5.6	10.5

volumes of these two protoplast suspensions, pipetting gently and slowly to ensure complete mixing.

2. Drip the mixed protoplast suspension (0.1–0.2 ml) onto a cover slip (22 × 22 mm^2) in the center of a 6-cm petri dish and leave it for 5–10 min so that the protoplasts sediment onto the cover slip.
3. Drop 0.2 ml PEG solution (Table 14) carefully and slowly onto the protoplast drops and leave it for 5–20 min until the protoplasts are in close contact and are sticky from the PEG. The appropriate duration of PEG treatment can be determined by viewing under the microscope.
4. Carefully add 0.4 ml eluting solution (high Ca and high pH; see Table 14) for 5 min to enhance the fusion frequency and to dilute the PEG.
5. Using a total volume of 10-ml washing liquid, wash 4 or 5 times at 5-min intervals, adding the washing liquid by pipette at one side of the cover slip and absorbing it from the other side.
6. Wash twice with medium in the same way at intervals of 5 min.
7. Add approximately 5 ml medium to form a thin liquid layer and to resuspend the treated protoplasts, seal the petri dish, and use the microscope to record the fusion frequency.
8. Put the petri dish into the culture room at 26°C under weak light.

Protoplasts of different colors or with morphological variations are frequently used in fusion experiments. For instance, leaf-derived (green and filled with chloroplasts) protoplasts are used with culture-derived protoplasts (white, with an obviously increased number of cytoplasmic filaments in the central vacuole) or white root protoplasts. With such combinations, the heterokaryons are easily identified under the microscope.

Many experiments have demonstrated that the PEG method results in a high frequency of heterokaryons. Under appropriate conditions, somatic hybrid plantlets can be formed from cultured heterokaryons. However, this method is rather complicated and takes much time. In recent years, the method of protoplast electrofusion has been developed. Protoplasts were place in contact with each other under an electric field and fused by adding electric pulses. As compared with the PEG method, electrofusion offers the following advantages: rapid and highly synchronized fusion; control of process and quantity of fused protoplasts, since the electric field and electric pulse intensity are tightly controlled; prevention of the effects of chemicals on the protoplasts; achievement of a higher fusion frequency. Although this method has not yet been used in experimental protoplast fusion of perennial crops, plantlets have been regenerated from tobacco cells fused through this method by Kohn et al. (1985). The method is as follows:

EXPERIMENTAL MATERIALS. Two complementary nitrate reductase-deficient suspension cell lines of *Nicotiana tabacum* were used. Hybrid cells were obtained after electrofusion. The complementing hybrid cells could directly use nitrate as the sole nitrogen resource and develop after culture into somatic hybrid plantlets.

The isolated protoplasts were washed with 0.6 M mannitol solution and then suspended in 0.6 M mannitol solution on ice at a density of 10^4 protoplasts/ml for fusion treatment.

ELECTROFUSION APPARATUS. The fusion chamber consisted of a bonded glass and plastic slide with two parallel platinum wires (300 μm apart) placed between them. The center of the plastic slide had a 1.0-cm-wide cylindrical cutout which fit over a section of the platinum wires (Fig. 3). This cylindrical well held the protoplasts dispersed in mannitol solution for the electrofusion treatment. The fusion chamber was connected to a digital display generator (Hewlett-Packard Model 4204 A) and pulse generator (Hewlett-Packard Model HP 214 B). The former generated the AC field across the electrodes for dielectrophoretic collection of the protoplasts; the latter supplied the single high-intensity DC square wave pulses necessary for protoplast fusion. The electrical output signals of the oscillator and pulse generator were monitored with an oscilloscope.

ELECTROFUSION PROCEDURE. Two drops (approximately 0.1 ml) of each protoplast preparation was introduced into the fusion chamber and subjected to a weak nonuniform electric AC field at a frequency of 0.9 megahertz (MHz) to align the protoplasts parallel to the field lines. After 20 sec, the AC field was reduced to 4% of its original value and the high-intensity DC square wave pulse was applied. Ten seconds after pulse application the residual weak AC field was removed. After 3 min, the fusion products were transferred to a plastic dish (60 mm in diameter). The fusion chamber was rinsed with mannitol solution, which was added to the plastic dish. This ensured complete removal of protoplasts

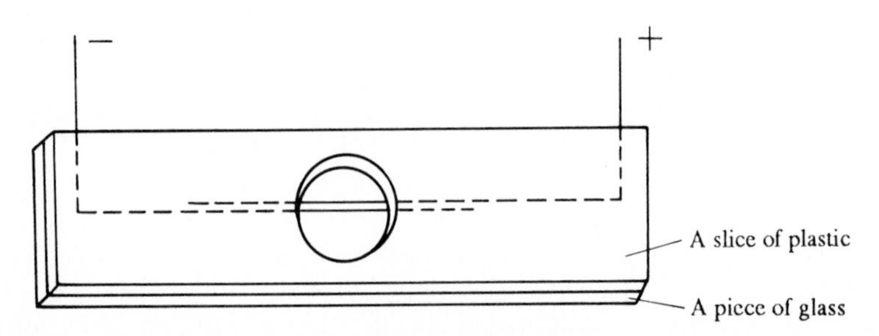

Figure 3. Fusion chamber with parallel platinum wire electrodes.

and fusion products from the fusion chamber. Then the previously enriched medium was added in a ratio of 1:1 v/v to the suspension of protoplasts and fusion products and gently mixed for culture.

DISCUSSION. The dispersed protoplasts were polarized under the influence of the 0.9-MHz AC field and formed "pearl chains" as they gravitated in the direction of higher field strength, randomly aligning with each other parallel to the electric field lines. The minimum-intensity electric field for the formation of pearl chains was 200 V. The protoplast membranes might be torn if the electric field intensity increased to more than 400 V.

The AC was reduced to 4% of the original value, to maintain pearl chain formation, by switching a resistor (1000 Ω) into the line immediately before application of the DC pulse.

Pulse intensity and pulse duration affected fusion rates. When the pulse duration was kept constant at 50 μsec, the fusion rate increased with increasing pulse intensities from 1 to 1.2 kV/cm; the maximum fusion rates were found between 1.2 and 1.5 kV/cm. When the pulse intensity was kept at 1.5 kV/cm and the pulse duration gradually prolonged (10–80 μsec), fusion rates increased.

In fusion experiments, the fusion of two protoplasts is most desirable, but the number of multifusions often increased and the bicellular fusion rate decreased in the treatments with higher pulse intensity and longer duration. In the present experiment, the optimum pulse intensity and pulse duration were 1.2–1.5 kV/cm and 50 μsec, respectively. The fusion of two protoplasts occurred at a rate of 5–10%.

Efficient electrofusion could be achieved by some technical improvements such as modifications to the fusion chamber and electrode distance and use of a uniform alternating electrical field. Protoplast fusion techniques could be further improved by combining electrofusion with chemical fusion (Kohn et al., 1983).

After fusion treatments, there were products of fusion between cells of the same line and between different parents as well as the unfused parental protoplasts. Screening the hybrid cells is very important. In herbaceous plants, screening depends either on different sensitivity of the protoplasts to environmental conditions (temperature, illumination), nutrients, or drugs or on complementation of mutants or albinos after fusion. These methods are quite critical in selecting parental materials and have limited the application of cell fusion techniques. Sun Yongru et al. (1982) used morphological differences of callus for primary screening in protoplasts fusion experiments with various *Nicotiana* spp. Along with protoplast fusion between two tobacco species, parental protoplasts were cultured separately as controls. After culture, calli were formed both by fusion and by separate cultures. Protoplast-derived calli from the parents were remarkably different in color and in texture. Only some of the calli which formed after fusion treatment were of intermediate type as expected if they developed from hybrid cells. After regeneration into plantlets, they demonstrated hybrid cytological, morphological, and biochemical characteristics. This screening method was complicated but reliable, especially when there was no specific selection system.

Identification of a regenerated plantlet as a somatic hybrid should be performed continuously. At present, the following identification methods are used:

Morphological Identification. The stems, leaves, and flowers should be observed to determine whether the hybrid plantlet displays parental traits. If distant parents are used, the morphological character of the hybrid plant is not always the intermediate between the parents but inclines to one or the other. It is necessary to observe the variations, even very slight variations, such as the distribution and density of hair on the surface of the stem and leaf, as well as the size of stomata and their distribution on the lower leaf epidermis (Li Wenbin et al., 1983).

Cytological Identification. Somatic hybrid plantlets often have the sum of the parental chromosome numbers. Because the chromosomes of one parent are usually eliminated to a certain extent, numerous aneuploid cells are usually observed. Their chromosome numbers are often higher than those of either parent. For example, root tip chromosomes from the somatic hybrid between *Nicotiana glauca* ($2n = 24$) and *Petunia* × *hybrida* ($2n = 14$) numbered 32–38 (Sun Yongru et al., 1982).

Isozyme Identification. The isozyme zymogram of the somatic hybrid should be different from those of the parents. In most cases, it has the bands of both parents; sometimes it has additional new bands. Somatic hybrid plantlets of *Nicotiana glauca* with *Petunia* × *hybrida* each had different peroxidase isozyme bands, but also had the bands of the parents (Sun Yongru et al., 1982).

Identification at Molecular Level. Nucleic acid fragments from the parents, if discovered in the hybrid, demonstrate that the hybrid was produced by fusion. After protoplast fusion the chromosomes of one parent are often lost during cell division; some fragments of DNA on one chromosome may be integrated into another chromosome. The hybrid character could then be identified at the molecular level.

For instance, RUBPCase of higher plants contains two subunits, large (LS) and small (SS). LS is maternally inherited and located on the chloroplast DNA. SS shows mendelian inheritance controlled by the nucleus. RUPBCase is synthesized in the ribosomes and transferred to the chloroplast. It can be used as a phenotypic marker of chloroplast DNA and the nuclear genome in analyzing the genetic constitution of the somatic hybrid. Genetic recombination might be proved if the LS or SS of both parents were found in RUBPCase of the hybrid plantlet (Komarnitsky and Gleba, 1981).

Other Genetic Manipulations of the Protoplast. In addition to protoplast fusion, which can combine two chromosome sets into one cell, other genetic manipulations of protoplasts, including transfer of genetic information of one plant into the cell of another species, are possible.

Protoplasts can take up exogenous nuclei, chromosomes, organelles, and DNA fragments. This might be accomplished by cloning genetic sequences and using vectors such as plant DNA viruses or the *Agrobacterium tumefaciens* Ti plasmid. Protoplasts might be also transformed by exogenous DNA.

Even in the case of herbaceous plants, genetic engineering is only at the preliminary stage; it has not been performed in perennial crops at all.

CONCLUSION

In recent years, much progress has been made in perennial crop tissue culture. The lag behind research on herbaceous plants is diminishing. Protoplast research of perennial crops is receiving increased attention. The fundamental characteristics of perennial crops' long life cycle, size, and high lignification lead to difficulties in studies of genetics and breeding and thus encourage researchers to work with protoplasts.

Although protoplast studies of perennial crops have just begun, they could provide wide possibilities for perennial crop improvement—growth rate, stem form, and wood texture, as well as cold, frost, disease, and pest resistance—and the yield and quality of specific metabolites could all be manipulated.

REFERENCES

Ahuja, M. R. 1983. Developmental potential of mega and normal protoplasts in *Populus*. In: 6th International Protoplast Symposium. Basel (Potrykus, I., et al., eds.), *Experientia Suppl.*, **45**:28–29.

———. 1984a. Protoplast research in woody plants. *Silvae Genet.* **33**(1):32–37.

———. 1984b. Short note: Isolation and culture of mesophyll protoplasts from mature beech trees. *Silvae Genet.* **33**(1):37–39.

Binding, H., Nehls, R., and Jorgensen, J. 1982. Protoplast regeneration in higher plants. In: *Plant Tissue Culture* (A. Fujiwara, ed.) pp. 575–578. Japanese Association for Plant Tissue Culture, Tokyo.

David, A. and David, H. 1979. Isolation and callus formation from cotyledon protoplasts of pine (*Pinus pinaster*). *Z. Pflanzenphysiol.* **94**:173–177.

Dorion, N., Godin, B., and Bigot, C. 1983. Isolation and culture of leaf protoplasts from *Ulmus* sp: Preliminary report. 6th International Protoplast Symposium. Basel (Potrykus, I. et al., eds), *Experientia Suppl.*, **45**:8–9.

Gamborg, O. L., Shyluk, J. P., and Shahin, E. A. 1981: Isolation, fusion, and culture of plant protoplast. In: *Plant Tissue Culture Methods and Applications in Agriculture* (T. A. Thorpe, ed.), pp. 115–153. Academic, New York.

Hakman, I. C. and von Arnold, S. 1983. Isolation and growth of protoplasts from cell suspensions of *Pinus contorta* Dougl. ex Loud. *Plant Cell Rep.* **2**:92–94.

Huhtinen, O., Honkanen, J., and Simola, L. K. 1983. Ornithine- and putrescine-supported divisions and cell colony formation in leaf prototoplasts of alders (*Alnus glutinosa* and *A. incana*). *Plant Sci. Lett.* **28**:3–9.

Kao, K. N. and Michayluk, M. R. 1974. A method for high frequency intergeneric fusion of plant protoplasts. *Planta* **115**:355–367.

Kohn, H., Schieder, R. and Schieder, O. 1985. Somatic hybrids in tobacco mediated by electrofusion. *Plant Sci.* **38**:121–128.

Komarnitsky, I. K. and Gleba, Y. Y. 1981. Fraction I protein analysis of parasexual hybrid plant *Arabidopsis thaliana* + *Brassica campestris. Plant Cell Rep.* **1**:67–68.

Krikorian, A. D., Cronauer-Mitra, S. S., and Fitter Corbin, M. S. 1988. Protoplast culture of perennials. *Sci. Hortic.* **37**:277–292.

Lakshmi Sita, G. and Shobba Rani, B. 1983. Preliminary studies in isolation and culture of protoplasts from sandalwood (*Santalum album*). In: 6th International Protoplast Symposium. Basel (Potrykus, I. et al., eds), *Experientia Suppl.* **45**:4–5.

Li, H., Li, W., and Huang, M. 1982. Plant regeneration from intergeneric hybrid cell between B_6S_3 and *Petunia* W43 and expression of LpDH. *Sci. Sin. (B)* **3**:223–228 (in Chinese).

———, Yen, C., and Li, W. 1978. Preliminary studies on protoplast fusion of wheat and *Petunia. Proc. Plant Tissue Cult. 1978*, Science Press, Beijing, 351–355.

Li, W., Sun, Y., Huang, M., and Li X. 1983. Further observation of somatic hybrid plants between *Nicotiana glauca* and *Petunia hybrida. Acta Genet. Sin.* **10**(3):194–196 (in Chinese, English abstract).

Niizeki, M., Hidano, Y., and Saito, K. 1983. Callus formation from isolated protoplasts of apple (*Malus pumila* Mill.) *Jap. J. Breed.* **33**(4):369–374.

Orozco, F. J. and Schieder, O. 1983. Isolation and culture of coffee leaf protoplasts. *6th Int. Protoplast* Symp., (Potrykus, I., et al., eds.), *Experientia Suppl.*, **45**:52–53.

Othman, R. B. and Paranjothy, K. 1980. Isolation of *Hevea* protoplasts. J. Rubb. *Res. Inst. Malaysia* **28**(2):61–66.

Rao, P. S. and Bapat, V. A. 1984. Regeneration of somatic embryos and plantlets in protoplast culture of sandalwood tree (*Santalum album* L.). *International Symposium on Genetic Manipulation in Crops, Oct. 22–26, 1984, Beijing, China*, p. 93 (abstract).

Rona, I. P. and Grignon, C. 1972. Obtention de protoplastes a partir de suspensions de cellules d'*Acer pseudoplatanus. C. R. Acad. Sci. D.* **274**:2976–2979.

Saito, A. 1980a. Isolation of protoplasts from mesophyll cells of *Paulownia*

growth cycles are long and their cultivation difficult, the concerns mentioned previously are particularly pressing. Some woody medicinal plants die after their medicinal components are harvested. For example, the medicinal part of *Eucommia ulmoides* Oliv. is the phloem, and the plant rarely survives when it is harvested. The propagation of woody medicinal plants and breeding of new varieties with high yield, high efficacy, and good vigor are urgent. There has been much research on tissue culture of herbaceous and woody medicinal plants, and it is clear that tissue culture techniques will be crucial for protection of resources and breeding of new varieties in medicinal perennial crops.

Tissue Culture of Genus *Lycium*

The fruit of *Lycium chinense* is a tonic in traditional Chinese medicine. It is used as an aphrodisiac, a means of improving vision, and a tonic for liver and kidney. Large quantities are used in various pharmaceuticals. At the same time, because it is not very expensive, unlike ginseng and Chinese caterpillar fungus, the demand is high. The tissue culture of *Lycium* is quite easy, and many scientists are currently involved in research on *Lycium*. Many recent advances have been made in this field.

TISSUE CULTURE OF LEAVES AND SHOOT TIP OF *LYCIUM*. (Cheng and Guo, 1980; Niu, 1983) Disease-free leaves were taken from a plant that had never bloomed. They were surface sterilized by immersion in 0.02% mercuric chloride in 70% alcohol for about 20 seconds, 5% antiformin solution for 4–5 min, and washed in sterile water three times. The leaves were cut transversely into 1-cm segments and inoculated into MS medium with 87.6 mM sucrose, 9.3 μM KIN, 9.0 μM IAA or 18.6 μM KIN, and 18μM IAA. Callus gradually grew at the cut surfaces; it was white and compact after 1 month of culture. The callus was transferred to MS medium with 2.2 μM BA or 1.1 μM BA and 1.1 μM 2,4-D for another month, and a white compact mass of callus was formed. This callus was cut into 3–4-mm segments and transferred to MS medium with 2.2 μM BA or 2.2 μM BA + 1.5 μM IAA, on which buds differentiated. When the buds had grown to 1–2 cm in height, they were cut and transferred to MS medium with 0.5 μM IBA or NAA with sucrose reduced to 43.8 mM. After about 2 months, roots emerged from the bud bases. Plantlets which had grown in test tubes for about a month could be transferred into vermiculite soaked with MS macronutrient solution. If segments with axillary buds were cut from the plantlets and transferred to MS medium with 0.5 μM IBA and 29.6 μM thiamine, they rapidly produced roots and formed a new plantlet after another month. This resulted in a fivefold increase in plantlets every month. Many plantlets could be produced in a year. Yang (1980) placed leaf pieces from field and in vitro plants, as well as lower leaf epidermis (ca. 0.3 cm), of *Lycium chinense* on MS medium with 5.4 μM NAA, and after 3–5 days callus emerged at the cut surface of the leaves and the epidermal margins. After 12–30 days, green buds differentiated from the callus, and after 30–45 days 10–30 buds, but sometimes more than 80 buds, were produced per flask. The total theoretical propagation rate was thus (1×10^8)–(2×10^8)/year. When the differentiated buds were transferred to hormone-free MS medium, intact plantlets were formed within 2 weeks. They were transferred to pots with fertile soil or vermiculite and grew rapidly.

Niu et al. (1983) placed hypocotyls from sterile seedlings and stem tips from new *Lycium barbarum* shoots on MS medium. Plantlets were regenerated with high frequency and survived after transplantation. A suitable hormone concentration for inducing callus was 1.1 μM NAA. The induction frequency of callus from hypocotyl was 95.7%; from stem tip, 62.5%. Suitable hormone concentration for differentiation of hypocotyl callus was 0.88 μM BA or 4.4 μM BA + 2.7 μM NAA. The induction frequency was 100% for both treatments as every callus differentiated buds. A suitable hormone combination for differentiation of stem tip callus was 8.8 μM BA + 1.1 μM NAA, and the induction frequency was again 100%.

CULTURE OF ANTHERS AND UNPOLLINATED OVARIES. The aim of anther and unpollinated ovary culture is to obtain haploid plants. The importance of anther culture and haploid plants in breeding, especially of perennial crops, has been emphasized (Chen, 1984). These techniques are still not widely used with medicinal perennial crops.

Anther Plants from Lycium barbarum. (Gu, 1981) Flower buds at the middle to late-middle uninucleate stage of pollen development were cut from the plant at the end of May or beginning of June. The buds were first soaked in 70% alcohol for several seconds, transferred into a 10% antiformin solution for 7–8 minutes for surface sterilization, and washed three times with water. The buds were then inoculated. The cultures were at a temperature of 28 ± 1°C, illuminated for 10 hr/day. The basal medium was MS medium containing 87.7 mM sucrose. The optimum combination of hormones was 4.4 μM BA and 0.54 μM NAA. The anthers immediately produced embryos or formed callus which differentiated into buds. Examination of root tip cells indicated a haploid chromosome number of 12 in plants regenerated from anther culture.

Regeneration of Plants from Unpollinated Ovaries of Lycium barbarum. (Niu et al., 1983) Ovaries from flower buds with pollen at the uninucleate stage or the tetrad stage of development were used for inoculation. The flower bud was surface-sterilized, and the ovaries were removed and inoculated onto MS or modified MS medium (NH_4 doubled; NO_3 decreased to one-half). A suitable combination of hormones was 4.4 μM BA and 2.7 μM NAA. The frequency of callus formation was 80–83%. Callus differentiated into buds immediately in the initial culture medium. After the buds were transferred to hormone-free MS medium, roots grew rapidly. Five ovary-derived plants were examined, of which one was $2n = 24$ (diploid), and the other four were all $2n = 48$ (tetraploid). No haploids were found. The ploidy of ovary plantlets should be studied further, but the large number of tetraploid plants obtained from ovary culture will nonetheless be very useful in breeding. We considered that these tetraploids might be completely homozygous, because the material for inoculation was stored at 2–4°C for 48 hr. Low temperature inhibits spindle formation during cell division, possibly resulting in chromosome doubling. The plants regenerated from materials which had not been treated at low temperature were all diploids.

Endosperm Culture. Endosperm cells are usually triploid. If triploid plants could be obtained from endosperm culture, it would be possible to obtain medicinal plants with large fruits with thin peels, thick pulp, and little or no seed, so the drug quality would be increased. Intact plants were obtained from *L. barbarum* endosperm culture by Wang et al. (1984). *L. barbarum* endosperm obtained after about 20 days of bloom was inoculated onto MS medium with 87.7 mM sucrose, 0.75% agar, 500 mg/l CH, 0.9 μM 2,4-D in the dark. White loose callus was induced with a frequency of 24% after 10 days. Callus was transferred onto MS medium with 0.88 μM BA and cultured in the light (2000 lux, 10–12 hr/day) for differentiation. Buds differentiated from the callus after 40 days. Each mass of callus differentiated up to three to four buds, and the overall differentiation frequency was 77%. Buds 1 cm long were cut and inoculated on MS medium without hormone. Roots were produced from the buds after about 10 days. The regenerated plants grew very well when transplanted. Cytological examination of six endosperm plantlets revealed that their chromosome number varied greatly. There were many aneuploid cells in these plants. The chromosome numbers in four of the plants ranged between 15 and 24 (close to the diploid number); chromosome numbers in the other two plants varied between 28 and 44.

Cell Culture. Isolation and culture of single higher plant cells, establishment of stable cell lines, and regeneration of plants from cell culture constitute a very useful experimental system. It may be used in breeding for mutant screening and is the basis for genetic engineering at the cellular level. Since cells are isolated from each other, the influence of chimeras can be prevented in mutation breeding. Hence, mutant cells can be selected immediately and used to regenerate plants. The establishment of cell lines and the regeneration of plants from single cell cultures have been achieved in *L. barbarum* by Niu et al. (1985). Sterilized seeds were germinated under aseptic conditions. When the hypocotyls grew to about 1 cm in length, they were cut into 5-mm segments and inoculated onto MS medium with 0.9 μM 2,4-D, 500 mg/l CH, and 87.7 mM sucrose. Callus formed after 10 days with an induction frequency of 95.7%. After callus was subcultured once in the initial culture medium, it was subcultured in liquid suspension culture to establish cell lines and then was frequently subcultured. The medium for suspension culture is identical to that for inducing callus, but without agar. After a few subcultures, the cells could be transferred back onto a solid medium, and callus would develop again. The callus was maintained by subculture. The fresh weight and volume of the suspended culture increased about fourfold after 6–7 days. The material entered into a dormant stage after 10 days and turned brown after 12–14 days. Subculture period longer than 10 days was, therefore, unsuitable. Normal diploids ($2n = 24$) are prevalent and stable in suspension cells. Suspension culture was reinitiated after 23 transfers. On the fourth day, it was filtered through a nickel filter of 100 mesh, and a single-cell suspension with average density of 4.5×10^4 cells/ml was obtained. More than 99% of the cells were single cells with only a few aggregates of 2–3 cells in the fluid. When the cell suspension was cultured on a rotation bed at 4 rpm cells began to divide after 2 days. On the third day cell aggregates formed. When these aggregates were transferred to solid culture medium, many of the aggregates grew into callus. The callus was subcultured once and transferred to MS medium with 1.32 μM BA, on which buds differentiated. The buds rooted and formed intact plants on MS medium with 1.1 μM NAA. Temperature could be used to control differentiation of cells in suspension. Suspended cells grew normally at a temperature of 24–26°C. When the temperature was kept at 30–32°C, embryos were produced in the liquid suspension. This discovery improved the experimental system, and mutant screening was more effective because the differentiation frequency of plantlet formation via embryos was increased.

Protoplast Culture. Protoplasts of *L. chinense* were cultured by Sun et al. (1982). Young leaves from young stem cuttings were cut into strips 3 mm wide and inoculated onto solid MS medium with 9.0 μM 2,4-D and 1.4 μM KIN. At a temperature of 25°C and a light intensity of 3000 lux, white, compact callus arose after about 10 days. CH (500 mg/l) was added to the culture medium and used for subculture. Callus was transferred once every 15–20 days. After about 4 months of subculture, the callus became very loose and soft, suitable for protoplast isolation. The enzymes used for isolation of protoplasts were 0.5% cellulase R-10

(Onozuka), 0.2% macerozyme R-10, 0.2% driselase, and 0.6 M mannitol. Seven to ten days after subculture, callus was transferred to the enzyme solution at a temperature of 25°C for 16 hours. Protoplasts were readily isolated. The enzyme solution in which the protoplasts were suspended was filtered with a nickel wire net (400 mesh) to remove small cell masses and cells which had not formed protoplasts. The protoplasts were washed twice with 3.5 mM $CaCl_2 \cdot 2H_2O$, 0.7 mM KH_2PO_4, and 0.55 M mannitol. Protoplasts were finally suspended at a density of 5×10^4/ml in D_2 medium. At 25°C and a weak light intensity (300 lux), the first division of regenerated cells could be seen after 4 days, and cell masses were formed after 25–30 days. At this time, a D_2 culture medium with 58 mM sucrose instead of glucose was added to the liquid medium containing the cell suspension in a 1:1 ratio. The cell masses grew rapidly. After about a week, the cell masses were transferred to a solid culture medium, and after a month callus was formed from the mass. However, organ formation was not successfully induced. This may be related to the long period of callus subculture.

Plant Propagation of Stem Segments of *Maytenus hookeri*

Maytenus hookeri is a wild tropical tree and the source of an anticancer drug, maytansine. Its seed germinates with great difficulty. Rapid propagation of this species via in vitro culture is thus very important.

Chen et al. (1984) removed leaves from unlignified young stems. The stems were washed in sterile water, rubbed with 95% alcohol, soaked in a 0.1% mercuric chloride solution for 10 minutes, then washed again thoroughly with sterile water. The sterilized stem was cut into 1-cm segments under aseptic conditions and inoculated onto MS medium with 8.8 μM BA, 0.9 mM 2,4-D, and 87.7 mM sucrose. The culture temperature was 25–30°C and the light intensity 2000 lux for 10 hr/day. Clustered buds were produced after 29 days of culture, with a frequency of 43%. The bud clusters were subcultured into the same medium once every 15 days, and many clustered shoots were obtained. They could be continuously proliferated. When the shoots were 2–3 cm long, they were cut and transferred onto MS medium with 1.1 μM NAA and 1.1 μM IAA. After 17 days, the shoots began to produce roots; the rooting rate was 83%. The plantlets could be transplanted about 1 month after rooting. The survival rate after transplantation was 79%. After transplanting to the field, the plantlets grew vigorously. Under optimum conditions, the plantlets could bloom and fruit in the test tube.

Regeneration of Plantlets from Hypocotyls of *Eucommiaceae*

Zuo et al. (1980) took shoots of *Eucommia ulmoides* bearing only two cotyledons, which had grown in soil for 25 days. The radical and cotyledons were removed and the shoots were sterilized with 0.1% mercuric chloride for 1.5 min, then washed four to five times with sterile water. The hypocotyls were cut into 0.5–1-cm segments under aseptic conditions and inoculated onto MS medium with different hormone combinations. After 20 days buds differentiated from hypocotyl callus. The explants nearest the end of the cotyledon had the highest rate of callus formation. When 13.2 μM BA and 5.7 μM IAA was added to the medium, the differentiation frequency was 75%; hormone combinations of 8.8 μM BA and 2.9 μM IAA or 8.8 μM BA and 5.7 μM IAA and 1.4 μM GA resulted in bud differentiation frequencies of 52% and 42%, respectively. When the buds grew to 2 cm in height, they were cut down from the callus and transferred to MS medium containing 4.9 μM IBA. After 15 days roots formed and intact plants were obtained.

In addition to *Lycium, Maytenus*, and *Eucommia*, intact plantlets have also been induced from young stems of *Ginkgo biloba* (Luo, 1985); cotyledons, shoot tips, and hypocotyls of *Forsythia suspensa* (Qin et al., 1985); and stem tips of *Daphne odora* (Song et al., 1984).

TISSUE CULTURE OF AROMATIC PERENNIAL CROPS

Many investigations have focused on the cultivation of aromatic plants and increases in the quantity and quality of their products. The economic value of aromatic plants is much higher than that for other cultivars. Plant perfume is widely used and has a very large world market in which demand persistently exceeds supply. The yields of woody aromatic plants are high, their management is simple, and the period of profitability is long, making them particularly attractive investments. An example is the aromatic vine *Vanilla planifolia*. Its product vanilla is used in the production of cigarettes, wine, tea, cake, butter, candy, ice cream, drinks, etc. It is a necessary raw material in perfume manufacture and is the most widely used flavoring. In medicine it is used for production of stimulants and treatment of hysteria, menoxenia, and pyreticosis.

These aromatic perennial crops are scarce and difficult to propagate. Many countries, including China, have to buy raw or semifinished products from regions where the aromatic perennial crops are endemic. In

China there are some aromatic plant resources in tropical areas. However, their yield is too small, for the demand and development of plantations requires large quantities of seedlings. For the purpose of rapid plant propagation, many workers have, therefore, studied propagation of aromatic perennial crops by tissue culture, and some advances have been reported.

Propagation of *Vanilla planifolia* via Tissue Culture

In Wu et al., 1982, young autumn shoots of *Vanilla planifolia* were cut into 3–5-cm segments after surface sterilization and were immediately planted in MS medium with 3% sucrose. After 20 days, the stem bases swelled and formed yellowish callus which grew continuously. Callus growth reached its peak after 15–20 days. Experiments showed that additional auxin inhibited formation of callus. The appropriate sucrose concentration of differentiation medium was 58.4 mM. After 10 days, the callus became compact and green regions appeared when placed on differentiation medium. Young leaves appeared after 30 days; they were followed by bud growth and the appearance of stem nodes. The effects of various hormones were examined. In culture media with BA, callus grew rapidly and formed many root primordia, after which roots were produced. In culture media with ZEA, callus grew slowly at first. Some of the calli soon turned dark green and sprouted green buds. It was also found that MS medium with 0.44 μM BA and 2.7 μM NAA induced bud primordia on stem segment to produce large quantities of strong young shoots. When segments from these shoots were cultured, they also produced young buds, which were transferred to MS medium (half mineral salts) with 2.9 μM IAA. After 10–15 days, thick roots emerged from their bases and intact plants were formed.

Kononowiez and Janick (1984) also studied in vitro propagation of *Vanilla planifolia*. The surface of stem segments from greenhouse-grown *Vanilla* were sterilized, and then segments were inoculated onto MS medium with 0.3 μM thiamine, 3 μM pyridoxin, 0.55 mM inositol, 40.6 μM nicotinic acid, 26.6 μM glycine, 87.7 mM sucrose, and 0.8% agar. When the BA concentration was 4.4 μM, the buds could be regenerated and propagated. BA concentrations of 2.2 or 4.4 μM gave the highest propagation rate of buds during subculture. However, at 2.2 μM BA the buds grew very large and aerial roots were produced. After five rounds of subculture, the culture still maintained a fivefold propagation rate in the medium with 2.2 μM BA.

In Vitro Micropropagation of *Rosmarinus officinalis*

Rosmarinus officinalis is also an important aromatic perennial crop (Chaturvedi et al., 1984). After the surfaces of stem tips and segments from elite individual plants of *Rosmarinus officinalis* were sterilized, the explants were soaked in 100 mg/l polyvinyl-pyrrolidone solution for a time to prevent oxidation of phenolic substances and then inoculated onto modified White medium with 0.49 μM IBA, 14.4 μM GA, 28.4 μM ascorbic acid, and 65 μM glutathione to induce shoots. Within a week, all stem tips and 80% of the stem segments inoculated turned brown, and after 15 days axillary buds grew out. After subculturing onto the same medium, healthy shoots were obtained.

Stem tips from sterile shoots were inoculated onto modified MS medium supplemented with 43 μM adenine, 0.57 μM IAA, and 50 mg/l malt extract, and 0.44, 0.88, or 2.2 μM BA. Many buds differentiated on the cut ends. When the BA concentration was 0.88 μM the greatest number of buds formed after 30 days. The effect of cytokinin on bud induction was determined largely by the concentration of mineral salts in the modified MS medium. When BA concentration was held constant, low concentrations of mineral salt, especially KNO_3 (2.0 mM), KH_2PO_4 (0.6 mM), and $MgSO_4$ (0.8 mM), were more favorable for bud production than high concentrations (14.8 mM KNO_3, 1.1 mM KH_2PO_4, 1.44 mM $MgSO_4$). When shoot apices were subcultured in the same medium with redifferentiated buds, the proliferation rate was very high.

When regenerated shoots were isolated and transferred to modified MS medium with 1.4 μM IPA (indole proprionic acid) or IAA, 80% of them produced roots, but root growth was faster and more normal on medium with IPA than with IAA. If the shoots were transferred to auxin-free medium just after the roots appeared, the results were much better. When the plantlets reached 3 cm in height, they were transplanted to sterilized soil, with a survival rate of 60%. After transplanting for the first 10 days, a polyethylene cover increased plantlet survival.

Approximately 5000 plants could be produced from a single nodal explant in 1 year. Cultures of proliferating shoots provided about 10 shoots each month, and the mother cultures were also divided in two each time they were subcultured at monthly intervals. It took about 1 month to obtain a 3- to 4-cm-high plantlet in soil from aseptically rooted shoots.

Culture of Vegetative Buds of *Rosa rugosa*

Rosa rugosa is not only an ornamental but also an aromatic perennial crop with very high economic value (Huang et al., 1984). The essential oil extracted from flowers of *Rosa rugosa* is used as a raw material in high-grade cosmetics, flavorings, perfumes, and drugs. However, its flowers do not produce any seed, and vegetative propagation is very difficult and inefficient.

Lateral buds from perennial branches and terminal buds from annual branches were cut 3–5 mm long and used as explants. They were inoculated onto MS solid medium with 2.2–4.4 μM BA, 2.3–9.0 μM 2,4-D, 500 mg/l CH, 7.3 mM sucrose, and half the original quantity of mineral salts. The explants swelled on this medium but did not produce callus. They were light green in color. Thereafter clusters of buds were formed, each with at least 5 buds and up to a maximum of 20. Each cluster could be divided into smaller clusters and transferred to fresh medium once a month for continuous proliferation. Small shoots 3 cm long were cut and roots were induced after about 23 days. Normal plantlets appeared after 35 days.

These examples demonstrate that in vitro rapid propagation of aromatic perennial crops with high economic value has great potential. The advantages of this method are its low cost, efficiency, and economy of space and labor. In vitro propagation will play an important role in developing the cultivation and preserving the germ plasm resources of aromatic perennial crops.

KEY REFERENCES

Chen, Z. H. 1984. Advances in anther culture of woody plants. *Hereditas* **6**:34–37 (in Chinese).

Niu, D. S., Shao, Q. Q., Wang, L., Qin, J. S., Chen, S. P., Cao, S. F., and Wang D. Z. 1983. Callus induction and plant regeneration from hypocotyl, stem tip, and young ovary of *Lycium chinese* Mill. *Hereditas* **5**:24–26 (in Chinese).

———, Shao, Q. Q., Qin, J. S., Wang, L., and Wang, D. Z. 1985. Establishment of cell line and plant regeneration from single cell culture of *Lycium chinense* Mill. *Kexue Tongbao* **4**:296–298 (in Chinese).

Sun, Y. R., Li, W. B., Huang, M. J., and Li X. H., 1982. Callus from protoplasts of *Lycium chinense* Mill. *Acta Bot. Sin.* **24**:477–479 (in Chinese, English abstract).

REFERENCES

Chaturvedi, H. C., Misra, P., and Sharma, M. 1984. In vitro multiplication of *Rosmarinus officinalis* L. *Z. Pflanzenphysiol.* **113**:301–304.

Chen, Z. Y., Wan, J. L., and Men G. Y. 1984. Culture of stem segments of *Maytanus hookeri* Loes. *Plant Physiol. Commun.* **2**:38 (in Chinese).

Cheng, W. L. and Guo D. H. 1980. Callus induction and plant regeneration from leaf of *Lycium chinese* Mill. *Plant Physiol. Commun.* **6**:40–41 (in Chinese).

Fan, Y. H., Zang, S. Y., and Zhao, J. F. 1982. Anther culture and haploid plant induction on two species of genus *Lycium. Hereditas* **4**:25–26 (in Chinese).

Gu, S. R. 1981. Anther plant from *Lycium chinese* Mill. *Acta Bot. Sin.* **23**:246–248 (in Chinese, English abstract).

Huang, O. C., Dong, M. S., and Lin J. F. 1984. Culture in vitro of vegetative bud of *Rosa rugosa* Thunb. *Plant Physiol. Commun.* **3**:44 (in Chinese).

Kononowiez, H. and Janick, J. 1984. In vitro propagation of *Vanilla planifolia. Hortic. Sci.* **19**:58–59.

Lu, Z. J. 1985. Tissue culture of stem segments of *Ginkgo biloba* L. *Plant Physiol. Commun.* **1**:35–36 (in Chinese).

Qin, J., Wang, L., Chen, S. 1985. Induction of plantlets *in vitro* of *Forsythia suspensa* Vahl. *Plant Physiol. Commun.* **1**:36–37 (in Chinese).

Song, C. H. and Lian, W. 1984. Culture of stem tip of *Daphne odora* Thunb. *Plant Physiol. Commun.* **5**:39 (in Chinese).

Tian, L. X. and Hu, Z. H. 1981. Study on tissue differentiation from stem of *Eucommia ulmoides* Olive. *Xibei Bul. Bot. Res.* **7**:65–73 (in Chinese, English abstract).

Wang, L., Cheng, S. P., and Qin, J. S. 1984. Intact plant obtained from endosperm culture of *Lycium chinese* Mill. *Plant Physiol. Commun.* **2**:33 (in Chinese).

Wu, Z. P., Zhan, X. J., and Lu, L. F. 1982. Preliminary report of propagation by tissue culture on *Vanilla planifolia* Andr. *Subtropical Plant Commun.* **2**:26–27 (in Chinese).

Yang, N. B. 1980. Propagation test of plantlets by culture in vitro on medicinal plants of five species. *Plant Physiol. Commun.* **6**:44–46 (in Chinese).

Zhong, W. G., Liu, S. Z., and Yang, G. M. 1984. Preliminary report on tissue culture of *Magnolia officinalis* Rehd. et Wils. *Z. Hongcaoyo.* **15**:31–33 (in Chinese).

Zuo, C. F., Hou, Y. H., and Han, W. L. 1981. Preliminary report of tissue culture of *Eucommia ulmoides* Olive. *Z. Hongcaoyo.* **11**:474 (in Chinese).

CHAPTER 7

Tissue Culture of Perennial Ornamental Plants and Virus Elimination

*Wenda Qiu**

INTRODUCTION

Cultivation

Ornamental perennial crops are the major plants used for gardens and environmental decoration. Trees bring pleasure to people for many years once they are planted. Most ornamental perennial crops are highly adaptable and can be easily cultivated. Therefore, they are important in decorative gardening and other environments. Of the 500 species of cultivated ornamental plants in China, half are woody species.

Taxonomy

Ornamental perennial crops consist of foliage species and trees that flower, fruit, or bud. Included are 9 families of gymnosperm, such as

*English translation by Kaiwen Yuan.

Cycadaceae, Pinaceae, and Cupressaceae; 52 families of dicotyledons, such as Magnoliaceae, Ranunculaceae, Theaceae, Ericaceae, Rosaceae, Leguminosae, Thymelaeoceae, and Rutaceae; and 3 families of monocotyledon: Gramineae, Palmaceae, and Liliaceae. Therefore, most ornamental perennial crops are dicotyledon, some are gymnosperm, and only a few are angiosperm monocotyledons.

Drawbacks of Seed Propagation and Advantages of Tissue Culture

Seed propagation of ornamental perennial crops is generally limited to a few species of foliage trees. For flowering and fruiting trees with heterozygous genotypes, seed propagation does not maintain parental characteristics. Since seed propagation is time-consuming, the methods of cutting, layering, and grafting are used mostly for asexual propagation. However, propagation rates are low, and propagation can be difficult in some species. Furthermore, viruses can accumulate after many generations, leading to degeneration of specific characteristics. In most ornamental trees, cross-hybridization is difficult. Therefore, plant tissue culture techniques have great potential for propagating clones of ornamental trees, accelerating multiplication of rare germ plasm, creating virus-free plants, and developing new species.

Shoot tip culture and proliferation of lateral buds are the most widely used approaches of tissue culture for ornamental perennial crops. The propagated materials are genetically stable, but the quantity of material is limited (e.g., *Rosa* hybrids). Some investigators regenerate adventitious buds through organogenesis. This method produces large quantities of propagules but may yield variations that include polyploids and aneuploids. Adventitious buds are hard to regenerate in a number of species, e.g., *Rhododendron* sp. The best approach to plant propagation is embryo development. A large number of genetically stable plants can be obtained for certain species. Complete plants of *Vitis vinifera, Citrus*, and others have been grown by using this method.

Literature Review

Studies on tissue culture of ornamental plants are extensive for herbaceous flowering plants, such as *Cymbidium goeringii, Dendranthema morifolium, Gerbera jamesonii, Dianthus caryophyllus, Gladiolus gandavensis,* and *Tulipa gesneriana*. Studies on tissue culture of perennial crops were well under way by the late 1960s. Success has been achieved in many species (Table 1). Regeneration frequency depends on

TABLE 1. Tissue Culture of Ornamental Perennial Crops

Plant species	Genus	Explant	Investigator
Cycas revoluta Thunb.	*Cycas* L.	Female gametophyte	Norstag (1970)
Ginkgo biloba L.	*Ginkgo* L.	Stem	Luo Zijuan (1985)
Araucaria Cunninghamii Sweet	*Araucaria* Juss.	Stem from young plant	Hains et al. (1977)
Pinus strobus	*Pinus* L.	Embryo	Minocha (1980)
Cedrus deodara Loud	*Cedrus* Trew	Stem from seedling	Liu Min et al. (1983)
Codiaeum variegatum B1	*Codiaeum* A. Juss.	Endosperm	Chikkannaian et al. (1974)
Euphorbia pulcherrima Willd.	*Euphorbia* L.	Seed, leaf stalk, stem	Nataraja (1971)
Rhododendron sp.	*Erica*	Shoot tip	Anderson (1975)
Citrus microcarpa	*Citrus* L.	Seed	Rangaswamy (1961)
Fortunella crassifolia	*Fortunella* Swingle	Nucellus	Esan (1973)
Bougainvillea spectabilis	*Bougainvillea*	Shoot tip	Chaturved et al. (1978)
Nandina domestica Thunb.	*Nandina* Thunb.		Matsayama (1977)
Punica granatum L.	*Punica* L.	Leaf	Wang Huaiming (1982)
Daphne odora Thunb.	*Daphne* L.	Shoot tip	Tada Hoyu (1973)
Rosa chinensis Jacq.	*Rosa* L.	Shoot tip	Skirvin et al. (1979)
Prunus cerasifera Ehrh.	*Prunus* L.	Shoot tip	Garland et al. (1981)
Paeonia suffruticos Hndr.	*Paeonia* L.	Shoot tip	Li Yulong et al. (1983)
Camellia chrysantha	*Camellia* L.	Young stem	Zheng Ruoxian (1980)
Camellia japonica L.	*Camellia* L.	Young stem	Samartin (1984)
Michelia alba DC	*Michelia* L.	Stem segment	Liu Min et al. (1983)
Loropetalum chinense Oliv.	*Loropetalum* R. Br.	Shoot tip	Song Peilun et al. (1982)
Hamamelis vernalis	*Hamamelis* L.	Shoot tip	Zilis et al. (1977)
Hibiscus rosa-sinensis L.	*Hibiscus* L.	Stem segment	Tian Huiqiao (1984)
Syringa vulgaris L.	*Syringa* L.	Shoot tip	Hildebrandt (1983)
Ficus carica L.	*Ficus*	Shoot tip	Muriithi et al. (1982)
Vitis vinifera L.	*Vitis* L.	Shoot tip	Berlass et al. (1980)
Serissa Serissoides (DC.) Druce	*Serissa* Comm.	Leaf	Shu Yinglan (1985)

the species or varieties inoculated, composition of the culture medium, and culture environmental conditions. More than 20 species, mostly dicotyledons, have been cultured, but very few of the species have been used in commercial production.

TISSUE CULTURE TECHNIQUES

Culture Materials

Shoot tip tissues are generally used for tissue culture because they have already differentiated morphologically, and explants can be readily initiated without much contamination. The regenerated plants are genetically stable. For example, a shoot tip of 10 to 15 mm, taken from vigorously growing *Rhododendron* sp., was used in culture (Fordham et al., 1982); and in another study shoot tips of 5–10 mm taken from a 1.5- to 2-year-old *Syringa vulgaris* plant growing in a greenhouse were used as explants (Hildebrandt and Hainey, 1983). Hence, the explant materials for ornamental woody species are usually shoot tips (1–15 mm in length), excised from healthy and young plants growing outdoors or indoors. Shoot tips divide rapidly and are not usually contaminated.

Sources of shoot tips are limited. Therefore, other areas of the plant can be used in tissue culture, such as the flower bud of *Rhododendron* sp. (Martin et al., 1982). The flower bud is good explant material and can also be more easily sterilized than the shoot tip, which is hairy and often sticky.

Since the explants of mature perennial crops contain polyphenol oxidase, they easily oxidize and turn brown, impairing the development of cultures. This may be solved by transferring the explant tissue into fresh medium two to three times, inoculating the explants into a liquid filter paper bridge, or adding an oxidation inhibitor, such as cysteine, ascorbic acid, or citric acid, to the medium (Sondahl et al., 1977; Skirvin et al., 1979).

Culture Procedure and Medium

CULTURE PROCEDURE. Murashige (1974) proposed three steps for rapid propagation: first, establishment of aseptic culture of explants; second, proliferation of the cultures; and third, rooting, acclimation, and transplantation. One problem of the first stage is overcoming dormancy in lateral buds; this can be done by temperature or chemical treatments.

Contamination is another problem. Since ornamental woody species are perennials, donor plants carry bacteria which can cause severe contamination after inoculation. Therefore, explants must be collected from disease-free plants. Furthermore, good explant materials are available only for a limited period, since the regeneration frequency for meristems of perennial crops is highest in the spring. Explant collection and inoculation should be concentrated at this time. During the second stage the key factors are to select an appropriate medium, use containers to prevent the desiccation of the medium, and provide sufficient quantities of nutrients since the culture period for woody plants is longer than that of herbaceous plants. The most difficult point in the culture is the third stage, root initiation, especially when the materials are from mature trees. This difficulty may be overcome by microbud grafting (Huang et al., 1980).

MEDIA REQUIREMENTS FOR EACH STAGE OF CULTURE. The MS medium is the most widely used medium for plant regeneration in ornamental woody species (Table 2). Its components can be revised according to different plant species. Solid media are used generally, but liquid media may also be used for some species. For most species, the inorganic salt concentration of MS medium is too high, affecting root initiation and even harming cultures. Therefore, half or one-fourth of the inorganic salt concentration of MS medium should be used. Among coniferous species, pine is the most difficult to regenerate. Suitable media are White, Heller, and Gautheret supplemented with arginine and malt extract. Economou and Reed (1984) used modified MS to culture shoot tips from *Rhododendron* sp., achieving good results. Concentrations of NH_4NO_3 and KNO_3 were reduced, $(NH_4)_2SO_4$ was increased, the ratio of NH_4^+ and NO_3^- was adjusted to 1:1, KI was replaced with Na_2EDTA, and $FeSO_4$ was replaced with FeNa EDTA.

Norton and Boe (1982) cultured shoot tips from 12 species and varieties of Rosaceae. Their experiments indicate that the best shoot proliferation was in LS medium supplemented with 0.44–11.0 μM BA (Table 3). For root initiation, supplementation with 4.9–49 μM IBA worked well.

Growth regulator requirements in different species or different varieties of the same species are varied and are related to species, location of explants, and seasons in which explants are collected. Therefore, supplementation with appropriate types and concentrations of growth regulators is essential for the preparation of culture medium for ornamental perennial crops.

In shoot tip culture of *Rhododendron* sp., Economou et al. (1981) found that KIN and BA were not as effective as 2iP in primary culture. In subculture, KIN was much better than 2iP. This was demonstrated by the number of regenerated shoots and the size and quality of the shoots.

TABLE 2. Medium and Hormone Level Used for Plant Regeneration in Major Ornamental Perennial Crops

Plant Species	Explant	Basal Medium	Hormone (μM)		Investigator
			Shoot Regeneration	Root Initiation	
Ficus carica L.	Shoot tip	MS	BA 0.44, NAA 1.1, GA 0.1	NAA 2.7, IBA 2.5	Muriithi et al., 1982
Bougainvillea spectabilis	Shoot tip	MS	BA 0.88, IAA 5.7	IBA 0.49	Chaturvedi et al., 1978
Nandina domestica Thunb.	Bud	Modified MS	BA 4.4, IBA 0.49	IBA 4.9	Chen, 1984
Rosa chinensis Jacq.	Shoot tip	MS	BA 13.2, IAA 1.67	½or ¼MS + NAA 0.54	Hasegawa, 1979
Citrus microcarpa	Stem segment	MS	KIN 9.3, NAA 2.7	No hormone	Dong, 1979
Fortunella crassifilia	Lateral bud	H	BA 4.4–8.8, IBA 0.5	IBA 1.0–2.45	Huang et al., 1983
Camellia japonica L.	Shoot tip	MS	BA 4.4	IBA 2.45	Samartin et al., 1984
Daphne odora Thunb.	Shoot tip	H	KIN 4.6, BA 1.8	IAA 57, KIN 0.24	Hoyu, 1973
Syringa vulgaris L.	Shoot tip	MS	BA 33, NAA 0.54	IBA 0.49	Hildebrandt, 1983
Vitis vinifera	Shoot tip	MS	BA 4.4, NAA 0.54	NAA 1.1	Chee et al., 1982

TABLE 3. Effects of BA on Growth of Shoot Tip in Rosaceae*

Species or Variety	BA (μM)	No. of Shoots†
Spiraea Froebelii	2.2	38±5.9
Potentilla Sutter's Gold	2.2	27±3.9
Potentilla Coronation Triumph	0.44	26±10.7
Prunus tomentosa	11.0	15±1.4
Crataegus braehyacantha	4.4	10±2.2
Pyracantha coccinea	11.0	11±2.6
Chaenomeles japonica	11.0	7±1.0
Cotoneaster dammeri	4.4	7±1.3
Prunus cerasifera	2.2	6±0.8
Crataegus Toba	*4.4*	5±1.4
Malus Golden Homet	11.0	4±0.5
Malus Dainty	4.4	4±0.6

*All were cultured for 4 weeks.
†Mean of four replicates with SD.
(Norton and Boe, 1982).

Regenerated shoots may have reduced their requirement for cytokinin after subculture, or the shoots themselves may have synthesized cytokinin. Low concentration of cytokinin may be used to replace 2iP in subculture.

Li Zhou (1983) reported effects of phlorizin and its product, phloroglucinol, on development of cultures. In the culture of bud explants of *Rosa* hybrids, a supplement of 1–7 mg/l phlorizin in the medium stimulated the growth of callus and regeneration of a large number of shoots. Excised young stem segments of *Cedrus deodara* grew poorly in MS medium with 8.8 μM BA and 1.1 μM NAA. When transferred to medium containing 1–7-mg/l phlorizin, green calli developed and buds were regenerated. These results are similar to the results of the work on test tube rooting in *Malus pumila* reported by Jones (1976). This discovery has promoted the tissue culture of perennial crops.

The sucrose concentration of the culture medium is usually 88 mM, and agar concentration is from 0.6 to 0.7%. The pH value of the medium is about 5.7. Exceptions to this are *Rhododendron* sp. and *Camellia japonica*, which grow better at pH 5.0. To achieve this, MS medium can be modified by reducing NH_4NO_3 and KNO_3, increasing $(NH_4)_2SO_4$, changing the ratio from 1:2 to 1:1, and reducing the pH.

Sterilization and Inoculation Procedures

1. Carefully excise the materials from donor plants and rinse with running water for 10–30 minutes.
2. Soak the explants in 75% ethanol for 10–15 secs, and rinse twice with sterile distilled water.

3. Soak the explants in 10% commercial hypochlorite solution for 20 min or 0.1% mercuric chloride solution for 5–10 min. Since most ornamental perennials, especially hairy, oily, or waxy explants, carry bacteria, explants should be well sterilized. This can be done by adding 0.1% Tween 20 to the sterilizing solution. Captan 0.1% has also been used to sterilize explants of *Rhododendron* sp. (Economou et al., 1981).
4. Rinse the explants with sterile distilled water three to five times.
5. After sterilization, the explants are placed on aseptic filter paper to absorb water and then inoculated into solid medium. For *Rosa chinensis*, the young stem tissue is cut into small pieces (each piece should include at least one axillary bud) under sterile conditions and inoculated into the medium (Cai et al., 1984), so that the number of explants can be increased.

The inoculation of shoot tip explants vertically or horizontally does not affect the number or height of regenerated shoots, as found for *Syringa vulgaris* L. (Hildebrandt, 1983) in medium which included 33 μM BA and 0.57 μM IAA (Table 4). However, there were obvious effects, especially in shoot height, when shoot tip explants were inoculated invertedly.

Incubation Conditions

TEMPERATURE. A constant incubation temperature of 25 ± 2°C for culture, proliferation, and root initiation in woody plants is used by most workers. A temperature fluctuation (high during the day and lower at night) can be used for some species.

LIGHT. The most commonly used photoperiod is 10–16 hr/day at a light intensity of 1000–4000 lux. To stimulate plantlet growth during root initiation, the light intensity should be raised to 3000–10,000 lux. Norton and Bal (1982) reported a method in which cultures from 11 species and

TABLE 4. Effects of Different Orientation on Growth of *Syringa vulgaris* Shoot Tips

Orientation of Shoot Tip	No. of Shoots Produced in One Explant (± SE)	Length of Shoots (± SE)
Upright	9.4 ± 0.97	21.9 ± 0.48*
Horizontal	8.0 ± 1.00	14.4 ± 0.48

*$p < 0.05$.

HILDEBRANDT, V. AND HARNEY, P. M. 1983. IN VITRO PROPAGATION OF *SYRINGA VULGARIS* VESPER. *HORTIC. SCI.* **18**(4):432–434.

varieties were incubated in darkness for 1 week and then provided with illumination for 16 hr/day during the period of root initiation. This encouraged root development; however, if the dark incubation was continued, only callus was generated.

ESSENTIALS AND TECHNIQUES FOR VIRUS ELIMINATION

Significance of Regenerating Virus-Free Plants

Many types of viruses harm ornamental perennial crops (Table 5). More virus types continue to be discovered, and virus damage to plants becomes more severe. For example, in nut plants, there were only 5 types of viruses identified by 1930. By 1950, 48 virus types had been identified, and by 1976, 95 virus types were identified (Hildebrandt and Harney, 1983).

Virus pathogens are generally eliminated during seed propagation in most species; however, in legumes, virus infections are limited to only one generation. In most ornamental perennial crops, which are propagated asexually, viruses are transmitted vegetatively. Infected plants grow slowly, have small flowers, exhibit flower color deterioration, and produce flowers of poor quality and low value.

Unlike fungal and bacterial diseases, viral diseases cannot be prevented and controlled by a chemical bactericide or fungicide. Although there have been investigations of virus inhibitors, virus multiplication is closely associated with plant metabolism and known viral inhibitors are toxic to plants. Furthermore, the action of the inhibitor is localized so that when the inhibitor loses its ability, viruses undergo replication soon and return to their original concentration. Pesticides can kill vector in-

TABLE 5. Major Viruses Affecting Ornamental Perennial Crops

Plant species	No. of different viruses
Rosa chinensis Jacq.	10
Ficus carica	5
Paeonia suffruticosa Hndr.	2
Vitis vinifera L.	26
Daphne odora Thunb.	4
Syzygium aromaticum	2
Camellia japonica L.	1
Jasminum sambac	2
Rhododendron sp.	1
Hibiscus rosa-sinensis	1

sects, thereby reducing the spread of some viruses. However, some viruses are disseminated mechanically, and these types are not eliminated by use of insecticide. In addition, plants do not have a general immune system.

Breeding of virus-free plants is important in cultivation of ornamental plants. In the 1950s plant tissue culture was found to be an effective technique for virus elimination which improved plant yield and quality. Some countries have adopted the virus elimination technique as an essential procedure in conventional breeding of improved plant varieties and have established large-scale centers for producing virus-free plants. Some countries export virus-free plants.

In the past, virus infected plants were manually eliminated from the field. Tissue culture is an effective approach to virus elimination. Since chemicals are not necessary, environmental pollution is reduced.

Methods of Virus Elimination

HEAT TREATMENT. In 1889 it was discovered in Indonesia that *Saccharum officinarum* plants infected by wilt disease (now known to be a viral disease) could grow well when they were placed into hot water at 50–52°C for 30 min. This method has continued to be used.

Kunkel (1936) reported that yellow virus in *Prunus persica* plants loses activity after heating at a temperature of 34–36°C for 2 weeks. Studies of successful virus elimination with heat treatment were reported for *Catharanthus roseus* (Kunkel, 1941), *Solanum tuberosum* (Kunkel, 1943), *Fragaria ananassa* (Posnette, 1953), *Dendranthema morifolium* (Hitchborn, 1956), *Rubus* spp. (Chambers, 1954), *Cucumis sativus* (Kassanis, 1954), *Dianthus caryophllus* (Kassanis, 1954), and *Pyrus* spp. (Campbell, 1965). For perennial crops, Nyland et al. (1969) have reported that virus elimination by heat treatment has been applied to *Citrus, Malus pumila, Prunus persica, Pyrus communis, Vitis vinifera*, and *Prunus pseudoccrasus*, among others.

Two methods of heat treatment are currently used. The first is to use hot water at 50°C for 3–15 min. This method is suitable for sugarcane, perennial crops, and dormant buds. Another method, applying hot air of 37–38°C for 2–4 weeks, is effective for most plant species. The temperature of the hot air should be increased gradually to the final temperature. Humidity and light must also be properly controlled. The effect of virus elimination varies with type of virus. In general, heat treatment is effective only for sphere-shaped viruses, including the fan-shaped leaf virus of *Vitis vinifera* and mosaic virus of *Malus pumila*, and filament-shaped viruses such as X and Y viruses of *Solanum tuberosum*. Not all of the spherical viruses can be eliminated. The treatment is not as effective for the rod-shaped viruses, such as mosaic virus of *Aretium lappa* L. and the virus of *Gomphicna globosa* L.

Heat treatment eliminates viruses because the virus becomes unstable when heated and loses its activity. Kassanis (1954) believed that virus concentration within an infected plant reflected the balance between production and damage of virus granules. High temperatures caused severe damage to the virus and therefore reduced the virus concentration. Virus infection was eradicated with heating.

VIRUS ELIMINATION THROUGH SHOOT TIP CULTURE. White (1943) found that virus concentration was very low, or even zero, around the apical growing point of a *Nicotiana tabacum* plant infected by tobacco mosaic virus; the virus concentration was found to vary at other regions of the plant, depending on the age of the plant. Encouraged by this discovery, Morel et al. (1952) obtained plants regenerated from apical meristem (250 μM in length) of mosaic infected plants of *Dahlia pinnata*. The regenerated plants were determined to be virus-free by using seedling stock grafting. Since these discoveries, shoot tip culture has been an effective approach for plant virus elimination and has been used successfully in a number of species, including *Solanum tuberosum* (Worris, 1954), *Dendranthema morifolium* (Holmes, 1956), orchid (Morel, 1960), *Lilium* (Phillips, 1962), *Fragaria ananassa* (Miller, 1963), *Petunia* (Mori et al., 1964), and *Iris tectoru* (Barach et al., 1966). The offspring of regenerated plants are genetically stable. This approach has been widely used to regenerate virus-free plants. In addition, viroid, mycoplasma, bacteria, and fungi are removed (Walkey, 1978; Murashige, 1980).

Studies by White (1943), Holmes (1948), and Limasset (1949) show that shoot tip culture eliminates virus because of different distributions of viruses in the plant body. Virus concentration is high in older leaves and mature tissues and organs, but low in immature tissues and organs. The apical growing point (about 0.1–1.0 mm long) contains little or no virus. This is because the virus infection moves upward slowly, while meristematic cells multiply rapidly.

Hollings et al. (1964) studied relationships between different sizes of inoculated shoot tips and elimination of mosaic virus in *Dianthus caryophllus*. Shoot tips measuring 0.1, 0.25, 0.50, 0.75, 1.0, and greater than 1.0 mm in length were cultured. Frequencies of virus infection of regenerated plants were identified as 33, 60, 87, 89, 100, and 100%, respectively. It was evident that the smaller cultured shoot tip explants had greater tendencies to be virus-free. The size of shoot tips needed for virus elimination appears to vary with different plant species and virus types. The results of virus elimination and size of explants inoculated are usually inversely correlated, but survival rate and size of shoot tip explant are positively correlated. In practice, both factors should be considered. Therefore, shoot tips for culture are generally between 0.2 and 0.5 mm long, having one to two leaf primordia. Since shoot tips for culture are small, excision may be difficult. Fine manipulation techniques are needed. The excision should be conducted under a dissection microscope, using good excising tools. Some workers place the plants in an environment of 37–40°C for 15 days to 4 months and then excise shoot tips for inoculation, which eliminates virus.

Basal medium for tissue culture is usually MS or White medium. KCl added to the medium increased growth of the cultures. Additives included 5–10% CW and 0.44–5.7 μM of the growth regulators IAA, NAA, IBA, KIN, BA, etc. Active carbon is required by some plant species. Shoot tips are cultured in solid or liquid medium with filter paper bridges. Most existing media are not suitable for shoot tip culture and must be appropriately modified according to the needs of a given species.

Sasahara Koyuki et al. (1981) grew virus-free plants of *Vitis vinifera* by shoot tip culture. Infected plants were first treated at 5°C and then placed in water at 25°C, with an illumination of 16 hr/day and a relative humidity of 80%. When new shoots regenerated with tips measuring 0.2–0.3 mm in length, including one leaf primordium, they were excised and inoculated into ½ MS medium containing 4.4 μM BA, 2.3 μM KIN, and 0.54 μM NAA. The survival rate was 88.6%. For subculture, 0.54 μM NAA was replaced by 1.1 μM IAA in the medium; otherwise the developed green tips died. Later on, the concentration of BA was reduced to 2.2 μM, which stimulated growth of stem and leaf and enhanced formation and elongation of multibuds. For root initiation, medium with 0.54 μM NAA was adequate. Frequency of root initiation may reach 89.8% in 50 days.

ELIMINATION OF VIRUS BY MICROGRAFTING OF SHOOT TIPS. It is difficult to induce roots in some ornamental perennial crops. In such cases micrografting of the shoot tip may be used to eliminate virus. This procedure is as follows:

1. Sterilize seeds and inoculate the seeds into MS medium containing 1% agar to induce seedlings for grafting stocks.
2. Remove seedlings from test tubes; cut off tops to retain epicotyls measuring 1–1.5 cm; remove cotyledons and axillary buds; and cut root tips to retain roots 4–6 cm long.
3. Using a micromanipulator, remove shoot tips about 1 mm long with one to three leaf primordia as scions from mature trees growing in the field or greenhouse.
4. Graft the scions on the stocks using the inverted T-shaped method;

then inoculate the grafted seedlings in liquid medium using filter paper bridges.

5. Transplant the surviving plants into soil when there are at least two full leaves.

Survival rate of a graft is usually 30–50%, which is closely associated with skill of the individual doing the micromanipulation. Navarro et al. (1983) micrografted shoot tips of *Pyrus* spp., using 10–14-day-old test tube regenerated shoots of 0.5–1.0 mm with three to four leaf primordia. The survival rate was 40–70%. They grew *Prunus domestica* plants free of ring spot virus, wilt virus, and chlorosis leaf spot virus.

Identification of Virus-Free Plants

Plants grown from shoot tip culture must be certified as being virus-free prior to cultivation. Common methods for virus indexing employ indicator plants, antisera, and electron microscopy.

INDICATOR PLANT. Holmes (1929) found that withered spots produced in test plants infected with a virus inoculant could be used for identifying virus types; e.g., the host plant in which disease spots are partially produced is the indicator plant, or identification host. The host plant is used only to identify viruses infected by inoculation with a liquid fraction of the plant to be tested. In asexually propagated ornamental plants, grafting is commonly used for inoculation with the indicator plant as the stock and the plant to be identified as scion wood. The methods of cleft grafting, grafting by approach, and bud grafting are usually employed.

ANTISERUM IDENTIFICATION. Each virus type produces a specific antiserum. Therefore, antisera of known viruses can be used to identify unknown viruses. Because antiserum is highly specific, this method is fast (a test can be accomplished within several hours or even several minutes) and simple. The antiserum method is one of the most useful means of identifying a virus. Precipitation reactions are the standard assays for virus identification.

For antiserum testing, the first step is to prepare the antigen. This procedure includes propagating the virus in the host plant; grinding infected leaves; separating the fractions, which includes purifying the virus suspension; and precipitating the virus. Highly purified antigen is necessary to produce the antiserum. Good-quality antiserum is prepared in rabbits. The following methods may be used to detect antigen-antibody reactions: precipitation, immunodiffusion, immunoelectrophoresis, fluorescent antibody, or enzyme-linked immunosorbent assay.

ELECTRON MICROSCOPE IDENTIFICATION OF VIRUSES. Unlike the indicator plant and antiserum methods, the presence or absence of virus can be observed directly by electron microscopy. In addition, size, shape, and structure of virus granules can be determined. Because these characteristics are stable, they are important for virus identification. Because electrons have low penetration ability, the prepared slides should be very thin: 10–100 millimicrons (slides are usually 20 millimicrons).

Prevention of Reinfection of Virus-Free Plants

When virus-free plants are grown, they should be well maintained by separate planting. Virus-free plants can be maintained for 5–10 years if stock materials are well preserved. Regenerated virus-free plants should be planted in an isolated net house with mesh size of 0.4–0.5 mm to prevent aphids from entering. Bedding soil should be sterilized, surroundings should be clean, and insecticide should be used from time to time. Cultivation of virus-free plants should strictly exclude virus infections. If possible, it is best to plant virus-free plants in isolated areas in cool weather and periods of low insect infestation, which favor growth and multiplication of the stock materials. When virus-free plants are in production, it is important to prevent virus reinfection. Planting soil should be sterilized. In new planting plots and small-scale planting areas, plants take more time to become reinfected by viruses; in areas of longtime cultivation, repeated planting, and large-scale production, virus-free plants can be reinfected within a short time. Plants are not readily reinfected in high and cold areas, where aphids are not usually found. In summary, a variety of methods should be used to prevent reinfection. If plants become reinfected, they should be replaced by virus-free plants.

TRANSPLANTATION AND GENETIC STABILITY OF PLANTLETS

In ornamental perennial crop tissue culture, it is difficult to induce roots from regenerated shoots. Root initiation frequency of *Syringa vulgaris* in the test tube was 58% (Hildebrandt, 1983) when shoots were placed directly into the substrate for root initiation. When *Syringa vulgaria* shoots measuring 31 mm were placed in vermiculite, root initiation frequency increased to 81%. When *Rhododendron* sp. shoots were inserted in substrate mixed with peat, vermiculite, and perlite, the rooting frequency approached 100% (Read, 1982).

Soil, pots, and other tools used for cultivating virus-free plants must be sterile. This may be done by autoclaving. If high-temperature treatment is used for soil sterilization, time should be limited to prevent damage to organic substances.

When test tube rooted plants have grown normal leaves and roots of 3 cm, they may be transferred to pots. Because plantlets have poor viability in dry environment, they must first be hardened by removing tube plugs for several days. Then plantlets are removed, the roots are rinsed, and plantlets are transferred into pots containing vermiculite or perlite and covered with glass vessels or transparent plastic sheets. Relative humidity should be 90–100%. It is advantageous to provide spray. Potted plants should be kept in the shade at first, placed in sunlight after 2 weeks, and transplanted into the ground after 4 weeks.

Genetic variation is often observed in regenerated plants, especially when the plants are grown from callus. This is related to the genotypes of the plant species and type of plant part inoculated, such as the leaf, root, or flower. In addition, culture of long duration, composition of the medium, hormones, and incubation environments induce variations. Genetic stability is required for rapid propagation. In plant breeding, variation provides opportunity for selection of desired characteristics. Genetic variation can be caused by polyploid, aneuploid, and chromosomal structural changes.

KEY REFERENCES

Bhojwani, S. S. and Razdan, M. K. 1983. Clonal propagation. In: *Plant Tissue Culture: Theory and Practice*. pp. 313–372. Elsevier, Amsterdam.

Luo, S. W. 1978. Plant tissue and cell culture: Recent advances and applications. *Acta Phytophysiol. Sin*. **4**:91–112 (in Chinese).

Qiu, W. D. *Tissue Culture of Horticultural Plants*. Shanghai Science and Technology Press, Shanghai in press (in Chinese).

Yang, N. B. 1982. List of test tube plants. *Plant Physiol. Commun*. **4**:61–80; **5**:59–71 (in Chinese).

REFERENCES

Bressan, P. H., Kin, Y. J., Hyndman, S. E., Hasegawa, P. M., and Bressan, R. A. 1982. Factors affecting in vitro propagation of rose. *J. Am. Soc. Hortic. Sci*. **107**:979–990.

Cai, J., Cai, M. Y., and Qian, D. L. 1984. Induction of multibuds and fast propagation of clones of *Rosa chinensis. Plant Physiol. Commun*. **5**:37–38 (in Chinese).

Chee, R. and Pool, R. M. 1982. The effects of growth substances and photoperiod on the development of shoot apices of *Vitis* cultured in vitro. *Sci. Hortic*. **16**:17–27.

Economou, A. S. and Reed, P. E. 1984. In vitro shoot proliferation of Minnesota deciduous azaleas. *Hortic. Sci*. **19**(1):60–61.

Hildebrandt, V. and Harney, P. M. 1983. In vitro propagation of *Syringa vulgaris* Vesper. *Hortic. Sci.* **18**(4):432–434.

Rhosh-Khui, M. and Sink, K. C. 1982. Micropropagation of new and old rose species. *J. Hortic. Sci.* **57**:315–319.

Li, Y. L., Wu, D. Y., Pan, S. L., Xu, S. L., Wei, Z. M., Xu, Z. H., and Li, X. J. 1984. In vitro propagation of *Paeonia suffruticosa. Kexue Tongbao* **12**(29):1675–1678.

Li, Z. 1983. The effect of phlorizin on in vitro cultures of *Begonia feastii, Rosa hybrida* and *Cedrus deodara* organs. *Plant Physiol. Commun.* **4**:37–39 (in Chinese).

Liu, M. and Shu, J. S. 1983. Tissue culture of *Sedum ergthrostictum, Michelia alba*, and *Cedrus. Plant Physiol. Commun.* **6**:38–39 (in Chinese).

Moore, J. N. and Janick, J. 1983. *Tissue Culture, Methods in Fruit Breeding*. pp. 124–135. Purdue University Press, West Lafayette, Ind.

Muriithi, L. M. 1982. In vitro propagation of fig through shoot tip culture. *HortScience* **17**:86–87.

Norton, M. E. and Boe, A. A. 1982. In vitro propagation of ornamental rosaceous plants. *HortScience* **17**:190–191.

Slack, S. A. 1980. Pathogen-free plants by meristem-tip culture. *Plant Dis.* **64**(1):14–17.

Song, P. L., Peng, Y. Y., and Zhang, T. X. 1982. Callus formation and organogenesis in tissue culture of *Loropetalum chinense. Plant Physiol. Commun.* **4**:33 (in Chinese).

Tian, H. Q. 1984. Tissue culture of *Hibiscus rosa-sinensis* stems. *Plant Physiol. Commun.* **2**:35 (in Chinese).

Wang, H. Z. 1983. Afforestation and tissue culture. *Sci. Silvae Sin.* **19**(3):292–301 (in Chinese).

Zimmerman, R. H. and Broome, O. C. 1980. Apple cultivar micropropagation. *Proc. Conf. Nursery Prod. Fruit Plants Tissue Cult.—Applic. Feasibil.*, USDA, SEA, ARR-NE-11, Beltsville, Md., 54–58.

PART B
Timber and Cork Trees

CHAPTER 8
Poplar and Willow

*Jingfang Lin**

INTRODUCTION

The trees of the family Salicaceae are economically important to human life. All members of the family are full-seed trees except some shrubs in the genus *Salix*.

Geographical Distribution

The Salicaceae have a wide distribution in the Northern Hemisphere, extending from the Arctic Circle to 3 degrees north latitude. Some species of the family also grow in the Southern Hemisphere (International Poplar Commission, 1979).

The cultivation and selection of Salicaceae have long been important in a number of countries, and monographs on poplar breeding are available from Italy, France, the United States, and the Soviet Union. Artificial and natural hybrids with economical value have been selected. Salicaceae trees have been cultivated in China since ancient times. Scientific breeding work started in the 1950s, and some new varieties have been developed since then.

*English translation by Kaiwen Yuan.

Taxonomy and Economic Importance

The family Salicaceae has three genera, *Populus, Salix*, and *Chosenia*. The genus *Populus* has the highest economic value because of its fast growth, simple cultivation, and easily processed timber (Xu et al., 1958).

Most species of Salicaceae are dioecious. Wide variation exists among individuals of Salicaceae species in nature. Because of dioecism, the complicated procedure of emasculation is avoided during controlled hybridization. Many species can also be propagated by cuttings. Clones can thus be readily formed from superior individuals obtained from artificial hybridization, and heterosis of F_1 hybrids can be utilized directly.

However, there are still some species which are difficult to propagate by cuttings, such as *Chosenia macrolepsis*, the only species of the genus *Chosenia; Populus diversifolia*, the only species under the section Turanga of the genus; and most of the species and hybrids under the section Leuce Duby of the genus *Populus*. These species have many superior characteristics which cannot be found in other species of the family.

Poplar breeders have made many interspecific crosses within the section Leuce Duby over the last several decades. Hybrids produced by some of the combinations, such as *P. tomentosa* × *P. bolleana*, and (*P. alba* × (*P. davidiana* × *P. simonii*)) × *P. tomentosa* (Jiang et al., 1980), have the superior characteristics of erect trunks, good plant form, and rapid growth. However, difficulty of propagating them by cuttings causes problems in asexual breeding and prevents the quick introduction of superior hybrids in production. Asexual propagation by tissue culture is a fast and effective approach enabling the timely use of superior genotypes.

LITERATURE REVIEW

Studies on tissue culture of poplar started in the late 1960s (Winton, 1968). Success has been achieved not only in those species which are easily propagated by cuttings, such as *Populus nigra* var. Italica and *P. euroamericana* under section Aigeiros Duby (Winton and Huhtinen, 1976), but also in the hard to propagate species and hybrids of section Leuce Duby, such as *P. tremuloides* (Winton, 1970), *P. tremula* (Winton, 1971), and *P. canescens* (Winton and Huhtinen, 1976).

In China the in vitro culture of poplar began in the 1970s, first concentrating on anther culture and later extending to somatic cell culture. The species used for culture included *P. adenopoda, P. alba, P. tomentosa, P. hopeiensis, P. grandidentata* (Lin et al., 1980; 1983), *P diversifolia* (Zhou and Gui, 1983), and hybrids of *P. davidiana* × *P. bolleana* (Li et al., 1979), *P. alba* × *P. bolleana* (Ren, 1984), and the composite hybrid, Superior 741, of (*P. alba* × (*P. davidiana* × *P. simonii*)) × *P. tomentosa*. Intact plants have been regenerated from these mentioned materials.

There are various species in the genus *Salix*. Individuals vary in size from bushes to large trees. Very few papers on tissue culture have been reported, possibly because most of the species in this genus can be propagated by cuttings. There has been only one report on *Salix babylonica* tissue culture up to now (Brown and Sommer, 1982). Letouze (1977) has obtained plantlets using lateral buds of *S. babylonica* as explants.

The genus *Chosenia* has only one species, *Chosenia macrolepsis*, growing in the provinces of Heilungjiang, Jilin, Liaoning, and the Inner Mongolia Autonomous Region of China and also found in Korea, Japan, and the USSR. It is a deciduous tree that can reach 30 m in height with a diameter at the base of more than 1 m. The tree has a beautiful form, and the crown is deep red when leaves are falling in autumn, so it is ideal as an ornamental tree in gardens. Its timber, which is light, soft, and easy to process, is good for use in building construction and furniture making. *Chosenia* has, therefore, aroused attention from forest workers in recent years. Use of this species has, however, not expanded because of difficulties in conventional vegetative propagation and segregation in sexual propagation (Zhao, 1982). The use of tissue culture not only accelerates propagation but also produces genetically uniform plant lines from superior individuals. *Chosenia macrolepsis* was cultured successfully in our laboratory recently, using segments of young twigs from mature trees as explants. Complete plants have been regenerated from these cultures (Lin et al., 1984) (Table 1).

CULTURE TECHNIQUE

Culture Procedures

1. Explants are first inoculated into a differentiation medium for induction of embryos or adventitious buds. Explants taken from the field should be sterilized. If the explants are taken from test tubes, they can be inoculated directly into the medium.

The medium developed for induction of adventitious buds from explants of poplar trees in our laboratory is ½ MS medium (with half-quantity of its major elements) supplemented with 1.32 μM BA, 0.27 μM NAA, 73.3 mM sucrose, and solid medium with 0.5% agar. The pH value is 5.8.

2. Callusing should be avoided at the root stem transition zone when roots are induced from plantlets. The rooting medium is ½ MS supplemented with 0.11 μM NAA (or 0.98 μM IBA), 44 mM sucrose, with 0.8% agar.

TABLE 1. General Information on Tissue Culture in Salicaceae

Species	Explant Sources	Developmental Stage	Date	Author
Populus tremuloides	Stems	Plants	1968, 1970	Winton
P. tremula	Stems	Small trees	1971	Winton
P. nigra	Stems	Small trees	1973	Brown et al.
P. canadensis	Shoot tips	Small trees	1972	Brown et al.
P. euroamericana var. robusta	Lateral buds	Small trees	1974	Brown et al.
P. canescens	Shoot tips, cambium	Plants	1974	Brown et al.
P. euroamericana	Twig tips, cambium	Plants	1974	Brown et al.
Salix babylonica	Lateral buds	Saplings	1977	Brown et al.
P. davidiana × *P. bolleana*	Twigs, axillary buds	Saplings	1979	Li et al.
P. hopeiensis	Twig segments	Saplings	1980	Lin et al.
P. tomentosa	Twig segments	Saplings	1980	Lin et al.
P. alba	Twig segments	Saplings	1980	Lin et al.
P. adenopoda	Twig segments	Plants	1983	Lin et al.
P. grandidentata	Segments of axillary branches and roots, stems, and leaves of test tube plants	Plants	1983	Lin et al.
P. diversifolia	Twig segments	Test tube plants	1983	Zhou et al.
P. alba × *P. bolleana*	Buds, shoot tips	Plants	1984	Ren
Chosenia macrolepsis	Young stem segments	Plantlets	1984	Lin et al.
(*P. alba* × (*P. davidiana* × *P. Simonii*)) × *P. tomentosa* Superior 741	Lateral buds and stem segments	Plants	Unpublished	
P. davidiana × *P. alba*	Axillary branches	Plantlets	Unpublished	

3. The whole induction process is conducted at 25° ± 1°C in an incubation room illuminated for 13 hr/day.

Selection of Different Organs as Explants

Explants of Salicaceae trees used for culture are usually shoot tips, stem segments, axillary buds, and leaf stalks. Explants can also be taken from the roots, stems, and leaves of test tube plantlets. Segments of young shoots and leaf stalks from five species of the section Leuce Duby have been used for culture in our experiments (Lin et al., 1980; 1983). Zhou and Gui used stem segments of the annual *Populus diversifolia* as explants (Zhou and Gui, 1983). Chen et al. used stem segments with axillary buds as primary explants to regenerate test tube plantlets and then took leaf stalk pieces (with part of blades) from the plantlets as secondary explants (Chen et al., 1983).

In 1980, the Chinese Academy of Forest Sciences sowed seeds of *P. grandidentata* introduced from the United States, but only three seedlings successfully germinated. In order to propagate this species, we collected axillary shoots from the three seedlings, cut them into pieces, and cultured them successfully. Because of the limited explant resources, we used stems, leaves, and roots of the regenerated plantlets as secondary explants for plantlet regeneration. This method is very effective for increasing and propagating species when explant resources are limited. Generally speaking, roots are seldom used in tissue culture, but we have achieved success in propagating *P. grandidentata* using roots as explants. Since every part of a test tube plantlet can be used, this provides a long-lasting source of aseptic explants. Figure 1 illustrates the feedback process between explants and test tube plantlets.

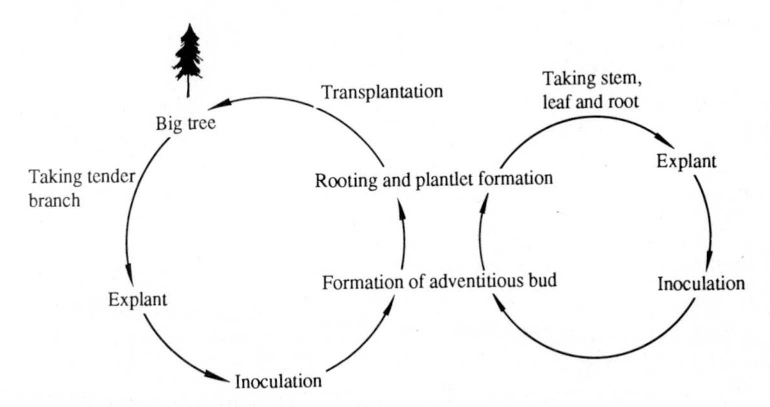

Figure 1. The feedback between explants and test tube plantlets.

TABLE 2. Comparison of Induction Frequency for Adventitious Buds Regenerated from Different Locations of Stem in *P. hopeiensis*

Shoot No.	Location	No. of Segments Inoculated	Frequency of Bud Differentiation (%)
No. 1	Upper part	21	100.00
	Lower part	15	60.00
No. 2	Upper part	22	95.24
	Lower part	15	60.00
No. 3	Upper part	22	90.91
	Lower part	14	71.42

The sources of the explant affect bud differentiation. With *P. hopeiensis*, it has been shown that varying differentiation frequencies were obtained when different parts of the same shoot were cultured under the same conditions (Dong et al., 1983). Shoot tips of 7 cm in length were taken from a mature tree. The leaves were removed, and explants were cut into two parts 4 cm below the apical point. Leaves were separately cut into pieces of 0.3 to 0.5 cm in length and inoculated into differentiation medium with the same composition. The results indicated that the differentiation frequencies of adventitious buds varied significantly between the two parts. The differentiation frequency of the upper part was more than 90%; that of the lower part was only 60–70%. The differentiation speed and growth of adventitious buds of the upper part were also faster than for the lower part (Table 2).

DIFFERENCES IN INDUCTION FREQUENCY BETWEEN MALE AND FEMALE TREES. In our experiment with *P. tomentosa*, we found that more adventitious buds were differentiated for each piece of explant from a female tree than for each explant from a male tree. Sexual differences have also been observed in other woody species. In *Phellodendron amurense*, for example, explants from a male tree regenerated at a higher frequency than those from a female tree (K. X. Wu, personal communication).

Sterilization

Two methods are usually employed for sterilizing Salicaceae explants. The first method is sodium hypochlorite sterilization, in which explants are soaked in 70% ethanol solution for several seconds, immersed in a solution containing 10% sodium hypochlorite for 5–10 minutes, and then rinsed with aseptic water three to four times. The second method is mercuric chloride sterilization, in which explants are soaked in the 70%

ethanol solution for several seconds, sterilized in a solution containing 0.1% mercuric chloride for 5–10 min, then rinsed several times with sterile water. The sterilization time should be in accordance with the lignification level of the explants; if young tissues are used, the time must be reduced. Mercuric chloride is much less damaging to the explants than sodium hypochlorite. Sodium hypochlorite is difficult to handle as it can easily damage young tissue, but contamination occurs if the sterilization time is shortened.

ANALYSIS OF SOME FACTORS IN THE CULTURAL PROCESS

Medium

BASAL MEDIUM. Many researchers use the inorganic salt components of MS medium as the first medium for the tissue culture of poplar, e.g., *P. davidiana* × *P. bolleana* (Li et al., 1979), *P. diversifolia* (Zhou and Gui, 1983), and *P. alba* × *P. bolleana* (Ren, 1984); some researchers use media with low levels of macronutrients. For instance, Chen et al. (1983) employed Miller medium for tissue culture of *P. japonica*. We used ½ MS medium for culture of *P. hopeiensis, P. tomentosa, P. alba, P. adenopoda, P. grandidentata*, hybrids of two combinations of section Leuce Duby, and *Chosenia macrolepsis*. Fewer adventitious buds developed in the complete MS medium than in ½ MS medium (Lin et al., 1980).

GROWTH REGULATORS. The cytokinin to auxin ratio appears to be critical for induction of organogenesis. In cultures of *P. hopeiensis* (Dong et al., 1983) the best results of adventitious bud induction were obtained when the BA/NAA ratio was 6 or the ZEA/NAA ratio was 10–20. The absolute concentrations of the hormones should not exceed a certain value. For example, when NAA was increased to 0.54 μM, bud differentiation was inhibited (Table 3).

We have found that an appropriate medium for adventitous bud induction is ½ MS containing 1.32 μM BA, 0.27 μM NAA; for rooting, ½ MS with 0.11 μM NAA. These media have lower levels of growth regulators than media commonly used for other perennial crops.

SUCROSE. Sucrose concentrations of 29, 44, 59, 73, 88, and 103 mM were experimentally added to the medium for adventitious bud induction from stem segments in three species of section Leuce Duby. High sucrose concentration (103 mM) made the explants black. Although a few buds regenerated under an appropriate growth regulator regimen, they turned yellow 1 month later. On the medium with 73 mM sucrose, the

TABLE 3. Effects of Cytokinin-Auxin Ratio on Adventitious Shoot Differentiation

BA/NAA (μM)	Ratio BA/NAA	Frequency of Shoot Differentiation (%)	ZEA/NAA (μM)	Ratio ZEA/NAA	Frequency of Shoot Differentiation (%)
1.3/0.27	4.8	82.61	1.4/0.27	5.2	84.91
2.2/0.27	8.1	50.94	2.3/0.27	8.5	91.67
3.5/0.27	13	48.00	3.7/0.27	13.7	93.55
4.4/0.27	16.3	38.00	4.6/0.27	17	96.55
1.3/0.54	2.4	80.00	1.4/0.54	2.6	66.67
2.2/0.54	4.1	65.00	2.3/0.54	4.3	61.36
3.5/0.54	6.5	49.33	3.7/0.54	6.8	46.15
4.4/0.54	8.1	42.42	4.6/0.54	8.5	82.14

explants and adventitious buds were all a normal healthy green. When the sucrose concentration was 29 mM, the explants grew very slowly and turned blue-green 1 month later. After 3 months, their growth was still very slow although they were transferred to fresh medium. A sucrose concentration of 44 mM appeared to be the best for the rooting medium (Lin et al., 1980).

AGAR. A low concentration of agar has been found to be suitable for primary cultures of stem segments in our work. The explants should rest firmly on the surface of the medium but not sink into it. In *Picea abies*, Von Arnold (1979) found that when the agar concentration was reduced from 0.7% to 0.3%, the bud induction frequency was increased. However, the culture sometimes suddenly stopped growing in such a medium. Von Arnold considered that this was because the medium texture was too soft and the organs sank into the medium. Therefore, she proposed that 0.5% agar was the most suitable concentration (Von Arnold, 1979). We have obtained similar results.

In rooting medium, more agar, at least 0.8%, should be added to hold the adventitious buds in place. If the medium is too soft, the adventitious buds will not grow straight.

Light and Temperature

Photoperiod significantly affects the speed and frequency of adventitious bud differentiation. Cultures of *P. hopeiensis* were tested under three different light treatments in our lab (Dong et al., 1983): (1) 13 hr/day under fluorescent light at 2000 lux for 5 weeks; (2) incubation first in darkness for 2 weeks, then the first treatment for 3 weeks; (3) incubation in darkness for 5 weeks. The best results were achieved with the first treatment (Table 4). Similar results were observed in *P. adenopoda* and *P. grandidentata*.

Temperature has a remarkable affect on rooting under consistent light conditions. Four different temperature treatments were conducted on November 3, 1981 (Dong et al., 1983): (1) 25 ± 1°C; (2) 22 ± 1°C; (3) 20 ± 1°C; (4) 18 ± 1°C. Illumination was 13 hr/day at 2000–2500 lux. The average temperature at night was about 17°C. Observations began on the 7th day. Green plants began to develop roots on the 8th day in the first two treatment groups, with a rooting frequency of 100%. The cultures of the first treatment grew faster, developing secondary roots on the 15th day; those in the second treatment initiated secondary roots on the 17th day. The rooting frequency of the third treatment also reached 100%, but the cultures rooted very slowly, developing primary roots on

TABLE 4. Effect of Light on Bud Differentiation of Stem Explants in *P. hopeiensis**

Treatment	No. of Explants Inoculated	Bud Differentiation		Differentiation % on Different Dates					Growing State of Culture
		No.	%	June 13	June 15	June 20	June 27	June 29	
I	98	90	91.8	21.4	56.1	82.7	83.7	91.8	Medium in size, compact, light green, multiple buds, and uniform growth
II	59	50	84.7	0	20.0	40.7	62.7	84.8	Medium in size, compact, light green, fewer buds, and nonuniform growth
III	37	30	81.1	0	0	51.3	78.4	81.1	Small and loose, pale yellow, and a few yellow buds

*Alll explants were inoculated on May 25 , 1981.

the 9th day and secondary roots on the 19th day. In the fourth treatment, roots appeared even later (on the 12th day) with a frequency of only 71.48%, and secondary roots did not appear at all. In addition, the plant structures above the surface of the medium showed abnormalities. Stems and leaves of the plants in the first three treatments grew normally; in the fourth treatment, shoot tips gradually became dormant buds, and the leaves rolled and turned yellow.

Culture Requirements in Different Species

The ½ MS medium was suitable for most of the species in section Leuce Duby, but not in *P. davidiana*, in which no success has been achieved in 2 years. However, adventitious buds could be induced in the related composite hybrid 741. The composite hybrid 741 had a wide adaptation to media. Stem explants from the hybrid of *P. davidiana* × *P. alba* have also been cultured successfully in the ½ MS medium for both differentiation and rooting.

Adventitious buds and roots of *Chosenia macrolepsis* could be induced in the same ½ MS medium used for section Leuce Duby. But 10 days after formation of an intact plantlet, basal leaf began to turn yellow and the entire plant died within 20 days. This phenomenon may be caused by phenolic toxins. Normal plantlets have been grown in wide-spectrum rooting medium supplemented with an appropriate quantity of polyvinyl polypyrrolidone (PVP) (Lin et al., 1984).

Characteristics of Morphogenesis

MODES OF MORPHOGENESIS. Both somatic embryos and adventitious buds have been grown from the same piece of stem explant in *P. hopeiensis* (Lin et al., 1984), arising spontaneously after 20 days in culture. In general, the somatic embryos were not completely separated from the explants. Occasionally, intact plants were regenerated from somatic embryos in medium containing cytokinin but not auxin. In most cases, however, both adventitious buds and roots grew from one piece of explant. Such buds were cut for root initiation in test tubes when they grew to a sufficient height. Somatic embryos alone arose from explants taken from pollen-derived plants of *Populus simonii* × *P. nigra* (Zhang et al., 1981).

In the composite hybrid 741, somatic embryo formation was the main type of organogenesis. In addition to normal differentiation of stems and leaves, roots could also be initiated in a differentiation medium without NAA.

In *Chosenia macrolepsis*, adventitious buds were formed in differentiation medium. Roots were not differentiated even when the shoots grew up to 5 cm long.

ENHANCEMENT OF ROOTING ABILITY DURING SUBCULTURE. When explants were transferred to differentiation media for subculture after adventitious buds were initiated, the buds continued to proliferate after 20 days in subculture, so that a group of adventitious buds could be produced every 20–30 days. Explants from one flask can be divided among three to four flasks after one round of subculture. In our experiments, the rooting ability of adventitious buds produced by repeatedly subcultured explants has been enhanced. For instance, in the composite hybrid 741, roots could be induced in ½ MS medium with only 1.32 μM BA after five subcultures. Cuttings taken from 1-year-old plants derived in test tubes were more easily rooted than those taken from big trees, favoring factory production of plantlets. Rugini also found that rooting ability was enhanced by repeated subculture of *Olea europaea sativa* (Rugini, 1984).

TRANSPLANTATION OF PLANTLETS AND PLANT PERFORMANCE

Transplantation of Test Tube Plantlets

The factors that affect the survival of transplanted plantlets are complex. The following points must be considered during transplantation:

1. There must be no callus tissue at the root stem transition zone because it will block free connections between the stem and root tissues and cause difficulties with transport of nutrients and water for growth of the plant. Therefore, if such plantlets arise in tubes, the rooting medium should be modified.
2. Before transplantation, the test tube plantlets should be hardened and lignified by exposing them to daylight for 15–20 days, adapting them to outside environments.
3. The soil mix should be sterilized, usually with a solution containing 0.4% ferrous sulfate. Soil with a good texture can be made by mixing equal amounts of loam, peat, and sand (Dong et al., 1983). Vermiculite soaked with the major elements of MS medium can also be used (Chen et al., 1983).
4. Do not bury the root collar in the soil. Plants grew best when they were planted with soil just under the stem-root union.
5. Humidity is a very important factor. In our experiments, trans-

planted plants were covered with beakers for 10–20 days to achieve a high survival rate.

6. The appropriate temperature for the transplanted plants is 15–28°C. When the temperature is below 12°C in the winter, plantlets cannot survive, and some surviving plants transplanted in autumn will be in a dormant state. Under appropriate environmental conditions, the survival rate can reach 100% (Dong et al., 1983) (Table 5).

Observation of Offspring

Offspring derived from the explants taken from the same plant were morphologically uniform in *P. hopeiensis, P. tomentosa*, and *P. alba*. Transplants of *P. hopeiensis* and *P. tomentosa* showed the same growth pattern, and their tree and leaf forms showed no differences. Plants regenerated from three annual seedlings of the introduced species *P. grandidentata* showed three different phenotypes.

PROSPECTS

Each explant of *P. hopeiensis* could produce at least 1 million buds within 1 year, and about 80,000 intact plantlets could be obtained. In 1983, the regenerated *P. hopeiensis* plants which had grown for 2 years in the nursery were planted on roadsides. They are now growing normally.

Chen et al. (1983) have also achieved good results propagating *P. tomentosa* by tissue culture. They first induced intact plants by using dormant axillary buds from mature trees as explants and then regenerated plantlets by using leaves (lower half blades with leaf stalks) of the test tube plants. These plants were transplanted in the nursery in 1983. More than 3000 plants over 3 m in height have been obtained.

This technique now can be used for commercial production of plantlets and for rapid propagation of introduced rare strains or superior hybrids. Survival will be enhanced if the greenhouse has facilities for controling temperature, light, and intermittent misting.

At the 17th Meeting of the International Poplar Commission held in Ottawa, Canada, in October 1983, Frohlich and Weisgerber (1984) reported that superior strains of *P. tremula* had been asexually propagated by tissue culture on a large scale in the Federal Republic of Germany. The regenerated plants have been used for forest culture in Hesse and other locations in Germany. Fifty thousand plants were obtained from a triploid clone Astria by tissue culture for the first time in 1981. Three multiclonal varieties composed of 49 clones were further proliferated in 1983, and about 150,000 plantlets were obtained within the year. One

TABLE 5. The Results of Transplantation of Seedlings of *P. hopeiensis*

Transplanting Date (Day/Month)	No. of Plants Transplanted	Plant Age* (Days)	Days for Light Hardening	Days with Covers	Survival		Remarks
					No. Plants	Percentage (%)	
10/11	16	56	10	15	15	93.8	
20/11	30	76	15	10	28	93.3	
24/11	22	78	10	15	22	100.0	
1/12	8	76	30	15	8	100.0	Greenhouse temperature
4/12	21	79	20	20	20	95.2	controlled at 12–27°C
6/12	21	82	30	20	21	100.0	by central heating from
25/12	35	56	20	15	33	94.3	November to April, next
10/1	25	66	15	15	21	84.0	year
30/1	50	74	18	15	46	92.0	
17/3	49	120	13	15	42	85.7	
17/4	30	102	32	15	24	80.0	
18/5	22	91	28	20	20	90.9	Temperature at 15°–30°C
19/6	30	90	19	15	23	76.7	Greenhouse was covered
20/7	28	115	15	15	19	67.9	with reed mats, with about 20% light leaking
18/8	15	84	18	15	6	40.0	through at 20°–30°C

*Number of days from transfer to rooting medium to transplantation into pot.

year later, these plants had grown to 187–257 cm in height with basal diameters of 1.3–2.5 cm. The German workers also tested disease resistance and stress tolerance of asexual lines in section Aigeiros and section Tacamahacca (*P. balsamifera*) by artificial infection on test tube plants. A reliable technique for breeding plants resistant to bacterial canker has been established.

It is also possible to overcome incompatibility in distant hybridization by culturing ovules in vitro. Li et al. (1985) have regenerated plantlets from young ovules of the intersectional hybrid (*P. simonii* × *P. diversifolia*). This cross normally would abort (Li Wentian et al., unpublished data). Tissue culture techniques of Salicaceae clearly have broad prospects not only in commercial tree production but also for basic research purposes.

KEY REFERENCES

Dong, M., Huang, Q., and Lin, J. 1983. Propagation of *Populus hopeiensis* through tissue culture. *Collect. Res. Pap. Forest Res. Inst. Chin. Acad. Forest. Sci.* **1**:50–59 (in Chinese).

Lin, J., Dong, M., and Huang, Q. 1980. Tissue culture of three species of Section Leuce Duby (C. Genus *Populus*). *Sci. Silvae Sin.* **16** (suppl.):58–64 (in Chinese, English abstract).

———, Dong, M. S., and Huang, Q. C. 1983. Tissue culture of *Populus adenopoda*, and *P. grandidentata*. *Plant Physiol. Commun.* **2**:137 (in Chinese).

———, Huang, Q. C., and Dong, M. S. 1984. Tissue culture of *Chosenia macrolepsis*. *Plant Physiol. Commun.* **2**:39 (in Chinese).

Xu, W., Ma, C., Jiao, Q., Lin, J., and Tu, H. 1958. *Poplar Trees*. China Forest Press (in Chinese). Beijing.

REFERENCES

Brown, C. L. and Sommer, H. E. 1982. Vegetative propagation of dicotyledonous trees. In: *Tissue Culture in Forestry*. pp. 109–149. Martinus Nijhoff, The Hague.

Chen, W., Guo, D., Yang, S., and Ciu, C. 1983. Rapid propagation of *Populus tomentosa* by tissue culture. *Bot. Res.* **1**(1):135–138 (in Chinese, English abstract).

Frohlich, H. J. and Weisgerber, H. 1984. Research on in vitro techniques within the framework of poplar breeding—results and future trends. Presented at the meeting during the 17th International Poplar Commission in Ottawa, Canada, October 1–4, 1984, p. 14.

International Poplar Commission. 1979. *Poplar and Willow*. FAO of United Nations, Rome (in Chinese).

Jiang, H., Han, Y., Zhao, H., Gu, W., Yang, C., Fan, K., and Xing, X. 1980.

Breeding of the superior hybrid clone 741 of Poplar. *Hebei Linye Keji* **1**:72–75 (in Chinese).

Li, C., Zhang, L., Guo, F., and Zhang, G. 1979. Bud culture in vitro of a new variety of poplar—*Populus davidiana* Dode × *P. bolleana* Louche. *Hereditas* **1**(4):27–28 (in Chinese).

Lin, J., Dong, M., and Huang, Q. C. 1984. Morphogenesis in tissue culture of *Populus hopeiensis. Collect. Res. Pap. Forest Res. Inst. Chin. Acad. Forest. Sci*. **1**:64–65 (in Chinese).

Ren, Y. 1984. Induction of test tube plants from *P. alba* × *P. bolleana. Plant Physiol. Commun*. **2**:33–34 (in Chinese).

Rugini, E. 1984. In vitro propagation of some olive (*Olea europaea sativa* L.) cultivars with different root ability, and medium development using analytical data from developing shoots and embryos. *Sci. Hortic*. **24**(2):123–124.

Von Arnold, S. 1979. Induction and development of adventitious bud primordia on embryos, bud and needles of Norway spruce (*Picea abies* (L). Karst) grown in vitro. *Report of Institute of Physiological Botany*, University of Uppsala, Sweden **51**:1–32.

Wang, H. 1982. Plantlet regeneration from pomegranate leaf. *Plant Physiol. Commun*. **6**:36 (in Chinese).

Winton, L. 1968. Plantlets from aspen tissue cultures. *Science* **160**:1234–1235.

———. 1970. Shoot and tree production from aspen tissue culture. *Am. J. Bot*. **57**(6):904–909.

———. 1971. Tissue culture propagation of European aspen. *Forest Sci*. **17**(3):348–350.

——— and Huhtinen, O. 1976. Tissue culture of trees. In: *Modern Methods in Forest Genetics*. pp. 243–264. Springer-Verlag, Berlin.

Zhang, L., Zhang, Y., Wang, F., and Liu, S. 1981. Induction of somatic embryoid formation of poplar. *Sci. Silvae Sin*. **17**(4):426–427 (in Chinese).

Zhao, J. 1982. Techniques of sowing and seedling raising of *Chosenia macrolepsis. Jilin Linye Keji* **2**:37–39 (in Chinese).

Zhou, L. and Gui, Y. 1983. Organogenesis and plantlet formation of *P. euphratica* cultured in vitro. *Bot. Res*. **1**(1):125–126 (in Chinese, English abstract).

CHAPTER 9

Poplar: Anther Culture

Zhihua Lu and *Yuxi Liu**

INTRODUCTION

Populus L. was the first tree species used for in vitro production of haploids. Anther culture has been carried out using *Populus maximowiczii* Aenty., *P. deltoides* Marsh, and (*P. siebldii* × *P. grandidentata*) hybrids. Plantlets have been differentiated from callus, but their origins were uncertain because of lack of haploids in chromosome examinations (Toru, 1974). Since this early work, anther culture in poplar has been performed successfully in China and a series of papers on the induction of haploid plants in different species and hybrids of poplar have been published (Table 1).

Poplar anther culture had been studied in our college for many years, dating back to 1974, when pollen plantlets were first induced. The regenerated plants were successfully transplanted in 1975. Pollen plants have been induced from 10 species and hybrids: *P. simonii* Carr., *P. pseudosimonii* Carr., *P. berolinensis* Dipp., *P. simonii* Carr. × *P. nigra* L., *P. pseudo-simonii* Carr. × *P. pyramidalis* Roz., *P. harbinensis* Wang et Skv. × *P. pyramidalis* Roz., *P. berolinensis* Depp. × *P. pyramidalis*

*English translation by Zhou Xiaxian.

TABLE 1. Poplar Species and Hybrids in Which Pollen-Derived Plants Have Been Obtained

Species	Reference
Populus nigra L.	Wang et al., 1975
P. simonii Carr × *P. nigra* L.	Laboratory of Breeding, Institute of Forestry, Heilongjiang Province, 1975
P. ussuriensis Komar.	Heilongjiang Province, 1975
P. canadensis Moench × *P. koreana* Rehd.	Laboratory of Forest Breeding, Forestry College of Northeast China, 1977
P. harbinensis Wang et Skv. × *P. pyramidalis* Roz.	College of Northeast China, 1977
P. berolinensis Dipp.	College of Northeast China, 1977
P. alba L. × *P. simonii* Carr.	Zhang et al., 1978
P. pseudo-simonii Kitagawa × *P. pyramidalis* Roz.	Zhang et al., 1978
P. berolinensis Dipp. × *P. pyramidalis* Roz.	Ying Kou Institute of Populus, Liaoning Province, 1978
P. simonii Carr.	Ying Kou Institute of Populus, Liaoning Province, 1978
P. simonii Carr. × *P. pyramidalis* Roz.	Ying Kou Institute of Populus, Liaoning Province, 1978
P. pseudo-simonii Carr.	Ying Kou Institute of Populus, Liaoning Province, 1978
P. euphratica Oliv.	Zhu et al., 1978

Roz., *P. canadensis* Moench × *P. koreana* Rehd., and *P. pekinensis* (*L. Henrf*) Hsii. Plantlets from *P. simonii* Carr. × *P. nigra* L., *P. berolinensis* Dipp., *P. canadensis* Moench × *Koreana* Rehd., *P. berolinensis* Dipp. × *P. pyramidalis* Roz., *P. simonii* × *P. pyramidalis* Roz., and *P. pseudo-simonii* have been successfully transplanted and are now 7–9 years old. Eight-year-old pollen-derived trees of (*P. simonii* Carr. × *P. nigra* L.) hybrids reached 13.5 m in height, and *P. berolinensis* Dipp. reached 9.7 m (Fig. 1).

POPLAR ANTHER CULTURE TECHNIQUES

Culture Procedure and Main Manipulations

Native commercial varieties, imported varieties, and selected hybrids were used in this study.

CULTURE PROCEDURES

1. Induction of pollen callus. The optimal medium was MS basal medium supplemented with 9.3 μM KIN, 9 μM 2,4-D, and 0.1 M sucrose.
2. Differentiation of buds from pollen callus. The optimal medium was

Figure 1. Eight-year-old pollen trees (*P. simonii* Corr. × *P. nigra* L.) 13.5 m in height.

MS basal medium with half the amount of macroelements, 4.4 μM BA, 1.1–2.7 μM NAA, or 1.1–2.9 μM IAA, 0.06 M sucrose.

3. Induction of healthy plantlets. Same medium used as for differentiation supplemented with 4.6 μM KIN, 0.5–2.7 μM NAA, or 0.5–2.9 μM IAA, 0.6 M sucrose.
4. Rooting was achieved on differentiation medium plus 4.4 μM NAA, 1.1 μM IAA, and 0.06 μM sucrose to induce well-developed roots.

CULTURE METHOD

1. Flowering cuttings were taken from healthy male plants and grown indoors to stimulate germination of buds from late winter to early spring.
2. The pollen development stage was determined when the flower bud began to swell. Pollen size and development status were observed and recorded. The acetocarmine squash method was used for identifying pollen development stages.
3. The flower buds were taken when the pollen was at the mononucleate stage in which the nucleus was adjacent to the periphery.

The buds were soaked in 10% sodium hypochlorite solution for 10 min after the removal of scales and calyx, and then rinsed in sterile, distilled water two to three times under aseptic conditions.

4. Anthers were removed and inoculated in medium for callus induction, 100–200 anthers per flask.
5. Anthers were cultured in the dark at 25–27°C.
6. Callus 3 mm in diameter was transferred into the differentiation medium, the room temperature was about 23°C, and the temperature at night was not less than 16°C. The illumination was supplemented with fluorescent lamps for 8–9 hr per day.

Keys to Successful Haploid Induction

IDENTIFICATION OF SOMATIC CELL CALLUS AND POLLEN CALLUS. The time of callusing after inoculation of anthers differed among the species studied. Callus appeared in 5–6 days in (*P. simonii* Carr. × *P. pyramidalis* Roz.) hybrids; it appeared after more than 20 days in *P. berolinensis* Dipp. There were two types of callus: (1) Loose, soft, and semitransparent callus appeared 5–6 days after inoculation at the site between filament and anther or appeared later from the break in anthers. The loose callus grew rapidly and filled the flask after more than 20 days, but no organ differentiation was observed. (2) Callus that was compact and hard in texture, milky or milky yellow in color, grew slowly and turned green gradually after transfer into differentiation medium. This callus could differentiate into plantlets.

It seemed that callus of the compact type was pollen-derived, and that of the loose type was derived from somatic cells. Our cytological observations demonstrated that callus appearing within 20 days after inoculation was of the loose type and was from somatic cells. Cytological observation of the early compact callus demonstrated, however, that most of these cells were diploid, although compact callus that appeared 20 days after inoculation contained mostly haploid cells. We, therefore, concluded that most of the compact callus appearing 20 days after inoculation and later contained pollen-derived cells.

SELECTION OF OPTIMAL EXPLANTS. Callus induction frequencies were markedly different for different explants. Different poplar species inoculated in MS medium supplemented with 9 μM 2,4-D, 9.3 μM KIN, and 0.1 M sucrose showed remarkable variation in callus induction. For instance, *Populus davidiana* Dode, *Populus alba* L., and *Populus balleana* Louche of section Leuce produced no callus. Callus was more easily induced from hybrids between section Aigeiros and section Tacamahacca, as these calli showed differences in morphology and frequency of induction.

Table 2 indicates that the relative frequency of the two types of callus, depending on the poplar species. The total induction frequency for (*P. simonii* Carr. × *P. nigra* L.) hybrids was only 11.9%, but almost half of them (5.1%) produced compact callus. On the other hand, the total induction frequency of (*P. simonii* Car. × *P. pyramidalis* Roz.) hybrids was as high as 50%, but compact callus accounted for only 0.7%. It is evident in Table 2 that the total average frequency was 16.6% and the maximum was 50.6%. As for compact callus, the total average frequency and the maximum frequency were 2.6% and 5.0%, respectively. The variation of callus induction frequency in different species suggested that the medium and supplements were species-specific. It is, therefore, necessary to develop specific media for each species in order to enhance the induction frequency.

Variations in induction frequency also existed among individuals of the same poplar species at the same age. For instance, the induction frequencies for (*P. simonii* Carr. × *P. nigra* L.) hybrids studied by Northeast Forestry College, by Nenjiang Forestry Institute, and by Qiqihaer County and Botanical Garden of the Heilongjiang Province were 8.1%, 11.9%, and 18.8%, respectively. In the first two cases, one-third of the callus was of the compact type; in the later case, two-thirds was compact.

SELECTION OF SUITABLE MEDIA AND SUPPLEMENTS. *P. berolinensis* Dipp. was used to test different basal media for callus induction frequency. MS, H, and N6, and 20% potato media were used with 9.3 μM KIN and 9.0 μM 2,4-D. Very little transparent loose callus was formed on 20% potato medium. All callus formed on N6 medium, where the induction frequency was more than 30%, was of the compact type. Callus resembled cotton wool, was pink, stopped growing when 2 mm in size, turned brown, and died. The induction frequency on H medium was higher, nearly the same as on MS medium, but most of the callus was loose. In general, MS medium was the preferred medium.

The effects of different concentrations of 2,4-D and KIN on callus induction were studied by using MS medium. (*P. simonii* Car. × *P. nigra* L.) hybrids were used as starting material, and different concentrations (9, 18, 27 μM) of 2,4-D were added separately to MS medium. The highest induction frequency occurred when the concentration of 2,4-D was 9.0 μM, yet most of the callus was of the loose type. For a constant amount of 2,4-D (9.0 μM), the medium was supplemented with different concentrations (2.3, 4.6, 9.3 μM) of KIN (Table 3). The lower the concentration of the KIN added, the sooner callus was formed and the higher the induction frequency. This callus was mostly loose in type. The callus induction frequency with *P. berolinensis* Dipp. reached 80%, and

TABLE 2. Results of Callus Induction in Various Poplar Species and Their Hybrids

Species	No. of Anthers Inoculated	No. of Calli Obtained			Induction Frequency (Total, %)	Induction Frequency (compact type, %)
		Compact	Loose	Total		
P. simonii × *P. nigra* L.	14,200	717	977	1694	11.9	5.0
P. simonii × *P. pyramidalis* Roz.	1740	12	868	880	50.6	0.7
P. berolinensis Dipp.	4250	147	231	378	8.9	3.5
P. berolinensis Dipp. × *P. pyramidalis* Roz.	5850	209	689	898	15.4	3.6
P. pekinensis (L. Henrf) Hsii	1500	63	87	150	10.0	4.2
P. canadensis Moench × *P. koreana* Rehd.	3500	54	147	201	5.8	1.5
P. harbinensis Wang et Skv. × *P. pyramidalis* Roz.	3450	29	158	187	5.4	0.8
P. pseudo-simonii Kitagawa × *P. pyramidalis* Roz.	1300	23	315	338	26.0	1.8
P. pseudo-simonii Carr.	900	16	127	143	15.9	1.8

TABLE 3. Effect of Different Concentrations of KIN on Callus Induction

Species	Concentration of KIN (μM)	No. of Anthers Inoculated	Loose Type Calli		Compact Calli		Date of Appearance of Callus after Inoculation (Day)
			No.	%	No.	%	
P. berolinensis	2.3	1000	800	80	25	2.5	22nd
Dipp.	4.6	400	200–240	50–60	10	2.5	25th
	9.3	1600	241	15	88	5.5	27th
P. simonii Carr. ×	2.3	400	360–380	90–95	13	3.3	8th
P. nigra L.	4.6	1000	500–600	50–60	3	0.3	25th
	9.3	1000	5	0.5	8	0.8	25th

that of (*P. simonii* Carr. × *P. nigra* L.) hybrid was as high as 90–95% on a medium with a KIN concentration of 2.3 μM. Most of the loose callus appeared as foam. This callus stopped growing when it was as large as a bean. Growth of this callus originated from the external anther wall. The cells were easily separated, and compact callus was seldom found. The foamlike callus disappeared on medium with 4.6 μM KIN, where there was much more loose callus which grew quickly after transfer and did not differentiate. In medium with 9.3 μM KIN, loose-type callus was greatly decreased and compact callus increased.

The remarkable effect of KIN concentration on the rate of abnormal division of pollen, development after division, and callus texture was observed cytologically. Ten percent of the pollen divided into bicellular structures very quickly on medium containing 2.3 μM KIN after 3 days of culture. However, 4 days later, pollen at bi- and multicellular stages quickly disintegrated and decreased, and a large number of somatic cell clusters were formed. When 9.3 μM KIN was used in the medium, only 2% of the pollen reached the bicellular stage after 3 days of culture. However, pollen at the bi- and multicellular stages increased to a certain extent after 7 days of culture. Disintegration of pollen cells was rare, and very little somatic cell callus was formed. It appeared that at 2.3 μM KIN, pollen underwent normal division, but callus of somatic cells developed quickly, too, and at late culture stages, a great number of abnormally dividing pollen aborted quickly. When the supplement contained 9.3 μM KIN, though, the abnormal pollen division was inhibited to some extent, but the growth of somatic cells was inhibited simultaneously, so that the percentage of pollen at bi- or multicellular stages increased gradually and only a few of them aborted.

OPTIMAL TIME FOR COLLECTING FLOWERING CUTTINGS. Ten collections of flowering cuttings were gathered between January 24 and June 28, 1975, to ascertain the optimal inoculation date. Significant variation in the texture and frequency of callus formation was detected among explants from different cuttings. In general, the closer in time to the natural flowering period (early May) that the flowering cutting was taken, the higher the callus induction frequency and the higher the compact callus rate. It is probable that the anthers are fully developed only during the flowering period.

To prolong the period of explant availability, cuttings were gathered in the season near thaw when they were still frozen and then kept at 0–3°C. The flowers blossomed normally after water culture from March 20 to June 30 in Heilongjiang District. The frequency of callus induction in anthers from cuttings which had been exposed to frost was raised markedly from 20% to 60–80%, but the proportion of compact callus did not increase.

Enhancing the Frequency of Callus Differentiation into Plantlets

DETERMINATION OF SUITABLE GROWTH REGULATOR CONCENTRATIONS. A high ratio of KIN to auxin in the medium is necessary for buds to emerge before roots. If roots were formed first, it was difficult subsequently to obtain buds. In the experiments described below, MS media with different combinations of KIN, BA, NAA, and IAA were compared for development of calli of hybrids of (*P. simonii* Carr. × *P. nigra L*).

Table 4 shows marked differences in plant regeneration frequencies between media supplemented with KIN and BA. Of 138 calli transferred to MS medium with KIN, 24 turned green and formed plantlets. A small proportion of the remaining callus formed roots, and the rest turned brown and died. The overall plant regeneration frequency was 17.4%. Among 74 calli transferred to MS medium supplemented with BA, 51 calli differentiated into plantlets (69%). The effect of BA on differentiation was clearly better than that of KIN. Desirable results were also obtained from the addition of 4.4 μM or less BA (Table 4).

BA and KIN have specific effects on differentiation. In medium with BA, numerous buds differentiated with high density and were clustered, yet the plantlets were thin, weak, and poorly lignified. With KIN, fewer buds were differentiated, but they usually grew into single strong plantlets with good lignification (Fig. 2). A two-step differentiation

TABLE 4. Effect of MS Medium Supplemented with BA or KIN on Plantlet Differentiation

Medium with Supplements (μM)	No. of Calli Transferred	No. of Shoot-Forming Calli	Shooting Frequency (%)
MS + KIN 13 + IAA 4.4	12	1	8.3
MS + KIN 9.3 + IAA 1.1–4.4	67	15	22.4
MS + KIN 8.3 + IAA 3.3	28	2	7.1
MS + KIN 5.5 + IAA 1.1–2.9	31	6	19.4
	Total 138	Total 24	Mean 17.4
MS + BA 8.8 + NAA 1.1–2.7	35	23	65.7
MS + BA 5.3 + NAA 1.1–2.7	37	26	70.3
MS + BA 4.4	2	2	100.0
	Total 74	Total 51	Mean 68.9

Figure 2. A well-lignified plantlet in flask.

method was adopted to improve the frequency and quality of plantlet differentiation, and favorable results were obtained. The method consisted of two steps: (1) Callus was transferred onto MS medium with 4.4 μM BA for bud differentiation, where one callus could give tens or hundreds of buds; (2) callus with buds was transferred onto MS medium + 9.3 μM KIN + 1.1–2.9 μM NAA (or IAA). Some of the buds grew as healthy plantlets, and the rest were discarded.

OPTIMAL TEMPERATURE FOR CULTURE. The temperature for callus differentiation should not be too high. If the temperature reached 28°C, callus generally stopped differentiating; when the temperature dropped to about 23°C, growth and differentiation began again.

Rooting of Plantlets

SUITABLE CONDITIONS. Root formation could not occur at the time of shoot differentiation from callus. So, rootless plantlets with well-formed stem and leaves were excised from callus and transferred into a specific medium for root induction. The time of transfer to such medium is critical for survival. The upper part of the shoot and the leaves may dehydrate when the root system is formed if the plantlet is sufficiently mature. The plantlet should be more than 4 or 5 cm in height and be at a fairly high level of lignification for root formation to be successful.

SUITABLE AUXIN CONCENTRATION. Good results have been obtained when NAA and IAA were used together in medium without KIN. It has been repeatedly demonstrated that MS medium with half macroelements containing 4.4 μM NAA and 1.1 μM IAA gave good results in rooting. White root tips were formed within 8–10 days, and an intact root system was formed in 20 days, after transfer of the rootless plantlets to this medium. A period of 5 months usually elapsed from anther inoculation to plantlet transplantation. For instance, in the case of (*P. simonii* Carr. × *P. nigra* L) hybrids, inoculation took place in February, and plantlets were ready to be transplanted to pots in July.

Transplantation of the Plantlets into Pots

The plantlets were very easily infected when they were transplanted from flasks to pots. On occasion, 6 or 7 days after transplantation, the leaves withered, and the stems dehydrated because of loss of water. On another occasion, the basal or upper part of the stem turned partially black and the leaves withered, apparently because of *Coryneum populinum* Bres. Lodging and *Helicobasidium* infection were also observed. Survival rates were enhanced as follows:

1. Using healthy plantlets with well-formed roots, a main stem, and good lignification for transplantation
2. Transplanting while the root system was white and fresh
3. Sterilizing the soil before transplantation and avoiding excess irrigation of transplants
4. Culturing at an appropriate room temperature (18–22°C) with sufficient light and moisture and good ventilation
5. Transplanting the plantlets during the growth season: in summer or autumn, or in spring of the next year
6. Transferring the surviving plantlets for dormant culture under cold conditions immediately after full lignification and after growth had stopped
7. Keeping the plantlet together with the soil when it was transferred from the pot into the nursery

POPLAR ANTHER CULTURE

Callus could be grown from both somatic cells and pollen in anther culture. The study of microsporogenesis in nature, as well as androgenesis during anther culture, is necessary to determine callus origin.

Anthers of different development stages were fixed in ethanol and acetic acid (3:1) for 4–24 hr and kept in 70% ethanol. Intact anthers were stained in a propionic acid–anhydrous trichloroacetaldehyde–iron haemotoxylin solution for 48–72 hr or longer. Anthers were then soaked in alkaline tap water for a few minutes to make the stain blue, and 45% propionic acid was used for color differentiation. Anthers could be stained with PIC CH after preparation of a smear for microscopic examination.

Induction of callus from cultured anthers should mainly produce pollen callus. Research should emphasize development processes in androgenesis and factors that affect the origin and development of abnormal microspores under natural conditions.

Types of Microspore Development

Poplar microspores began to divide after 2 days in culture. One route of development has been called Path B by Sunderland. (See Handbook Vol. 1, Chapter 6 for detailed description of paths) Two daughter cells are formed by an equal division of monokaryotic pollen and continue to grow through successive divisions. The other route is Path A. Two daughter cells are formed through an unequal division of monokaryotic pollen. The vegetative cell then divides further, while the generative cell is forced to the side. No pollen development further than six cells resulted from unequational division except for one case of unequal eight-cell pollen, which aborted. As shown in Table 5, Type B was the more common type of microspore development.

Effect of Different Development Stages on Pollen Dedifferentiation

The potential for callus formation varied with the developmental stage of the pollen. If anthers, which are mostly at mononuclear stages (including the mid- and late-mononuclear stages), were used for inoculation, a higher induction frequency would be obtained. Histological examination showed that microspore development varied for different anthers. Anthers in which all of the pollen aborted accounted for one-third or one-half of all anthers observed. Some of the anthers had aborted pollen and only a small amount of viable pollen, but others contained much more viable pollen, even though more than one-fifth of the pollen still aborted.

Inspection of anthers with more viable pollen revealed that the first mitosis in cultured mononuclear pollen occurred within 2–3 days. Multicellular pollen was observed on the 7th day of culture and increased thereafter. Individual multicellular pollen broke out of the other wall on the 15th day, and callus began to emerge from the anther on the 20th day. However, large-scale callus formation did not occur until 30–40 days after inoculation.

Mononuclear and binuclear giant pollen was observed in pollen after in vitro culture. The giant pollen grains were thick in exine and filled with

TABLE 5. Development of Microspores Inoculated at Mononuclear Stage in *P. berolinensis* Dipp.

Days after Inoculation		Mono-nucleus	Two nuclei (Equal Division)	Two nuclei (Unequal)	Bicells (Equal)	Bicells (Unequal)	Multi-nuclear Pollens	Multi-cells (Equal)	Multicells (Unequal)	Aborted	Total
3	Number of pollen	261	100	68	18	5	10	—	—	550	1012
	%	25.79	9.8	6.8	1.79	0.49	0.99	—	—	54..34	100
7	Number of pollen	232	112	94	37	11	5	5	1	640	1137
	%	20.40	9.85	8.27	3.25	0.97	0.44	0.44	0.09	56.29	100
10	Number of pollen	29	17	8	39	7	5	25	—	750	880
	%	3.30	1.93	0.91	4.43	0.80	0.57	2.84	—	85.22	100
15	Number of pollen	10	6	—	8	—	—	22	—	340	386
	%	2.60	1.55	—	2.07	—	—	5.70	—	88.08	100
30	Number of pollen	2	1	—	11	5	—	46	3	480	548
	%	0.36	0.18	—	2.01	0.91	—	8.39	0.55	87.60	100

starch. Starch could be removed by staining with PIC CH and heating until boiled so that internal changes of the mononucleus and unequational and equational binuclei were visualized. As the culture progressed, most of the giant pollen grains stained indistinctly until the pollen aborted.

The results indicated that poplar pollen callus was induced mainly from mononuclear stage pollen, and other stages were not suitable. It is notable that variation existed in development of anthers from different parts along the flower spike. If the anthers from the basal part were at the late mononuclear stage, those from the middle and upper parts might be at the midmononuclear stage or tetrad stage. It was, therefore, quite difficult to choose anthers at a definite stage accurately for inoculation.

CHROMOSOME VARIATION AND SPONTANEOUS DOUBLING IN POLLEN-DERIVED POPLAR

Most species in the genus *Populus* are natural diploids ($2n = 2x = 38$), with only a few triploids and tetraploids (Storova and Kits, 1976). Artificially induced haploids and polyploids generally consist of mixoploids with various chromosome numbers. Mixoploids were produced from young branches treated with low concentrations of colchicine (Larsen et al., 1970). In *P. trichocapa*, a "possible haploid" with a chromosome number of 17–22 could be induced by parthenogenesis. In this plant, few cells with a chromosome number less than 28 and more diploid cells with chromosome numbers of 34–38 were found (Stettler and Bawa, 1971). Variation in chromosome number also existed in haploids induced through parthogenesis in aspen, where some were haploids, some were mixoploids composed of haploid and diploid cells, and others were diploids (Illies, 1976). The progeny of parthenogenic plants of *P. tremuloides* were mixoploids consisting of haploid and diploid cells (Winton and Einspahr, 1970).

The somatic cells of poplar plants derived from anther culture in (*P. simonii* Carr. × *P. nigra* L.) hybrids or *P. berolinensis* Dipp. were mixoploids to a certain extent. It is possible that haploid, diploid, polyploid, and various aneuploid cells could coexist in the same plant, branch, and root tip; where haploid cells were dominant at an early stage, diploid cells may predominate later.

Root tip squashes were prepared for chromosome inspection. The primary root was used from plantlets cultured in flasks; the root that emerged from a water-cultured shoot was taken for examination. The root tips were pretreated by keeping them at low temperature (0–3°C) for 24 hr or soaked in saturated solution of dichlorobenzene for 2.5–4 hr. The root tips were fixed in ethanol and acetic acid (3:1) for 4–24 hr and stored in 70% ethanol. The materials were dissociated with 1 N HCl at 60°C for 10–15 min. Two staining methods were used: phenol-magenta preparation II and iron-hematoxylin (Zhu, 1982). The former was simple to use and was employed for general observations and counts of chromosome numbers, and the latter was used for photomicroscopy. Iron-hematoxylin was used as a mordant for 15–24 hr, then stained with 0.5% hematoxylin for 0.5–1.5 hr, differentiated with 45% acetic acid for 1–3 hr, and finally squashed.

Karyotypes observed in this study were classified into four groups: (1) haploidlike cells with chromosome numbers less than 22; (2) diploidlike cells with chromosome numbers of 30–42; (3) aneuploidlike cells with chromosome numbers of 23–29; and (4) polyploidlike cells with chromosome numbers over 50.

Chromosome Inspection of Plantlets Cultured in Flask

Roots from 11 flask-cultured plantlets of (*P. simonii* Carr. × *P. nigra* L.) hybrids were chosen at random for chromosome examination. The results are summarized in Table 6. In 10 of 11 plantlets, 88% of somatic cells on average were haploid. These plantlets were clearly generated from differentiated pollen. In 1 plantlet 80% of the cells were diploid and 20% were aneuploid. Probably it also arose from pollen but its chromosome number doubled at an early stage. In general, the plantlets in culture flasks were mixoploid with the predominant number of cells (88%) haploid, and with 12% diploid and 8% aneuploid cells (Fig. 3).

TABLE 6. Root Tip Chromosomes from Pollen Plantlets Cultured in Flask

		Percentage of cells (%)		
Plantlet No.	Number of Cells Observed	Haploid Somatic Cells	Diploid Somatic Cells	Aneuploid Somatic Cells
1	36	80.6	8.3	11.1
2	37	100.0	0	0
3	59	86.4	5.1	8.5
4	10	0	80.0	20.0
5	10	80.0	10.0	10.0
6	39	89.7	7.7	2.6
7	14	100.0	0	0
8	17	100.0	0	0
9	52	90.4	0	9.6
10	46	63.0	10.9	26.1
11	29	89.7	10.3	0

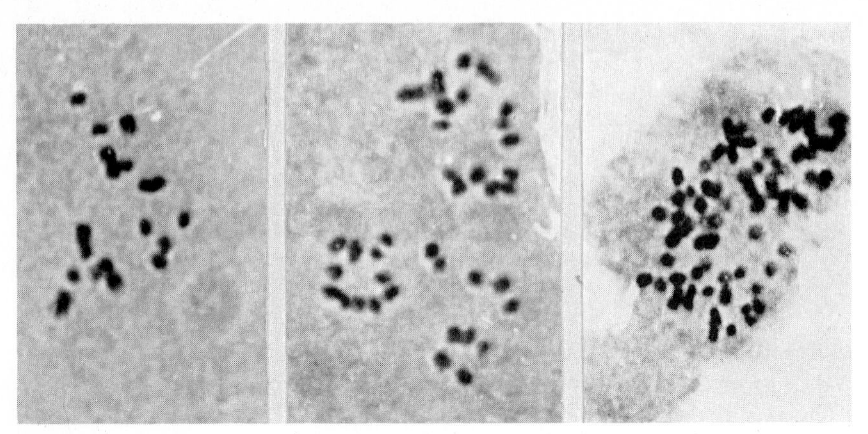

Figure 3. Variation in chromosome number in somatic cells of pollen-derived trees (haploid [$2n$ = 19], diploid [$2n$ = 38], triploid [$2n$ = 57]).

Chromosome Inspection of Anther Culture Plants Growing in the Nursery

Five plants were taken from each species for chromosome analysis in 1979, 1981, and 1983 after they had been transplanted to garden plots (Table 7 and Table 8). The conclusions were as follows:

1. Once the pollen plants had been transplanted in soil, their chromosome numbers began to stabilize. The frequency of haploid cells gradually decreased and diploid cells concomitantly increased as the plants grew. The process reflects the tendency to spontaneous doubling. The phenomenon held true not only in the *P. simonii* × *P. nigra* and *P. berolinensis* Dipp. plants examined, but also in most other species. For instance, Table 7 indicates that diploid cells increased in (*P. simonii* Carr. × *P. nigra* L) hybrids after aging, from 5% in 1976 to 50% in 1979, 85% in 1981, and 91% in 1983. A similar process occurred in *P. berolinensis* Dipp. However, no consistent changes were found in the percentage of diploid cells in control plants, the proportion fluctuating between 90% and 100%. The change in proportion of haploid and diploid cells is shown in Fig. 4.
2. The rate of spontaneous doubling seems to proceed more rapidly at early stages and more slowly at late stages. The doubling process may be completed at the age of 7–8 years, when homozygous diploid levels were reached.

Figure 4 shows that the percentage of diploid cells in pollen plants of the two species increased rapidly from 1976 to 1981. In the case of (*P.*

TABLE 7. Results of Three Chromosome Inspections for Anther-Derived Plants during Growth in Nursery

Materials	Year of Inspection	Number of Cells Observed		Percentage of Cells with Various Chromosome Numbers							
				Haploid		Diploid		Aneuploid		Polyploid	
		1*	2†	1	2	1	2	1	2	1	2
Anther-derived plants	1979	181	190	24	26	49	50	26	20	1	4
	1981	581	569	13	5	75	85	12	8	0	2
	1983	516	632	1	4	92	91	3	4	4	1
Controls	1979	30	26	0	0	100	100	0	0	0	0
	1981	109	52	0	0	100	91	0	0	0	9
	1983	95	68	0	8	100	92	0	0	0	0

*_P. berolinensis_ Dipp.
†(_P. simonii_ Carr. × _P. nigra_ L.) hybrid

TABLE 8. Comparison of Chromosome Numbers between Anther-Derived Plants and Cuttings in *P. berolinensis* Dipp.

Plant No.	Origin	Tree Age (Years)	No. of Cells Examined	Percentage of Different Groups of Cells		
				Haploids	Diploids	Others
1	Anther-derived	7	82	15.9	73.1	11.0
	plant cutting	2	130	6.9	81.5	11.6
2	Anther-derived	7	107	14.0	76.4	9.6
	plant cutting	2	135	10.3	79.3	10.4
3	Anther-derived	7	105	15.3	68.5	16.2
	plant cutting	2	130	10.8	79.1	10.1
4	Anther-derived	7	140	13.6	69.3	17.1
	plant cutting	2	125	6.4	78.4	15.2
5	Anther-derived	7	99	14.1	76.8	9.1
	plant cutting	2	113	5.3	84.1	10.6

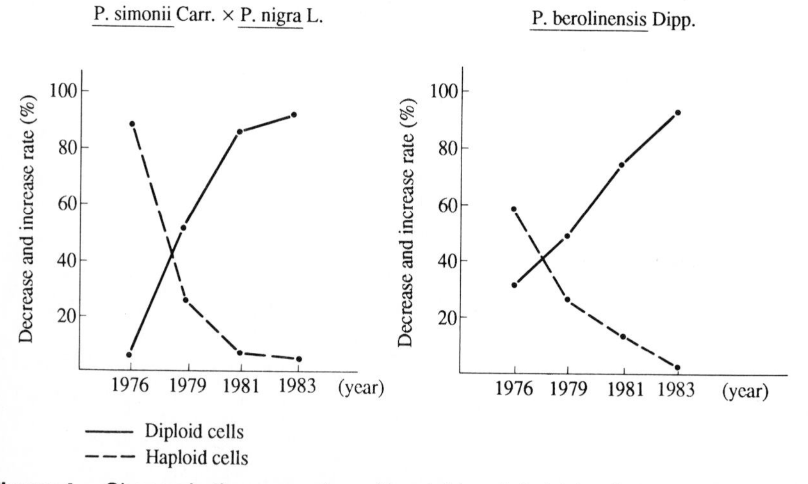

Figure 4. Change in the proportion of haploid and diploid cells in growing pollen-derived plants.

simonii Carr. × *P. nigra* L.) hybrids, diploid cells increased an average of 17.5% per year; in *P. berolinensis* Dipp., they increased 13% per year. However, from 1981 to 1983, the increase in diploid cells for both species was sharply reduced. It increased only 3% and 3.5% per year for (*P. Simonii* Carr. × *P. nigra* L.) hybrid and *P. berolinensis* Dipp., respectively. These figures indicate that once a large number of diploid somatic cells accumulate, the doubling process is slowed. The time from plantlet differentiation to completion of spontaneous doubling is approximately

7–8 years. This is particularly interesting because poplar trees of such an age are just coming to their flowering period and may, therefore, be able to flower, bear fruit, and be crossed for breeding in homozygous diploid condition.

Spontaneous Chromosome Doubling in Plants from Anther Culture after Reproduction by Cuttings

In 1980, some of the anther-derived plants were reproduced by cuttings for comparison in the field. Five plants and cuttings of *P. berolinensis* Dipp. were used for chromosome examination. It is evident in Table 8 that the percentage of diploid cells in each plant derived from cuttings increased in comparison with that of its original plant. The percentage of diploids from cuttings varied from 78.4% to 84.1% and averaged 80.5%, while the original plant varied from 68.5% to 76.8% and averaged 72.8%. This means that the increase of 7.7% on average resulted from increased spontaneous doubling during the cutting process. The experimental results suggest that homozygous diploids could be obtained quickly through the technique of repeated propagation from cuttings.

Mechanism of Variation in Chromosome Number

Variation in chromosome number in poplar plants derived from anther culture showed that a particular culture had specific characteristics. Not only did plants exhibit a high level of mixoploid cells, but they also showed a tendency to spontaneous doubling.

RELATIONSHIP BETWEEN MIXOPLOIDY AND PLANT ORIGIN. This study indicates that the majority of cells in mixoploid plants were haploid prior to transplantation from culture flasks. One year after transplantation into the nursery, the mixoploidy became even more prevalent, as more and more diploid cells emerged and some aneuploid cells appeared. In a study of poplar anther culture, Tabatskaya (1981) reported that *P. balsamifera* anther callus consisted of haploid, diploid, andaneuploid cells, of which aneuploids accounted for 64% of the cells. This was considered to be due to abnormal division of haploid cells. Stettler et al. (1969) obtained mixoploids with monoploid and diploid cells in a study of haploid induction by parthenogenesis in *P. deltoides* Marsh. They assumed that the phenomenon could be easily explained by the plants' origin from haploids with chromosome doubling in part of the tissue. Winton et al. (1970) also assumed that the mixoploid phenomenon demonstrated that haploids had been produced but their chromosomes had doubled prior to plantlet examination. In discussing whether anther-derived plants arose from pollen, Cen (1981) suggested, "If all or most of the early embryoids were either haploids or mixoploids with a large haploid component, it would be confirmed that most of them originated from pollen."

On the basis of similar assumptions, we propose that mixoploids in poplar anther-derived plants reflect a state of transition due to spontaneous chromosome doubling of haploid cells. The fact that most of the plantlets at early stages were mixoploids composed largely of haploid cells demonstrated that anther culture plants were derived from pollen.

OCCURRENCE OF MIXOPLOIDS. Many investigators have suggested that it is easy for cells to undergo endomitosis, endoreduplication, karyogamy, multipolar mitosis, chromosome nondisjunction, etc., in in vitro culture. These mitotic abnormalities are the source of mixoploid chromosome numbers and other types of chromosomal variation (Hu, 1978). Hu suggested that multipolar mitosis may be one of the factors which induce aneuploidy. Our chromosome observations indicated that multipolar mitosis, which would produce aneuploid cells with decreased chromosome number after cytoplasmic division, occurred. It was quite possible that in the process of proliferation, pollen callus or plantlet somatic cells produce cells with various chromosome numbers caused by multipolar mitosis and endomitosis leading to mixoploidy. However, the mechanism is no doubt very complicated, and more studies are needed for a full explanation.

SPONTANEOUS DOUBLING. Stettler (1971) has pointed out that natural haploids exist in some forest tree species and haploids should, therefore, be inducible. But there remains the tendency toward chromosome doubling. Both Stettler and Winton induced mixoploid poplar plants (containing haploid and diploid cells) by means of parthenogenesis, and they attributed the production of mixoploids to spontaneous doubling of chromosomes. In reporting chromosome doubling in an anther-derived *P. simonii* × *P. nigra* plant which had grown for 3 years, Kexian Wu mentioned that diploid cells increased from 33% to 64% at the age of 4 months to 3 years. We thus assume that spontaneous chromosome doubling in forest trees is common.

Two years of observation on chromosomes of anther-derived *P. berolinensis* Dipp. plants demonstrated that spontaneous doubling did occur, and the doubling process could be completed within 7–8 years. This spontaneous chromosome doubling is safe and reliable, produces few nuclear aberrations, occurs in perennial wood plants only, and may be advantageous to the haploid breeding of forest trees.

FUTURE PROSPECT

Poplar anther culture has been studied in China for 10 years. The oldest living pollen-derived trees are 8–9 years old. Repeated chromosome observations on anther-derived plants demonstrated that the plants originated from pollen. These studies documented the occurrence of mixoploidy. Thus a series of reliable techniques for induction and establishment of poplar pure lines has been developed. If desirable cross-combinations and hybrids among pure lines are to be selected, numerous pollen-derived plants are needed. It is therefore necessary to expand the number of poplar varieties studied in vitro and induce pollen plants on a large scale. Two-thirds of our pollen-derived trees have blossomed, and all the flowers are male flowers, showing that the trees which have blossomed are male trees. The sexuality of the remaining trees is unknown to date. If all the pollen plants produce only male flowers, new questions about sex determination in poplars will need to be asked.

Morphological observations on pollen plants from hybrid poplar indicated that segregation had occurred. Remarkable variation was seen in certain traits such as bark color, rust resistance, plant height, and trunk width. These results imply that it is possible to select more desirable homozygous individuals from anther culture of excellent hybrids.

Since anther-derived plants are mixoploids they can be used to obtain single cell lines with various chromosome numbers. Such material is useful for studies of genetics and for breeding.

REFERENCES

Cen, M., Chen, Z., Qian, C., Wang, C., He, Y., and Xiao, Y. 1981. Investigation of ploidy in the process of another culture of *Hevea brasiliensis*. *Acta Genet. Sin*. **8**:169–174 (in Chinese, English abstract).

Group of Tree Breeding, Northeast Forestry College. 1977. Induction of haploid poplar plantlets from pollen. *Acta Genet. Sin*. **4**:49–54 (in Chinese, English abstract).

Hu, H., Hsi, T., and Chia, S. 1978. Chromosome variation of somatic cells of pollen calli and plants in wheat (*Triticum aestivum* L.). *Acta Genet. Sin*. **5**:23–29 (in Chinese, English abstract).

Illies, Z. M. 1976. XVI World Congress, Division II. 267–271. Institute of Forestry, Academy of Forestry, Heilungkiang Province, 1976. Haploid plants of poplar induced from anther culture *in vitro*. *Acta Genet. Sin*. **3**(2):145–149 (in Chinese, English abstract).

Liu, Y., Zhang, J., and Lu, Z. 1979. Microsporogenesis in *P. berolinensis* Moeitzburg. *J. Northeast Forest. Coll*. **2**:1–4 (in Chinese).

Lu, Z., Zhang, J., and Liu, Y. 1979. Chromosomal observation of somatic cells in poplar pollen plants. *J. Northeast Forest. Coll*. **2**:14–21 (in Chinese).

Muhle-Larsen, C., 1970. Recent advances in poplar breeding. In: Ramberger, J.

A. and Mikola, P., eds., *International Review of Forestry Research*, Vol. 3, pp. 1–67, Academic Press, New York.

Stettler, R. F., Bawa, K. S., and Livingston, G. K. 1969. Experimental induction of haploid parthenogenesis in forest trees. In: *Induced Mutation in Plants*. pp. 611–619. International Atomic Energy Agency, Vienna.

——— and Bawa, K. S. 1971. Experimental induction of haploid parthenogenesis in black cottonwood (*Populus trichocarpa* T. & G. ex Hook). *Silvae Genet.* **20**:15–25.

Storova, N. V. and Kots, S. P. 1976. 16th International Forestry Res. Organization World Congress, Division II, pp. 246–257.

Toru, S. 1974. Callus induction and organ differentiation in anther culture of poplars. *Jap. Forest. Coc.* **56**:55–62.

Wang, C., Chu, Z., and Sun, C. 1975. The induction of *Populus* pollen plants. *Acta Bot. Sin.* **17**:56–59 (in Chinese, English abstract).

Winton, L. L. and Einspahr, D. W. 1970. The use of heat treated pollen for aspen haploid production. *Forest. Sci.* **14**:406–407.

Wu, K., Xu, M., and Xie, Q. 1980. Studies on chromosome spontaneous doubling and artificial doubling in poplar pollen plants. *Forest. Sci. Technol.* **3**:6–10 (in Chinese).

Ying Kou Institute of Populus, Liao Ning Province. 1978. A study on the induction of populus pollen plants and the techniques of plantation. *Forest. Sci. Rep.* **1**:6–7 (in Chinese).

Zhang, J., Liu, Y., and Lu, Z. 1979. Cytological observation on microsporogenesis in poplar pollen culture. *J. Northwest Forest. Coll.* **2**:5–13 (in Chinese).

Zhu, C. (ed.). 1982. Chromosome in plants and chromosomic techniques. *Sci. Press* 82–84 (in Chinese).

Zhu, X., Wang, R., and Liang, Y. 1980. Induction of *Populus* pollen plants. *Forest. Sci.* **3**(1):190–197 (in Chinese).

CHAPTER 10
Poplar: Ovary Culture

*Kexian Wu**

INTRODUCTION

Poplar belongs to the family Salicaceae under the genus *Populus* L. Poplars are usually dioecious. There are about 110 primitive species of poplar in the world, of which more than 30 species grow in China (Forest Service of Kangle County of China, 1976). There are also numerous artificial hybrids.

As poplar is a cross-pollinated plant, it is highly heterozygous. The use of haploid induction in poplar breeding would, therefore, greatly enhance the efficiency and accuracy of selection of superior trees. The in vitro culture of ovaries or ovules and the induction of female gametogenesis are important techniques for obtaining haploids in poplar (Wang and Kuang, 1981; Zhu and Wu, 1979; Karnosky, 1981).

LITERATURE REVIEW

The in vitro culture of unpollinated ovaries or ovules has been investigated widely. The literature indicates, however, that unpollinated ovary

*English translation by Kaiwen Yuan.

culture has succeeded in only a few herbaceous plants, such as barley (Wang and Kuang, 1981; San Noeum, 1976), tobacco (Zhu and Wu, 1979; Zhu et al., 1981), rice (Zhou and Yang, 1980), maize (Ao et al., 1982), and lily (Gu and Zheng, 1983). In perennial crops, Mullins and Srinivasan cultured unfertilized ovules of grape in 1976, and plantlets were regenerated from the callus of nucellar cells. Niu et al. cultured unpollinated ovaries of *Lycium chinense* in 1983, and tetraploid plants were obtained from low-temperature pretreated material. In poplar, maternal haploid plants have been obtained in the author's laboratory by culturing unpollinated ovaries (Wu and Xu, 1984).

CULTURE TECHNIQUES

Culture Procedures and Media

CALLUS FORMATION. Plantlets derived from ovary culture were obtained in two ways: by differentiation of callus tissue and by direct development from the split ovary. The dedifferentiation medium used for the first approach was BN medium (Bourgin and Nitsch, 1967). The components are $Ca(NO_3)_2 \cdot 4H_2O$ 3 mM, KNO_3 1.2 mM, $MgSO_4 \cdot 7H_2O$ 0.5 mM, KH_2PO_4 0.9 mM, $MnSO_4 \cdot 4H_2O$ 0.1 mM, H_3BO_3 0.16 mM, $ZnSO_4 \cdot 4H_2O$ 40 μM, $Na_2MoO_4 \cdot H_2O$ 1 μm, $CuSO_4 \cdot 5H_2O$ 0.1 μM, ferric EDTA 90 μM, inositol 0.55 mM, glycine 27 μM, pryridoxal chloride 2.4 μM, thiamine chloride 1.4 μM, nicotinic acid 41 μM, folic acid 1.1 μM, and biotin 0.2 μM, supplemented with 4.6 μM KIN, 9 μM 2,4-D, and 0.3 M sucrose. Callus could be induced and plantlets could be regenerated directly from MS medium supplemented with 9.3 μM KIN, 0.54 μM NAA, and 0.1 M sucrose.

SHOOT DIFFERENTIATION. The callus began to turn green 15 days after it was transferred to the differentiation medium. Shoots emerged 25 days after transfer; leaves grew out and rootless plants formed 10 days later. Basal MS medium with 9.3 μM KIN, 2.2 μM BA, 2.7 μM NAA, and 0.15 M sucrose was used for shoot differentiation.

SHOOT ELONGATION. To accelerate stem growth, the callus on which clustered shoots had already differentiated were transferred, every 25–30 days, to fresh MS medium (or medium with half the major element concentration of MS medium) with 9.3 μM KIN, 4.4 μM BA, 2.9 μM IAA, and 0.06–0.1 M sucrose.

ROOT INDUCTION. Young plantlets were cut with scalpels and placed in rooting medium when they grew to 3 cm in height. The appropriate depth of placement in the medium was about 0.5 cm. BN or ½ BN medium

Figure 1. Haploid plant with roots, stem, and leaves.

with 2.9 μM IAA, 2.7 μM NAA, and 59 mM sucrose was suitable for root induction. The cuttings started rooting in 6–7 days at a room temperature of 23–25°C, and complete plantlets were formed (Fig. 1). The rooting frequency of the plantlets was 78.79%. The pH of the root induction medium was adjusted to 6.0; 0.45% agar was added to the medium. The medium was then poured into 50- or 100-ml Erlenmeyer flasks and heated under a pressure of 1 kg/cm^2 for 20 min.

Manipulation Procedures

SELECTION OF MATERIALS. Experimental materials were from *Populus* × *simonigra* Chon-Lin. Pistillate flower stems were taken from trees in early April and grown in water indoors to accelerate flower growth. They were ready to be inoculated when the buds began to swell. The inoculation was carried out when the flower scales were about to split open. At this time, the pollen was at the uninucleate stage.

STERILIZATION. Prior to inoculation, the flower buds were immersed in a dilute solution of commercial bleach for 15 min and washed twice with sterile distilled water; bud scales were aseptically removed; and the ovaries were picked out with small tweezers and inoculated in the medium.

CULTURE CONDITIONS. The cultures were put on an incubation shelf, illuminated by fluorescent lights for 12–14 hr/day. Room temperature was controlled at 20–25°C.

TRANSFER OF CALLUS AND PLANTLETS. Callus was transferred to differentiation medium for organ formation when it reached 3–4 mm in diameter. When young plantlets were 2–3 cm in height, they were cut at the basal part of the stem and placed in rooting medium. They were transplanted to pots containing sandy soil after roots were well developed.

Keys for Successful Plant Regeneration from Embryo Sac

RESTRAINING THE GROWTH OF SOMATIC CALLUS AND STIMULATING THE DEVELOPMENT OF HAPLOID CELLS IN THE EMBRYO SAC. The ovaries began to swell 3–5 days after inoculation. In most of the media, somatic callus emerged from breaks in the basal part of the ovary, especially in Number 1–4 media (Table 1). Callus was loose in texture and translucent. Callus grew rapidly and covered the whole ovary after 2 weeks in culture. The growth of the ovary decreased until

TABLE 1. Effects of Different Media on Callus Formation

Medium No.	Composition	No. of Ovaries Cultured	No. of Ovaries Callusing	Percentage of Ovaries Callusing
1	BN + 4.6 μM KIN + 2.3 μM 2,4-D + 0.15 M sucrose	432	7	1.62
2	BN + 4.6 μM KIN + 9 μM 2,4-D + 0.15 M sucrose	305	8	2.62
3	BN + 4.6 μM KIN + 18 μM 2,4-D + 0.15 M sucrose	317	5	1.58
4	BN + 4.6 μM KIN + 36 μM 2,4-D + 0.15 M sucrose	374	3	0.80
5	BN + 4.6 μM KIN + 9 μM 2,4-D + 0.3 M sucrose	97	36	37.11
6	BN + 4.6 μM KIN + 9 μM 2,4-D + 0.45 M sucrose	237	12	5.06
7	BN + 9.3 μM KIN + 2.7 μM NAA + 0.15 M sucrose	392	4	1.02
8	BN + 9.3 μM KIN + 2.9 μM IAA + 0.15 M sucrose	211	8	3.79
9	BN + 9.3 μM KIN + 0.15 M sucrose	350	2	0.57
10	MS + 9.3 μM KIN + 1.67 μM IAA + 0.1 M sucrose	210	27	12.86
11	MS + 9.3 μM KIN + 0.54 μM NAA + 0.1 M sucrose	243	52	21.40
12	MS + 4.6 μM KIN + 0.57 μM IAA + 0.1 M sucrose	119	9	7.56

it turned brown and died. Excision of somatic callus was ineffective, for it would grow out again 3 or 4 days after removal. Most of the ovaries split vertically along their axes, and some opened to fan shapes as the inside callus increased. Some ovaries split at one end and several white protuberances, which were recognized as swelled ovules, emerged. Horizontal splitting was occasionally observed. The walls of some of the incubated ovaries became thinner after the ovaries swelled until round breaks, from which spherical-shaped callus protruded, appeared. Regenerated plantlet from ovary callus was usually haploid.

Three factors inhibited proliferation of somatic cells: the concentrations of inorganic salts, sucrose, and hormones. Experiments were carried out in our laboratory to determine the best composition of these three factors. The results are summarized in Table 1. In Number 9 medium without auxin, somatic callus grew slowly, but most ovaries could not grow normally, turned brown, and died quickly. The increase of somatic callus was significantly inhibited in BN medium with high sucrose concentration. In the medium containing 293 mM sucrose, more than 80% of calli regenerated from somatic tissue associated with ovaries turned yellow and atrophied; about one-third of the ovaries grew normally, became green, split after 25 days in culture, and finally formed callus. A 440 mM sucrose concentration was too high, as both proliferation of somatic callus and ovary growth were inhibited. High concentrations of inorganic salts in MS medium could both inhibit the proliferation of somatic callus and stimulate the normal growth of some ovaries.

Table 1 shows that high sucrose concentration or high level of inorganic salts could inhibit proliferation of somatic callus and favor the further development of female gametophytes. Among the tested medium compositions, that of Number 5 medium was the best, being composed of BN + 4.6 μM KIN + 9 μM 2,4-D + 0.3 M sucrose. The frequency of callus induction in this medium was up to 37.11%. This suggests that an appropriate increase of osmotic pressure in the medium favors the development of haploid cells in the embryo sac. Good results were also achieved in Number 11 medium with an induction frequency of 21.40%. Its components were MS + 9.3 μM KIN + 0.55 μM NAA + 0.1 M sucrose. The high concentrations of inorganic salts and KIN in this medium inhibited the proliferation of somatic cells and resulted in a high frequency of callusing ovaries.

SELECTING APPROPRIATE KINDS AND CONCENTRATIONS OF HORMONES FOR SHOOT DIFFERENTIATION. Different hormones were tested for shoot regeneration in our laboratory (Table 2). KIN combined with BA gave better results than KIN alone, and NAA was more effective than IAA. However, all regeneration frequencies were rather

TABLE 2. Effects of Hormones on Differentiation of Shoot from Callus

Medium No.	Composition	Number of Calli Cultured	Number of Calli Shooting	Percentage of Calli Shooting
13	MS + 13.8 μM KIN + 2.7 μM NAA + 0.15 M sucrose	35	None	None
14	MS + 13.8 μM KIN + 2.9 μM IAA + 0.15 M sucrose	42	None	None
15	MS + 9.3 μM KIN + 2.2 μM BA + 2.7 μM NAA + 0.15 M sucrose	51	12	23.53
16	MS + 9.3 μM KIN + 2.2 μM BA + 2.9 μM IAA + 0.15 M sucrose	42	5	11.90

low, possibly because of unfavorable cultural conditions. Most cultures died gradually in 30 days. The reason that ovaries died so early in culture is not yet clear. The cultural conditions required for each stage must now be investigated.

In Number 11 medium, shoots regenerated directly from swelled ovaries without callusing after 40 days in culture. Use of this medium not only simplified the culture procedure but also prevented variation in chromosome number during callus formation (Zhu and Wu, 1979). This direct shoot regeneration technique should be further studied.

PLANTLET TRANSPLANTATION AND EXAMINATION OF CHROMOSOME NUMBER IN ROOT TIP CELLS. Plantlets were transplanted when their root systems were well developed and their stems showed clear lignification. The test tube covers were removed 2 days before transplantation for hardening. The plantlets were then taken out, washed clean of the medium, and planted in pots with care. The greenhouse should be appropriately humid. The survival rate reached 65%. The vigorously growing plants could be transplanted into the nursery the following spring.

Ploidy levels in the regenerated plants were examined. Young root tips were cut, pretreated in 3 mM 8-hydroxyquinoline for 3 hr, fixed in Carnoy's fluid for 24 hr, dissolved under 1 M hydrochloric acid at 60°C for 12 min, stained with iron haemotoxylin, squashed, and examined under the microscope. Root tips from 12 plantlets were examined. Of the cells, 78–90% contained 19 chromosomes, indicating that these plants were haploids derived from the female gametophytes of maternal donors (Fig. 2).

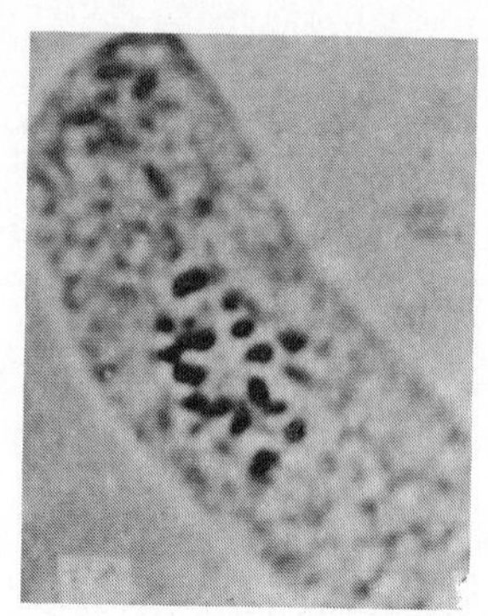

Figure 2. Chromosomes ($2n = 19$) of a root tip cell of plantlet derived from embryo sac.

For ovary culture to be an important part of poplar breeding, more experimental materials must be tested, so that ovary culture can be applied to more varieties and more haploid plantlets with different genotypes can be obtained. Ovary culture can also be combined with pollen culture techniques to regenerate plants for poplar breeding. Only by so doing can many different genotypes be combined in directed crossbreeding to create more new varieties. Increased numbers of pure lines will make it possible for us to utilize heterosis effectively and to increase timber production. The induction frequency of ovary-derived plants should be further increased in the varieties for which ovary culture has been successful. Similarities and differences between ovary-derived and pollen-derived plants should also be investigated to make better use of these materials.

KEY REFERENCES

Ao, G., Zhao, S., and Li, G. 1982. In vitro induction of haploid plantlets from unpollinated ovaries of corn. *Acta Genet. Sin.* **9**:281–283 (in Chinese, English abstract).

Karnosky, D. F. 1981. Potential for forest tree improvement via tissue culture. *Bioscience* **31**:114–120.

Wang, J. and Kuang, B. 1981. In vitro induction of haploid plants from female gametophytes in barley. *Acta Bot. Sin.* **23**:329–330 (in Chinese, English abstract).

Wu, K. and Xu, M. 1984. In vitro induction of maternal haploid plants from unpollinated ovaries of poplar. *Acta Genet. Sin.* **11**:47–51 (in Chinese, English abstract).

Zhou, C. and Yang, H. 1980. In vitro induction of haploid plants from unpollinated young ovaries of *Oryza sativa* L. *Acta Genet. Sin.* **7**:287–288 (in Chinese, English abstract).

REFERENCES

Bourgin, J. P. and Nitsch, J. P. 1967. Obtention de *Nicotiana* haploids a partir d'etamines cultivées in vitro. *Ann. Physiol. Veg.* **9**:377–382.

Gu, Z. and Zheng, G. 1983. In vitro culture of unpollinated ovaries and embryological observation in lily. *Acta Bot. Sin.* **25**:24–28 (in Chinese, English abstract).

Kangle County, Forest Service. 1976. *Cultivation of Poplars*. Ganshu People's Press, p. 5 (in Chinese). Lanzhou.

Mullins, M. G. and Srinivasan, C. 1976. Somatic embryos and plantlets from an ancient clone of the grapevine (cv. Cabernet-Sauvignon) by apomixis in vitro. *J. Exp. Bot.* **27**:1022–1030.

Niu, D., Li, J., Li, A., Zhou, Z., Jiang, X., Chen, Y., Shao, Q., Qin, J., Wang, L., and Wang, D. 1984. Tetraploid Chinese wolfberry from culture of unpollinated ovary. *Annual Report of the Institute of Genetics, Academia Sinica*, p. 81.

San Noeum, L. H. 1976. Haploids d'*Hordeum vulgare* L. par culture in vitro d'ovaries non fecondes. *Ann. Amelior Plantes*. **26**:751–754.

Zhu, Z. and Wu, H. 1979. In vitro production of haploid plantlets from the unpollinated ovaries of *Triticum aestivum* and *Nicotiana tabacum. Acta Genet. Sin.* **6**(2):181–183 (in Chinese, English abstract).

———, Wu, H., An, Q., and Liu, Z. 1981. In vitro induction of haploid plantlets from unpollinated ovaries of *Triticum aestivum. Acta Genet. Sin.* **8**(4):386–390 (in Chinese, English abstract).

CHAPTER 11

Poplar: Rapid Propagation

Yanlei Ba and *Jianshong Guo*

INTRODUCTION

Our experience indicates that 70,000 seedlings can be obtained in 8 years using conventional methods to propagate cuttings of *Populus simonii*. For *P. jrtyschensis*, only 62 seedlings were obtained after 3-year propagation; i.e., it takes 5 years to produce enough seedlings for a small-scale field trial. Usually it takes 10–15 years to release a new variety. Tissue culture is a practical technique for rapidly propagating a newly selected variety (Lao, 1983; Wang, 1983). There are several technical problems which have to be resolved before applying this tissue culture method for the rapid vegetative propagation of a variety (Oyuang et al., 1980).

First, techniques for regenerating plants via embryogenesis should be established to prevent a callus development stage. Only by doing so can large quantities of plants be obtained rapidly. Second, the techniques for rooting plantlets outside test tubes should be developed to meet the need for commercial production. Third, to provide more rootless plants for rooting, the differentiation frequency of subcultures and the coefficient of propagation should be increased.

Studies of these problems were carried out in our station using improved individuals of hybrids between section Aigeiros and section Tacamahacca and natural hybrids of section Leuce Duby. Rapid propa-

gation techniques in four improved individual plants have been developed since 1980, and the results will be described in this chapter.

CULTURE TECHNIQUES

Culture Procedures and Media

SELECTION OF MATERIALS. The following improved hybrid trees were selected for culture:

1. A natural-hybrid tree of *P. jrtyschensis* with characteristics of rapid growth, erect trunk, small crown, and cold resistance; code number 076-28
2. A hybrid tree of *P. simonii* × *P. italica* × *P. rassica* with characteristics of rapid growth, erect trunk, narrow crown, and cold resistance; code number 741-9-1
3. A natural hybrid tree of *P. simonii* with characteristics of erect trunk, round crown, drought tolerance, and cold resistance; code number 077-11
4. A hybrid tree of *P. alba* × *P. laurifolia* with characteristics of rapid growth and cold and disease resistance;
5. A natural hybrid tree of *P. alba*, code number 080-1

PRETREATMENT OF MATERIALS. The cuttings were first taken from the field and put into water culture to induce etiolation. The explant material taken after the etiolation treatment was young, shortening the culture duration and reducing contamination. The procedure is as follows: The lateral buds, young stems, or shoot tips of cuttings were covered with black paper bags and the lower openings sealed. They were cultured in water for 10–15 days at a room temperature of 18–22°C. Cuttings were cultured in water before the normal growing season. The new lateral buds and stems were taken for sterilization and inoculation when they became yellowish green or yellow.

MEDIA AND HORMONES FOR EACH STEP. The culture process was divided into two steps: embryo induction and plantlet regeneration. Two additional steps were required for subculture in order to produce plantlets continuously. The media used for each step are listed in Table 1.

We have tried different medium compositions in an effort to regenerate plants from a range of poplar varieties.

Manipulation Techniques

STERILIZATION. The etiolated materials were directly sterilized using aseptic conditions; the materials taken from the field were first washed

TABLE 1. Culture Media and Hormone Concentrations (μM) for Each Stage of Culture

			Subculture	
Variety	Embryo Induction	Plantlet Growth	Embryo Induction	Plantlet Growth
076-28 077-11 741-9-1	H + BA 4.4 + NAA 2.7	MS + BA 0.44 + NAA 0.11	MS or H + BA 0.88 H + BA 4.4 + NAA 2.7	MS + BA 0.44 + NAA 1.1
Clone of poplar	MS + 0.5 macro-elements + ZEA 2.3 + IBA 0.98	MS + BA 1.32 + GA 0.87 + NAA 0.54	MS + 0.5 macro-elements + ZEA 2.3 + IBA 0.98 or MS + BA 0.88	MS + BA 1.32 + NAA 3.3 or MS + GA 0.87 + BA 1.32 + NAA 0.54

several times then sterilized. The materials collected from trees of section Leuce Duby were in particular need of repeated rinsing. Their contamination rate was otherwise very high, as there are fine hairs on the young stems, buds, and shoot tips. The explants were soaked in a solution containing 0.1% mercuric chloride for sterilization for 10–15 min, rinsed four to five times with sterile water, and inoculated in medium.

CULTURE CONDITIONS. The culture temperature for embryo induction from the explants of 076-28, 077-11, and 741-9-1 was 26°C, and for plantlet regeneration 26–28°C. For poplar cultures, 28°C was required for embryo induction, and 24–26°C for plantlet regeneration.

OBSERVATION OF MORPHOGENESIS. To observe morphogenesis of embryos, embryos were fixed with FAA solution, embedded in paraffin, and stained with acetocarmine.

Key Points in Culture

EMBRYO MORPHOGENESIS. The characteristics of induction and development of embryos varied with species. After 10–15 days in culture, yellow or pale yellow calli with protuberances emerged at the point of excision of the explant taken from superior trees of section Aigeiros and section Tacamahacca. Spherical and stick-shaped embryos arose from protuberances in the calli after 30 days of culture. Many of the embryos were produced from bark lenticels of the explant (Chu, 1978; Zhou, 1981). The young shoot tips, stem segments, leaf stalks, and leaf veins produced most of the embryos. The embryos regenerated from the callus. In contrast, embryos that were not induced synchronously from epider-

mis of stem segments or shoot tips were not uniform in development but were spherical, heart shape, torpedolike, and cotyledonary. Plantlets were even found growing simultaneously in one explant or one tube. Similar results have been obtained by other investigators (Zhang et al., 1981). These embryos could sometimes produce complete plantlets when they were transferred to a plant regeneration medium.

Another approach to plant regeneration was via organogenesis. A large quantity of granular tissue emerged from callus and then differentiated into clustered green shoots. This form of shoot development was similar to that found with *Angelica sinensis* (Zhang and Zheng, 1982). The vigorously growing shoots could develop into healthy rootless plantlets when cultured in plant regeneration medium. There were also a few callus-derived plantlets, which usually produced only one leaf at first but no stem. After a long period of culture, the stem was developed. These plantlets derived via organogenesis often emerged before embryos or grew among them. They differed from embryo-derived plantlets in the following ways: the embryo-derived plantlets had small leaves, compact form, short leaf stalks, densely green and thick leaves, thick and strong purplish red stems, strong viability, healthy growth after cutting for proliferation, more generations of proliferations, and high frequencies of rooting. In contrast, plantlets derived via organogenesis had large leaves, loose plant forms, long leaf stalks, drooping leaves with rolled leaf margins, green-white stems, and weak growth. Their leaves turned yellow when the culture period was prolonged. They usually died of stem rot while rooting outside these tubes. The frequency of plant regeneration was very low.

The induction and development of embryos in *P. alba* asexual lines were different from those described above and were similar to that described by Zhu (1978). The embryo developmental process was as follows: the explants began to swell after 5 days of culture, a yellow or pale yellow cell mass emerged at the bud primordium of the cultured stem segment in 15 days, and a large number of embryos in sticklike, spherical, and cotyledonary shapes were produced in 25 days. Under the microscope, the cultures contained embryos in various stages of development. Many callus-derived plantlets were also produced. Hybrids of section Aigeiros and section Tacamahaca *P. alba* cultures differed in that no embryo was found to emerge from epidermal cells of the explant. The tissues that readily produced embryonic cell masses were shoot tips, young stems, and abscission layer in leaf stalks.

INDUCTION OF LARGE NUMBERS OF EMBRYOS. When embryos develop into rootless plants, they can be cut in 15 days and plantlets regenerated in another 15 days. If a plantlet is cut into three segments, the

number of plantlets doubles every month. Therefore, it is possible to get 1 million plantlets from one piece of explant per year. The size of plantlets varied greatly because of nonsynchronous embryo development. Of the plantlets cultured in flasks, 14% could be used for rooting. Plantlets 3–5 cm long accounted for 20% of this total, those 1–2 cm long for 20%, and those less than 1 cm long for 60%. As the plantlets used for pot rooting should be selected from those 3–5 cm long, over 60% of the plantlets were wasted. The utilization frequency of test tube derived plantlets was higher.

TECHNIQUES FOR ROOT INITIATION OUTSIDE TEST TUBES

Root initiation outside test tubes can simplify the culture procedure and cut the cost remarkably. Pot and flask root initiation methods have been used successfully in our experiments.

Root Initiation in Pots

SOIL PREPARATION. Humus and peat soil were selected as substrate and sterilized with 190 μM potassium permanganate or 0.2 M ferrous sulfate.

CLASSIFICATION AND SELECTION OF PLANTLETS. Plantlets were classified according to the following standard. Plantlets of Classes I–III were used for rooting, and the others were eliminated.

First-Class Plantlets. Class I plantlets were 3 cm long with erect and young stems 0.14 cm in diameter. The stems and leaf stalks were purplish red. Leaves were expanded, dark or deep-green, with visible leaf serrations.

Second-Class Plantlets. Class II plantlets were 2–2.9 cm long with erect and tender stems 0.10–0.12 cm in diameter. The stems and leaf stalks were purplish red or pink. Leaves were slightly rolled or curled and green or deep green. There were a few yellow leaf blades.

Third-Class Plantlets. Class III were 1–3 cm long. The stems were under 0.10 cm in diameter, weak, S-curved, and green or green-white. The leaves were curled and green. Leaf apices were usually yellow-green.

Fourth-Class Plantlets. Class IV plantlets were 0.5–1 cm long with strong stems and two whorls of leaves. The leaves were erect and green with unexpanded top leaves.

SELECTION OF POTS. Small flowerpots or specially made wooden pots (22 by 13 cm) were used. The plantlets were inserted into the soil at

a depth of 0.5–0.8 cm and covered with flasks after watering. The temperature, humidity, ventilation, and illumination should be properly controlled.

RESULTS OF POT ROOT INITIATION. The pot rooting experiment was carried out commercially in 1982 using plantlets regenerated from *P. simonii*. The experiment started on June 17. Altogether, 5144 test tube developed plantlets were planted on 13 different occasions. The survival rate was greatly affected by the planting time. Higher survival rates were achieved in the plantlets planted from late June to early August. The quality of the plantlets also influenced their survival rate. There were 1267 first-class plantlets transplanted with an average rooting rate of 30.5% (the most successful group reached 54.8%); 1578 second-class plantlets, with a 22.3% rooting rate; and 2299 third-class plantlets, with a rooting rate of 22.3%. The keys to successful pot rooting are to promote the plantlet growth and to choose the correct planting period. Fine, white roots appeared from the lower epidermal lenticels of the stem 20 days after potting. The number of roots increased, and secondary roots developed 30 days later. The survival rate then stabilized. The shoot tips above the ground almost stopped growing during root development, then started to grow again 30 days later.

Root Initiation in Bottle

1. Soil preparation: A mixture of three parts of humus and one part fine sand (0.25–0.5 mm) was placed into a 10–12-cm-high bottle, producing a 4-cm layer. In each bottle 20–30 seedlings were cultured. The cuttings were inserted about 0.3–1.2 cm into soil.
2. Nutrient solution for inducing root: MS medium with 5.7–29 μM IAA or 4.9–25 μM IBA, but without sucrose, was used.
3. Results of root initiation in bottle: Flask root initiation was performed with plantlets regenerated from *P. simonii* and asexual lines of *P. alba*.

At first, appropriate nutrient solutions were screened by using the plantlets of Classes I–III. A total of 1424 plantlets was planted and 824 plants rooted, for a survival rate of 57%. Further experiments were conducted from September 1984 to June 1985, with results shown in Table 2. The rooting rates of the two materials were more than 72%, more than three times higher than that of pot rooting.

High survival rates, 75–78%, were obtained using the hybrid tree of *P. alba* × *P. laruifolia*, which rooted with difficulty. The plantlets in

TABLE 2. Rooting of In Vitro Cuttings in Bottles (1984–1985)

		Date of	Date of	Totals		
Year	Variety	Cutting (Day/Month)	Observation	No. of Cuttings	No. Rooted	Percent Rooted
1984	077-11	18/9	11/10	400	290	72.5
1984	Hybrid of poplar	18/9	11/10	50	41	82.0
1985	Hybrid of poplar	29/4–18/6	26/6	624	469	75.2

flasks showed good development of roots. Rootlets can be induced in 15 days. After 30 days of culture, the length of roots was 8–12 cm.

About 60% of the green seedlings, whose length is less than 1 cm in the experiment where roots are initiated in pots, can be cultured in bottles to produce first- to third-class seedlings. The method of root initiation in bottles is cheap and easy to handle. It can be an ideal method for large-scale propagation of seedlings if the conditions of temperature, humidity, and oxygen are optimized.

Field Observation of Clonal Progeny

The desirable characteristics of fine hybrid varieties have to be maintained after in vitro propagation. This was proved in several superior varieties such as 077-11, 741-9-1, and *P. alba* × *P. laurifolia*. There is no significant difference in most of the morphological characteristics (leaf color, leaf tip shape, leaf petiole, angle between vein and secondary vein) between tissue culture-derived seedlings and conventionally vegetatively propagated seedlings.

Future Prospects

The research results of tissue culture in perennial crops during 1978 and 1984 resolved the following two problems:

1. It is practical to apply embryogenesis, in vitro cutting to induce rooting, and long-term subculture of propagules to propagate seedlings rapidly. The method of root initiation in pots can raise seedling production rate 179-fold. Further study of the procedure of large-scale propagation of seedlings is necessary.
2. Induction of seedlings from somatic embryo, induction of roots in test tubes, and capacity of plant regeneration of long-term subcultures can be applied to not only section Aigeiros but also section

Tacamahacca. For example, we established the preceding techniques in 080-1, a natural hybrid of *P. alba*.

REFERENCES

Loo, S. W. 1983. Application of plant tissue and cell culture and its prospect. *Plant Physiol. Commun.* **2**:1–5.

Ouyang, Q., Li, P., and Li, Q. Q. 1980. Somatic embryogenesis from callus in *Eucalyptus. Acta Plant Physiol.* **6**:429–432.

Wang, H. Z. 1983. Tree planting and tissue culture. *Forest. Sci.* **19**:292–301.

Wang, K. Q., Zhang, P. F., Ni, D. X., and Bao, C. H. Callus formation and organogenesis in tissue culture of several woody plants. *Acta Bot.* **23**:97–103.

Zhang, L. K., Zhang, Y. H., Wang, F. Q., and Liu, S. L. 1981. Induction of somatic embryogenesis in Poplar. *Forest Sci.* **19**:426–427.

Zheng, K. C. Plant regeneration in vitro in Chinese angelica. *Acta Bot.* **24**:512–517.

Zhou, J. Y. 1981. Somatic embryo derived from plant tissue culture. *Acta Plant Physiol.* **7**:389–397.

Zhu, Z. 1978. Somatic embryo in plant tissue culture. *Acta Genet.* **5**:79–88.

CHAPTER 12
Eucalyptus

Quan Ouyang and *Haizhong Peng**

INTRODUCTION

Economic Importance

Eucalyptus trees originated in Australia, Tasmania, New Guinea, and the South Phillipines. They have been introduced for cultivation and acclimatized in various regions of the world since the eighteenth century. According to a report from the UN Food and Agriculture Organization (Planting of Eucalyptus), in 1981 eucalyptus had been introduced for cultivation in 96 countries and districts throughout the world, and artificial forest farms covered an area of 6 million ha. This represents a fourfold increase in eucalyptus forests in 20 years for a timber yield of around 6000 m^3/year.

The eucalyptus, poplar, and bamboo are three genera famous for their fast growth. Eucalyptus has the following characteristics:

1. They are useful for a variety of forests: industrial timber, cash forests, shelter and windbreak forests, and scenic forests. They have hard, high-quality wood, and their wood, bark, leaves, and flowers have high economic value.
2. They reproduce easily, grow quickly, and have high yields. Young

*English translation by Jifang Huang.

plants have a high growth rate which greatly shortens the productive cycle of forest management, increasing the economic value.

3. Eucalyptus, poplar, and bamboo include various species and varieties suitable for cultivation in different climate zones: heat-tolerant or cold-tolerant, resistant to plant pathogens and destructive insects, adaptable to poor soils and barren lands. These three genera are favored by the forester. Eucalyptus timber is one of the best materials for papermaking, shipbuilding, and mine timber. Eucalyptus leaves may be used for producing oils, the eucalyptus bark is a source of resin, even the residues may be used for culturing edible fungi, and the final wastes may be used as fertilizer. The flowers of eucalyptus trees are a good source of honey in apiculture.

Classification

The genus *Eucalyptus* includes 455 species, 24 varieties, 8 subspecies, and 115 hybrids. About 100 of these are used for industrial and cash forests, and more than 40 kinds are widely cultivated. The most widely cultivated *Eucalyptus* species are *E. globulus, E. maideni, E. saligna, E. grandis, E. camaldulensis, E. tereticornis, E. maculata, E. citriodora, E. robusta, E. exserta, E. viminalis, E. diversifolia, E. resinifera, E. pilularis, E. delegatensis*, and *E. microcorys*. In China, eucalyptus has been cultivated since 1890, when it was introduced from Italy and France. Over 300 kinds of eucalyptus have been introduced and 211 kinds have been used to raise nursery stocks for afforestation. Eucalyptus forests are found in over 600 counties of 15 provinces and districts from south China (Hainan Island, 18°20′ north latitude) to north China (Ping-yangguan, Hanzhong District of Shanxi Province, 33°10′ north latitude); from the east of China (Taiwan, 22° north latitude) to the west (south Sichuan Province, 32°26′ north latitude). The introduced poplars in north China and the eucalyptus in south China are fast growing and high-yielding varieties which have been essential species for large-scale afforestation and for public planting in cities and countrysides.

Problems in Production

Because all eucalyptus species are cross-pollinated perennial crops, natural interspecific hybridization is frequent and results in complicated segregation patterns in their progeny. It is thus difficult to maintain the characteristics of eucalyptus by conventional sexual methods of propagation. The traditional vegetative propagation methods are also not effective for maintaining the characteristics, because of difficult proliferation of ad-

ventitious roots. Application of plant growth regulators to some traditional vegetative propagation methods, such as stem cutting, stem layering, and stem-to-stem and scion-to-stock grafting, to promote proliferation and growth of adventitious roots has done little to overcome the problem of propagating eucalyptus trees in quantity. Many laboratories have, therefore, attempted to use in vitro culture techniques to seek new approaches for fast propagation of good eucalyptus varieties and species.

History of Research

In 1965, Sussex obtained callus from *E. camaldulensis* seedling explants. Ahuja and Atal (1969) regenerated plantlets from stem callus of *E. citriodora* callus. De Fossard et al. (1974) obtained plantlets by using in vitro culture of internodes of eucalyptus. Edwards et al. (1975) grew callus from in vitro culture of *E. alba* hypocotyl, from which plantlets arose. De Fossard et al. (1976) produced plantlets as nursery stocks in large quantities from *E. ficifolia* using in vitro culture of stem nodes. Lakshmisita (1979) induced plantlets from *E. citriodora* hypocotyl and cotyledon cultures. Gupta et al. (1981) asexually reproduced *E. citriodora* plantlets by in vitro culture of young stem segments of an adult tree; they derived 100,000 plants for nursery stock within 1 year from the bud of an adult eucalyptus tree. Kondas reported that shoots could be regenerated from a 20-year-old hybrid eucalyptus tree.

The application of tissue culture to eucalyptus has advanced rapidly in the last few years in China. Laboratories to develop tissue culture techniques for mass propagation of eucalyptus have been founded in Guangxi Province, Guangdong Province, and Shanghai since 1983. Cooperative research organized by Shanghai Institute of Gardening Sciences and Fu Dan University on tissue culture in 12 eucalyptus species indicated that quantities of nursery stocks could be reproduced for *E. botryoides, E. cinerea, E. regnans*, and *E. citriodora* (Zhuang, et al., 1980). In the Institute of Forestry Sciences of Leizhou District, Forestry Bureau of Leizhou District, Guangdong Province, tissue culture of the eucalyptus variety Lei Lin No. 1 could be used to produce plantlets for transplanting (Xie Yinen et al., personal communication). The Institute of Forestry Sciences of Ganzhou District, Jiangxi Province, introduced test tube-cultured plantlets of *E. botryoides* into that district; propagated large quantities; and used them to afforest certain mountain regions. The Institute of Forestry Science of Quinzhou District, Guangxi Zhuang Autonomous Region, used young stem segments and sprouting buds to obtain calli from which clustered greenish shoots were regenerated in February 1979. Plantlets were obtained through tissue culture of these

greenish shoots, and some plants survived after transplantation. Embryos were obtained by culturing seeds and aseptic seedlings of Lei Lin No. 1 and these embryos differentiated into plantlets with normal roots and shoots. Clustered greenish shoots derived from calli of the F_1 hybrid seeds (*E. saligna* × *E. exserta*) and (*E. grandis* × *E. exserta*) were obtained and propagated in large quantities for trial afforestation (Ouyang et al. 1981). In 1981 tissue culture was used for rapid propagation of *E. saligna* × *E. exserta* F_2 hybrids in the Guangxi Institute of Forestry Sciences, Guangxi Zhuang Autonomous Region. The greenish shoots developed into green plants, and after transplantation, the surviving plantlets were sent to a mountain region as nursery stock for afforestation.

There are two ways for plants to form from eucalyptus tissue cultures: callus may form plantlets, or single cells may develop into embryos and then plantlets. Callus first differentiates into green shoots, then requires transfer to a root induction medium for root differentiation. The differentiation of shoots and roots is, therefore, not synchronized. Although callus in general exhibits a high frequency of differentiation, the low survival rate after transplantation has resulted in high production costs. In contrast, the embryo-derived plantlets showed three evident advantages: high survival rates, rapid growth, and complete organ development in the plantlets. The cell mass derived from a seed or an aseptic seedling could produce up to 10^{11} plantlets per year with continuous subculture. This method is a good foundation for commercialization of nursery stock production. The combination of embryo culture with cross-breeding could overcome the low reproductive rate that has resulted from cross-breeding and could be used to propagate nursery stock from varieties with hybrid vigor. All the eucalyptus nursery stock formed from tissue culture grew juvenile leaves with opposite phyllotaxy. They were easy to use in conventional stem cutting or grafting. It is also possible to produce new varieties which do not exist in nature by clonal screening of embryo plantlets derived from single cells.

CULTURE TECHNIQUES

E. ficifolia Culture Medium

De Fossard et al. developed a basic medium for initiation and proliferation of *E. ficifolia* stem segment cultures by screening a wide range of media. The basal medium for *E. ficifolia* is listed in Table 1 (De Fossard et al., 1976).

De Fossard established four large-sized clonal populations of different genotypes in *E. ficifolia*. His investigation advanced rapidly although

TABLE 1. Media for Initiation, Regeneration, and Rooting of Eucalyptus Hueng-hua

Compounds	Medium M	Medium R	Compounds	Medium M	Medium R
	mM		Hormones/Vitamins	µM	
Macroelements					
NH_4NO_3	10	5	IBA	5	5
KNO_3	10	—	BAP	2	—
NaH_2PO_4	1	1	Inositol	600	
KCl	—	1.9	Nicotinic acid	40	—
$CaCl_2$	2	1	Pyridoxine HCl	6	—
$MgSO_4$	1.5	0.5	Thiamine HCl	40	—
Microelements	µM		Biotin	1	—
H_3BO_3	50	150	Pantothenate	5	—
$MnSO_4$	50	100	Riboflavin	10	—
$ZnSO_4$	20	40	Ascorbic acid	10	—
$CuSO_4$	0.1	1.5	Chlorocholine chloride	10	—
Na_2MoO_4	0.1	1.0	L-Cysteine	120	—
$CoCl_2$	0.5	1.0	Glycine	50	—
KI	2.5	5.0		mM	
$FeSO_4$	100	100	Sucrose	120	60
Na_2-EDTA	100	100		g/l	
Na_2SO_4	650	—	Agar	8	8

decay appeared on the young stems and leaves of some plantlets after the roots were formed using the medium listed in Table 1.

Plantlet Differentiation from Callus of Adventitious Buds

Cooperative experiments by the Shanghai Institute of Gardening Sciences and Fu Dan University showed that in general differentiation of callus from eucalyptus adventitious buds could be divided into three steps.

CALLUS INDUCTION. Visible callus at various levels of development was induced from explants, such as young stem segments, which were taken from different vegetative organs of various species and cultured on different media for 20–30 days.

Different media were evaluated in Shanghai and an ideal procedure for organogenesis was determined. Callus was induced from several eucalyptus varieties and clones were obtained. The culture procedure was as follows: sterilized explants were inoculated on a medium of MS with 2.2 µM BA and 4.9 µM IBA. This medium resulted in the highest induction rates for eucalyptus callus: 97% in *E. tereticornis* and

95.5% in *E. cinerea*. The calli that were granular and compact in texture proliferated at the cut surface of the stem segments. Callus was identified with the following colors: pink, red, pale yellow, yellowish green, brown and black. The calli were derived from the cells between the cortex and cambium. Another second type of callus occurred at the surface of the stem segment away from the cut plane. This callus seemed tumorlike with a loose texture. It was derived from the same cells as the former callus but appeared to include dividing epidermal cells.

SHOOT REGENERATION. Cultures of *E. botryoides* transferred onto MS medium with 2.2 μM BA and 1.23 μM IBA had the highest rate of shoot regeneration. Less regeneration was observed in cultures of *E. gunnii* and *E. cinerea*, even less in *E. maideni* and *E. trabutii*; no shoots were observed in *E. alba* cultures.

Both tumorlike and granular calli produced from the different parts of cultured stem segments regenerated buds. Buds regenerated from the tumorlike calli were clustered or difficult to separate or developed into leafy shoots. Histological observations showed that the bud primordia occurred on the surface of calli and were difficult to separate as they were embedded within the callus. In contrast, large numbers of single compact buds regenerated from granular calli that were easy to separate in spite of their density. These shoots had a well-developed stem and expanded leaves. Histological observations of granular calli clearly revealed that the shoot tip meristem arose from meristematic cells on the callus surface, developed into bud primordia, and then developed into single compact buds. Such buds became separated from each other as they emerged from the callus, sprouted, and expanded their leaves.

ROOT INDUCTION AND PLANTLET FORMATION. When the regenerated shoots grew to 2–3 cm in height, they were transferred onto different media for root regeneration. Masses of callus occurred at the base of all the shoots. But only those of *E. botryoides, E. cinerea, E. citriodora, E. regnans*, and Lei Lin No. 1 regenerated adventitious roots and developed into plantlets on the following media: (1) White's containing 0.98–4.9 μM IBA, 296 μM adenine and 58.6 mM sucrose; (2) White's with 2.5 μM BA and 117 mM sucrose. According to histological observations, the meristematic region formed within the calli and grew with polarity in a single direction, developed into root primordia, and differentiated to form roots.

Induction of Shoots from Embryonic Cell Aggregates

Eucalyptus tissue culture was started in August 1978 in our laboratory. Young stem segments, stem tips, and leaf blades from young eucalyptus

trees (2–5 years old) or adult trees as well as young shoots sprouted from adult tree stumps were used for callus induction organ differentiation. Over 10 eucalyptus species produced callus. In Lei Lin No. 1, *E. saligna, E. citriodora*, and *E. exserta*, green shoots differentiated and developed into plantlets after subculture and survived after transplantation. B_5 and H media were selected from screened media by using a multiple-factor orthogonal design and a simple comparison method. Those media were not suitable for rapid propagation because roots and shoots could not be regenerated simultaneously from the callus and rates of survival were, therefore, low after transplantation.

Improved B_5 and H media were, therefore, developed. Aseptic seedlings or seeds were used as material for callus induction instead of nonsterile seeds. In 1980 embryonic cell aggregates were induced from calli. Embryos were differentiated and plantlets formed with simultaneous development of roots and shoots. In September 1980 10 cloned cell lines were established; they maintained vigorous growth, differentiation capacity, and similar seedling morphology for 50 subcultures, 4 years' growth. These lines demonstrated the advantages of rapid propagation of eucalyptus as they maintained the good characteristics of the cell clones. The lines were developed as follows:

CALLUS INDUCTION

1. Healthy seeds of eucalyptus were wrapped in wet gauze for more than 10 hr and soaked in 75% ethanol for 10–15 sec, transferred into 1% mercuric chloride for 10 min, and rinsed four to five times with aseptic water.
2. Water was blotted from the seed surfaces, and the seeds were inoculated onto medium solidified with 0.7% agar.
3. After 5–7 days, seeds germinated and developed into seedlings. The seedlings were cut into 0.2–0.3-cm segments after they grew to a height of 0.5–2 cm, and the shoot segments were inoculated onto an induction medium for continuous culturing. Aseptic seeds were inoculated directly onto induction medium. The young stems or the sprouted stem tips were also used for inoculum, being cut into segments 1 cm in length. The induction medium was as follows: B_5 medium or H medium or improved H medium with 3.7–4.6 μM KIN, 9–18 μM 2,4-D, 117–147 mM sucrose, and 0.6–0.7% agar. On this medium, the cultures developed callus in 3–4 days.
4. The culture temperature was 25–29°C. Fluorescent illumination was 12–14 hr/day at an intensity of 1000–1500 lux.

DIFFERENTIATION OF EMBRYONIC CELL AGGREGATES. After 20 days calli induced from the explants had undergone rapid proliferation and were red. They were then transferred to differentiation medium containing H or improved H medium with 0.88–4.4 μM BA, 2.7 μM NAA,

88–147 mM sucrose, and 0.6–0.7% agar. On this medium, calli proliferated slowly. After 1–2 weeks, their surfaces appeared smooth and pink, and their texture became compact. After 4–8 weeks the surface of the calli became smooth and moist, with a purple or light-red color. A portion of the calli appeared granular and uneven; this appearance indicated differentiation of embryonic cell aggregates from the calli. Such embryonic aggregates underwent further differentiation to produce embryos.

PLANTLET DEVELOPMENT. Solitary plantlets with roots and shoots developed in large numbers after 10–30 days' culture. Around the bases of the shoots and roots, clusters of newly arising cell aggregates and embryos were regenerated, indicating that the hypocotyl was the essential embryo-producing site. However, the differentiation of shoots and roots after embryo development depended on the auxin-cytokinin ratio.

Table 2 lists the five different media used. All five media produced a large number of green shoots, but roots were regenerated by using only three of the media. This was due to an unfavorable auxin-cytokinin ratio in the other two treatments. The results show that 0.88 μM BA and 2.7 μM NAA was the best for overall plantlet development. When the cultured callus regenerated more green shoots and fewer roots because of an unfavorable ratio, the cultures were transferred onto a medium composed of H medium or improved H medium with 2.5–4.9 μM IBA, 88 mM sucrose, and 0.6–0.7% agar, to improve root development. The cultures that developed too many roots but no green shoots as a result of excessive auxin in the medium failed to form plantlets upon subculture.

The selection of proper medium is critical for successful induction and differentiation of callus. Although the H medium and B_5 medium could be used for tissue culture in eucalyptus, they still needed to be improved for the induction of embryonic cell aggregates. In the improved H medium, the vitamins used were those used in B_5 medium. Nitrogen content was raised from 382.6 mg/l to 521.7 mg/l by adding 48.8 mg/l NH_4NO_3 and 41.6 mg/l KNO_3, changing the difference between the ammonium nitrogen value and nitrate nitrogen value from 131.7 to 173.3. This improved H medium promoted induction and differentiation of callus. It took more than 30 days to regenerate green shoots on the H medium and just over 10 days on the improved H medium. Temperature and illumination conditions were also important for the regeneration and differentiation of embryonic cell aggregates and plantlet development. With a room temperature of 25–29°C and illumination of 1500–1600 lux, the embryonic cell aggregates were scarlet, plantlets developed vigorously, and roots regenerated in 6–8 days. The adventitious roots regenerated to complete root systems at high rates. Under an illumination of 500–600 lux, aggregate development was less successful. Although the embryonic

TABLE 2. Numbers of Plantlets Induced from Embryogenic Calli with Different Hormones and Their Combinations (μM)

Media	No. of Test Tubes	No. of Green Plantlets Induced	No. of Green Plantlets Each Test Tube	Greenhouse No. of Rooted Plantlets	Field No. of Rooted Plantlets	Notes
H + BA 0.88 + NAA 2.7 + sucrose 88 mM + agar 0.65%	24	2000	84	840	35	Most green plantlets induced after 9 days, rooted after 27 days
Revised H + BA 0.88 + IBA 2.5 + sucrose 147 mM + agar 0.65%	18	3600	200	1200	200	Most green plantlets induced after 24 days, 1 of 3 with roots
Revised H + BA 4.4 + NAA 2.7 + sucrose 147 mM + agar 0.6%	9	1800	200	0	0	Most green plantlets induced after 15 days, no roots
Revised H + BA 2.2 + IBA 2.5 + sucrose 147 mM + agar 0.65%	24	4000	167	0	0	Most green plantlets induced after 26 days, no roots
Revised H + BA 4.4 + 2,4-D 2.3 + sucrose 147 mM + agar 0.7%	20	4000	200	1600	170	Most green plantlets induced after 36 days, 9 of 11 with roots

cell aggregates produced shoots and roots, calli also regenerated on a few stems and leaves of the plantlets, retarding plantlet growth and even causing plantlet death. The worst results were obtained from illumination of 50–60 lux: the embryonic cell aggregates turned dark brown, and the green shoots and weak root system developed slowly.

SUBCULTURE OF EMBRYONIC CELL AGGREGATES. After some time in culture, callus derived from embryonic cell aggregates needed to be transferred to fresh medium and subcultured. Otherwise, the cultures aged because of exhaustion of nutrients and loss of water from the medium as well as accumulation of undesirable metabolites.

In our experiments, the callus lost its capacity to differentiate after three generations of subculture. The embryonic cell aggregates were very different from the calli. They maintained vigorous growth and capacity to differentiate even after subculturing in differentiation medium at intervals of 1–2 months from 1980 to 1984. The essential features of the eucalyptus embryonic cell aggregates were as follows:

1. In addition to differentiating embryos, it was capable of continuous proliferation upon subsequent subculture. Within 15 days after subculture, the embryonic cell aggregates recovered from the effects of transfer; they underwent rapid proliferation in 15–35 days. The embryonic cell aggregates began to age 40 days after subculture.
2. The most vigorous cell division and the fastest proliferation in the embryonic cell aggregates occurred when they were 1–3 cm^3 in volume.
3. The best cell aggregates for subculturing were 0.1–0.2 cm^3. The recovery period from subculturing was longer when the cell aggregate was smaller than 0.05 cm^3, and cell aggregates larger than 0.2 cm^3 browned easily. Every 20 days was the optimal time for subculturing.

In general, an embryonic cell aggregate which grew to a size of 1–3 cm^3 after 20–30 days' subculture might be cut into 10–30 pieces for the next subculture. If proliferation continued in a geometric progression, 10^{10}–30^{10} pieces of embryonic cell aggregates might be obtained from 10 successive subcultures.

Since the embryonic cells did not regenerate synchronously, embryos, plumules, and plantlets in various developmental stages occurred together in the same embryonic cell aggregate. This condition was fundamental to long-term subculturing, providing different types of materials for culture. Green shoots could be removed with a small piece of embryonic cell aggregate attached for subculturing, or green plantlets could be recovered similarly. Embryonic cell aggregates alone could be removed for subculturing, or plantlets more than 2 cm in height with developed

Figure 1. Large number of eucalyptus plantlets that grew synchronously after several subcultures.

roots could be removed for transplantation and the remaining very young plantlets and embryonic cells could be transferred to freshly prepared culture medium. This successive transplantation over a long period raised the plantlet yield to high levels (Fig. 1). The differentiation of plantlets from embryonic cell aggregates was strictly controlled. Cell aggregates were transferred to subculture for about 20 days to obtain large quantities of embryonic aggregates. Plantlets regenerated from the cell aggregates were transplanted before afforestation season. If the subculture followed the preceding method, the survival rate of plantlets was quite high. More than 200 plantlets from a culture tube and over 2000 plantlets from a 1-l culture flask could be obtained.

Plantlets did not develop synchronously; large and small plantlets as well as the cell aggregates were removed from the culture tube or flask for transplantation at the same time. As a result of the failure to separate the different sizes, most of the smaller plantlets were drowned in watered soil, and only plantlets 2–4 cm in height grew as nursery stock. Therefore, only 10–20 plantlets were obtained from the cell aggregates in a single culture tube. Nevertheless, the cell aggregates in one tube still produced 10^{11}–20×30^{10} embryonic plantlets per year after 10 subsequent subcultures.

CHARACTERISTICS OF EMBRYO DEVELOPMENT. Ouyang Quan et al. were successful in the induction of embryonic cell aggregates from callus and differentiation of embryos from embryonic cell aggregates in 1979. The aggregates were fixed with FAA, sectioned (5–8 μM thick), and stained with a dual ferric alum hematoxylin and safranin-fast green stain. The preparations showed the whole process of plantlet formation from embryos: proembryo, spherical embryo (Fig. 2), heart-shaped em-

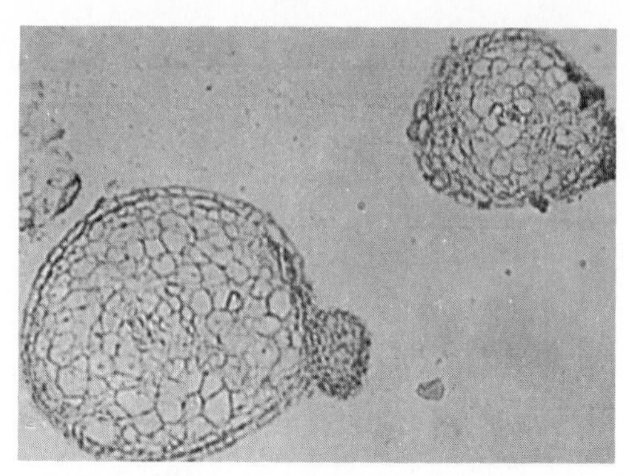

Figure 2. Histological section of globular somatic embryos of eucalyptus grown from embryogenic cell aggregates.

bryo, torpedo-shaped embryo, cotyledon-shaped embryo, to fully developed embryo plantlet with shoot and root system. Embryo morphogenesis was very similar to that of normal zygotic embryos.

According to Halperin et al. (1970), in carrot cell suspension cultures the embryo-forming cells could be distinguished several days after proliferation. Embryo formation was not due to direct proliferation of a single cultured cell but occurred from the surface cells of aggregates. Eucalyptus embryos were also formed from the epidermal cells of embryonic cell aggregates.

In embryonic cells, the newly formed daughter cell wall was always perpendicular to the mother cell wall, which made the cell aggregate become granular. In contrast, the cell aggregates which do not regenerate embryos enlarged irregularly and rarely showed definite orientation of the newly formed cell walls. After a period of time in culture, the embryo-forming cells grew into a compact cell aggregate with distinct histological characteristics. The outer layers of the cell mass showed characteristics of meristem regions; they were composed of small cells with concentrated protoplasm and many small starch grains. The inner region of the cell mass was composed of large cells with large starch grains, which were obviously vacuolated and seldom divided. The proembryo formed cell aggregates, after the process of embryo development, and grew with distinct polarity. It regenerated a plumule and radicle at opposite ends and then developed into an intact plantlets with well-formed shoots and roots.

Eucalyptus embryonic cell aggregates contain many embryos clustered together. The epidermal cells of embryos at the sphere-shaped stage were a little longer and were arranged regularly. Some embryos at the sphere-shaped stage possessed a suspensor and small proembryonic cells in a more compact group at the center. When these small cells divided to produce seven to eight layers of cells in different shapes, the embryo of the sphere-shaped stage progressively developed into a heart-shaped, and then a torpedo-shaped embryo. The bud and the cotyledon primordia of the embryo at the torpedo-shaped stage then began to emerge. The embryo elongated gradually and formed a cotyledon-shaped stage. In the mature embryo, the cotyledon primordium degenerated, and the two cotyledons were thick and grew fast, with the long vessels connected with the hypocotyl. The bud primordium was slender and small and was connected with the cotyledon. Cells in the bud primordium had dense cytoplasm and a slightly elongated nucleus. Cell division in the cells of the cotyledon and bud primordia was clearly active. Cells of the root primordia had dense cytoplasm, clearly distinguishable from the outer cells. The root primordium regenerated more slowly than the shoot primordium, while its cells divided less actively than those of the cotyledon and the shoot primordia. This effect may be due to high levels of KIN and an insufficiency of auxin in the medium, which inhibited root development. Most of the green plantlets developed roots when they were transferred onto medium without KIN but with 2.7 μM NAA or 2.5–4.9 μM IBA.

The diameter of the sphere-shaped embryos was 0.05–0.1 mm; the heart-shaped embryos were 0.15–0.6 mm by 0.2–0.5 mm; the torpedo-shaped embryos were 0.2–0.6 mm by 0.4–1 mm; the cotyledon-shaped embryos were 0.6–2 mm long, with cotyledons covering half the total length. Larger embryos produced plantlets more quickly than the smaller ones.

Embryo regeneration was completed in two ways: The embryos could be regenerated directly from cell aggregates derived from the calli of aseptic seeds or seedlings. Embryos could also be obtained from proliferated embryonic cell aggregates, which had been subcultured. The former gave low regeneration rates, around 70%, but when the culture was subcultured for a long period (398 passages), 100% regeneration was obtained. The subculture method was not affected by seasonal change or availability of explants since materials for inoculation were available throughout the year. In order to induce embryo formation from callus, successive changes of the medium, especially of the hormone contant, were important. The key was the decrease or elimination of auxin in the medium. Many experiments indicated that 2,4-D inhibited embryo formation, even if coconut water or cytokinin were added to the medium.

TRANSPLANTATION OF PLANTLETS

To enhance the survival rate of transplanted plantlets, the following points are important:

Selection of Nursery Bed Soil

An artificial soil mix (equal parts sandy loam or peat soil, burnt earth, and river sand) should be available for growing plantlets. This soil mixture must have a porous texture and the proper proportion of sand to earth. Good water holding capacity, good aeration, and permeability are favorable for the development of plantlet root systems. In our laboratory, nursery beds 1 m in width and 12–15 cm high were used for transplantation. In some cases, bamboo baskets, plastic bags, or paper bags were used. The baskets or bags were 15–20 cm high and 8–10 cm in diameter.

Selection of Plantlets for Transplantation

Strong plantlets 2–4 cm high with vigorous root systems were selected for transplantation. The cultures were placed in a hothouse with the lids of the culture tubes or flasks removed for 2–3 days to allow the plantlets to become acclimated. Clean water was poured into the culture tube to loosen the medium. For transplantation, the plantlets were taken from the culture tube with a small hook and placed in a basin with clean water. The plantlets were then rinsed to remove medium adhering to the roots. The roots of the plantlets were then dipped in mud, and they were transplanted in bunches into bamboo baskets or nursery beds. The distances between rows and between transplants in the nursery beds were 3 and 6 cm, respectively. The bunches were then separated into single plantlets which were put into cultivation. Another method involved taking plantlet clusters from the tube or flask, separating them into single plantlets, and transplanting them into baskets or nursery beds. The transplanted plantlets were thoroughly watered and protected with a glass cover for 10–20 days to maintain a humid atmosphere.

SHADING PLANTLETS. The plantlets taken from culture tubes were very delicate and needed protection against direct sunlight and hard rain. In our experiment, shades were erected and the light permeability was initially set at 30–40%, but gradually raised to 50–60%. After the glass covers were removed, the plantlets were cultivated under shade for 5–6 days. The plantlets were watered once or twice a day. Available moisture and drainage had to be closely monitored. A thin plastic cover was put over the shade when it rained heavily and the greenhouse was used as the temperature dropped.

AFFORESTATION USING CULTURED PLANTLETS AS NURSERY STOCK

An afforestation experiment was carried out to determine whether plantlets obtained from tissue culture maintain hybrid vigor under practical conditions, to investigate methods for screening clones for afforestation programs, and to observe variation in the cultured plantlets. The experiments were carried out on May 26, 1980. Tissue cultured plantlets and conventional seedlings of the same age were used. The average height of the plantlets was 51.9 cm for Clone A, 39.0 cm for Clone E, and 31.2 cm for Clone C; the average height of seedlings was 30.8 cm. The measurements were taken on December 31, 1982. Results are summarized in Table 3. The height and diameter of plantlets and seedlings set on August 27, 1981, was measured on October 7, 1982, and the results are summarized in Table 4. In these two experiments, the plantlets of Clone C were 21–36% taller and 34–38% larger in diameter than the seedlings. The plantlets derived from tissue culture had a smaller coefficient of variation in height and diameter than the seedlings (Table 5). With regard to general appearance, the forest of trees from seedlings varied in height, stem, and leaf size. The forest from the Clone C appeared uniform, with stems plump and erect, leaf blades almost identical in size and green in color. These forest trees grew vigorously and were uniform in hereditary character.

The survival rates of both seedlings and plantlets were high (Table 6). The afforestation was carried out when the plantlets and seedlings grew to a height of 30–40 cm after growing in nutrition bags for 3 months. Both the plantlets and the seedlings achieved 100% afforestation. Forest trees of all three kinds of stock grew rapidly. The first measurements were carried out in June 1982, 3 months after afforestation. The eucalyptus trees had reached an average height of 2.1 m in the plantlet Clone C, 1.7 m average height in Clone D, and only 1.65 m in seedlings. The second measurement, in June 1983, showed that the average height of Clone C (1-year-old) was 3.8 m with a diameter at ground level of 4.5 cm; in Clone D, the average height was 2.9 m with a diameter of 3.9 cm; the average height of seedlings was 3.0 m with a diameter of 3.9 cm.

Best results with cultured plants occurred when the plantlets' root systems were not damaged during transplantation. The plantlets should

TABLE 3. Comparison of Zygotic Seedlings and In Vitro–Derived Plantlets in Field Trials

Growth/Variety	Zygotic Seedlings	In Vitro Plantlets		
		A	E	C
Average height (m)	8.72	8.15	8.18	10.98
Average trunk diameter (cm)	6.82	6.39	8.3	9.81
Wood Yield (m^3)	3209	2633	4622	8361
Relative height (%)	100	93.5	97.3	125.9
Relative trunk diameter (%)	100	93.7	121.7	134.6
Relative wood yield (%)	100	82	144	260.5

TABLE 4. Comparison of Zygotic Seedlings and In Vitro Plantlets in Field Trials

	Average Height (m)	Average Chest Diameter (cm)	Relative Height (%)	Relative Chest Diameter (%)	Volume of Timber per m^3
Zygotic seedlings	3.594	2.61	100	100	0.2064
In vitro plantlets (Clone C)	4.898	3.66	136.3	138.3	0.5408

TABLE 5. Coefficient of Variation of Height and Trunk Diameter of Zygotic Seedlings and In Vitro Plantlets

	Coefficient of Variation			
		In vitro plantlets		
Parameter	Zygotic seedlings	A	E	C
Height of tree (%)	17.4	6.33	9.02	9.02
Chest diameter (%)	25.8	14.4	4.92	13.8

TABLE 6. Field Trials in Shan-kou Forest Station

	Height When Transplanted (cm)	No. Planted	No. Survived	Survival Rate (%)
In vitro plantlets				
Clone O	3–5	3810	3710	97.4
Clone C	3–5	3640	3535	97.1
Zygotic seedlings	3–4	9000	8743	97.1

be kept at the proper temperature when the plantlets are being transported. This may entail heating systems or insulation if the temperature is too low or well-ventilated shade if the air temperature is too high.

The Institute of Forestry of Qinzhou District has used these methods to transport embryo-derived plantlets to the Institute of Forestry of Guangxi Zhuang Autonomous Region for afforestation in May 1981. The plantlets traveled over 100 km and were planted as nursery stock with 100% survival rate. After 1½ years, the eucalyptus trees had an average height of 9.2 m and an average diameter at breast height of 7.67 cm. These results showed the rapid growth of plantlets from tissue culture.

PROSPECTS

Development of plantlets from embryonic aggregates provides the base for commercial production of plantlets in quantity.

In our investigation, mendelian segregation occurred in embryo plantlets which were derived from F_2 hybrids of Lei Lin No. 1 eucalyptus. However, the elite clones which were obtained through field selection gave the forest a uniform appearance, rapid growth, and the advantages of hybrid characteristics, showing great prospects for practical production.

There are numerous widely different species and varieties of eucalyptus trees. A first consideration for further research is to discover suitable culture conditions and to overcome problems arising from interspecific differences among the cultures. Well-chosen methods will make it possible to popularize good nursery stock of numerous clones.

It is still impossible to induce embryonic cell aggregates when stem, buds, or leaves of mature trees are used as explant material. This problem remains to be solved in research for clonal propagation in eucalyptus.

KEY REFERENCES

De Fossard, R. A. 1974. Tissue culture of *Eucalyptus*. *Aust. Forest*. **37**:43–54.

———, Nitsch, C., Cresswell, R. J., and Lee, E. C. M. 1976. Tissue and organ culture of *Eucalyptus*. *N. Z. J. Forest Sci*. **4**:267–278.

Ouyang, Q., Peng, H., and Li, Q. 1981. Studies on regeneration of embryoid from callus in Eucalyptus. *Forest. Sci*. **17**(1):1–7 (in Chinese, English abstract).

Zhuang, M., Bao, C., Liu, J., Ma, Y., Wang, K., Ni, D., and Zhang, F. 1980. Studies on induction of intact plantlets through in vitro culture of organs in *Eucalyptus*. *Forest. Sci*. **16**(2):151–153 (in Chinese, English abstracts).

REFERENCES

Aueja, A. and Atab, C. 1969. Plantlet formation in tissue culture from lignotubers of *Eucalyptus citriodora* Hook. *Curr. Sci. (India)* **38**:39.

Kitahara, E. J. and Caldas, L. S. 1975. Shoot and root formation in cell cultures of Eucalyptus. *J. Forest. Sci*. **26**:242–243.

Laboratory of Cytology of the Institute of Plant Physiology, the Chinese Academy of Sciences, ed. 1978. Cell differentiation and morphology. *Tissue and Cell Culture in Plant*. pp. 150–189. Shanghai Science and Technology Press, Shanghai (in Chinese).

Zhu, C. 1978. Embryoid in tissue culture of plant. *Acta Genet. Sin*. **5**(1):79–88 (in Chinese, English abstracts).

CHAPTER 13

Cork Tree

*Kexian Wu**

INTRODUCTION

Geographical Distribution and Economic Importance

Cork tree (*Phellodendron amurense*) Rupr., is a large tree species belonging to the family Rutaceae. It is a dioecious species, distributed in the north and northeast of China and also growing in Japan, Korea, Siberia, and Eastern Europe. It was introduced into the United States in 1986 (Forest Service, U.S. Department of Agriculture, 1974).

The wood of *Phellodendron amurense* is tenacious and elastic and has beautiful grain with specific luster. It is good material for making furniture, ships, and vehicles and also for use in aircraft and military industries. Its bark is a source of medicine, dyestuffs, essences, and other products. The cork tree is also nectarous. It is acclaimed as one of the three preeminent hardwood trees in Northeast China because of its wide use and high economic value.

Problems in Seeding Production

P. amurense trees are highly heterozygous because they are cross-pollinated. It is difficult to maintain superior characteristics by seed

*English translation by Kaiwen Yuan.

propagation, and the yield of seeds is not stable. If seed propagation is the only production method, male trees possessing superior characters cannot be propagated sexually. Furthermore, success is difficult to achieve by grafting because of the tree's poor regeneration ability. This has limited the development of artificial afforestation. Tissue culture techniques can solve these problems and save much space. Propagation by tissue culture can be carried out all year on a large scale and with a greater propagation coefficient. It can thus provide a large number of high-quality plantlets for propagating this precious woody species.

Studies of plant tissue culture and its techniques have developed rapidly since the 1940s, especially in the last 20 years. But up to now, only a few investigators have regenerated plants by tissue culture from hard broadleaf species growing in frigid zones, as most of them have a poor regenerating ability (Wang, 1983; Yang, 1982; Bonga and Durzan, 1982; Karnosky, 1981). As for tissue culture in the genus *Phellodendron*, Dong et al. (1984) have regenerated plantlets using cotyledons and hypocotyls of 2-day-old seedlings. Since 1983, systematic studies on tissue culture of *Phellodendron amurense* Rupr. have been carried out in our laboratory. Plantlets have been induced from shoot tips taken from trees 20–50 years old. Large numbers of test tube plants have been produced through 12–15 generations of subculture.

CULTURE TECHNIQUES

Selection of Explants

SHOOT TIP. From the end of April to early May each year, shoot tips were taken from trees when the overwintered buds had begun to develop but the bud scales had not yet burst. The donor trees were divided into three groups according to their age: young (2 years), middle-aged (20 years), and mature (40–50 years). Shoot tips taken from the mature trees were the best materials for culture. Shoot tips 2–3 mm in length were used for culture after removal of the bud scales and unexpanded young leaves.

COTYLEDONS AND HYPOCOTYLS FROM FRUIT SEEDS. Cotyledons and hypocotyls were taken from four different fruit development stages: (1) early stage: fruits are 5–7 mm in diameter, seedcoats are soft and white, and seeds are small and tender; (2) middle stage: fruits are close to 10 mm in diameter, seedcoats are white and slightly hard, seed size is similar to that of ripe seed; (3) late stage: fruits are about 10 mm in diameter, seedcoats are hard, and black seeds are close to maturity; and (4) mature seed: when using cotyledons or hypocotyls as explants, the pericarp

TABLE 1. The Effect of Hypocotyl Stages on Differentiation

Developmental Stage	No. of Inoculations	No. of Buds	Differentiation Rate (%)
Early stage	27	1	3.7
Middle stage	61	13	21.3
Late stage	26	4	15.4
Mature seeds	76	0	0

and pulp of the sterilized fruits were first removed. The seedcoats were then cut open with care and the cotyledons or hypocotyls picked out for inoculation. Among the four groups of materials, buds could be readily regenerated from hypocotyls at middle and late developmental stages, with the highest regeneration frequency achieved with middle stage (Table 1). The bud regeneration frequency of cotyledons was not as high as for hypocotyls. Fifty-four cotyledons at middle stage were inoculated with a differentiation frequency of 11.1%. Only 3.6% bud regeneration was achieved from 56 cotyledons of the late developmental stage, and those from early and mature seeds did not produce shoots.

Culture Procedures

INDUCING BUD DIFFERENTIATION. The medium used for induction of buds was MS medium with 8.8 μM BA, 1.64 μM NAA, and 58.7 mM sucrose. The pH was adjusted to 6.0, 0.45% agar was added, and medium was dispensed into 50–100-ml Erlenmeyer flasks and autoclaved. All explants were first sterilized in 70% alcohol solution for several seconds, then immersed in a saturated solution of bleaching powder and washed three times with sterile water. The cultures were incubated under 1–12 hours' illumination per day at an intensity of 1500 lux. Temperature was 25–28°C in the day and 18–20°C at night.

Shoot tips began to swell after 7–10 days' culture. Buds were regenerated and clustered in the swelled shoot tips after 20–25 days in culture. One bud cluster usually contained two to five buds (Fig. 1). Sometimes both buds and calli arose in one explant. Dissection revealed that all buds were derived from axillary buds, not from calli.

The cotyledons and hypocotyls enlarged and turned green after 8–18 days' inoculation and were 10 times their original size in 20–25 days. The cultured cotyledons became crooked, and the hypocotyls grew larger and longer; bud differentiation followed.

ROOT INITIATION. The rooting medium was MS medium (with only half-quantity of macroelements) with 1.64 μM NAA, 1.67 μM IAA, 161 μM H_3BO_3, and 58.7 mM sucrose. The rooting frequency was 31.6%.

Figure 1. A cluster of shoot buds that developed from a cultured cork tree shoot tip.

Root initiation was found to be difficult in regenerated shoots of *P. amurense*.

SUBCULTURE. Axillary buds of plantlets growing in tubes are not dormant. They grow rapidly with growth of the plantlet. Plantlets 4–5 cm in height can be cut at the internodes and subcultured. The axillary buds grow very quickly and can reach 4–5 cm in 1 month (Fig. 2). Histological observations showed that axillary buds do not grow for the first 10 days after subculture, then begin to sprout in 15–20 days. Axillary shoots are formed in 25 days.

The number of nodes in test tube plants varied between 3 and 10, with an average of 4.95 nodes per plant observed.

ESSENTIAL POINTS IN CULTURE

Appropriate Hormone Compositions

The results of bud regeneration from shoot tips of a 20-year-old cork tree under different concentrations of hormones and sucrose are shown in Ta-

Figure 2. Shoots that grew in 1 month from axillary buds on nodal segments.

TABLE 2. Effect of Hormones and Sucrose on Shoot Tip Regeneration (Basic Medium, MS)

Supplements in Medium (μM)					No. of Inoculated Explants	No. of Regenerated Buds	Regeneration Rate (%)
KIN	BA	NAA	IAA	Sucrose (mM)			
9.3		1.64		147	15	0	0
	8.8	1.64		147	17	1	5.88
	8.8		1.67	147	16	0	0
	8.8	1.64		58.8	20	2	10.00
4.6	4.4	1.64		147	20	0	0
4.6	4.4	1.64		58.8	17	1	5.88

TABLE 3. Effect of NAA, IAA, and Boric Acid on Rooting (0.5 Strength MS)

NAA μM	IAA μM	H_3BO_3 Supplement μM	Sucrose (mM)	No. of Tested Plantlets	Rooted Plantlets	Rooting Rate (%)
0.16	0.17	0	58.7	14	3	21.43
1.64	1.67	226	58.7	19	6	31.58
16.4	16.7	549	58.7	20	2	10.00

ble 2. BA was more effective than KIN in bud regeneration, and NAA gave better results than IAA. The optimum sucrose concentration was 58.7 mM. The combination of BA and NAA with a low sucrose concentration was favorable for shoot tip differentiation.

Root initiation experiments, shown in Table 3, indicated that half the macroelements of MS medium with low concentrations of NAA and IAA favored root differentiation, and the addition of boric acid stimulated root development.

INCREASING INDUCTION FREQUENCY BY USING MALE MATERIALS. Two groups of comparative tests were conducted with basal MS medium plus 8.8 μM BA, 1.64 μM NAA, and 147 mM or 58.7 mM sucrose using both male and female 20-year-old trees. Table 4 shows that the differentiation frequencies of shoot tip explants varied between male and female trees. Not only was the differentiation frequency of male explants higher, but the growth of regenerated plantlets was also faster and stronger than for females. Male trees are usually larger and more vigorous and, therefore, can be used as superior donors for propagation.

TABLE 4. Comparison of Shoot Tip Regeneration from Female and Male Plants

Sex	Sucrose (mM)	Inoculation No.	Bud Regeneration No.	Regeneration Rate (%)
Male	147	23	2	8.70
Female	147	26	1	3.85
Male	58.7	76	8	10.53
Female	58.7	74	3	4.05

TABLE 5. Effect of Age on Shoot Tip Regeneration*

No.	Age of Tree	Inoculation No.	Regeneration No.	Regeneration Rate (%)
1	2	24	1	4.17
2	20	56	7	12.50
3	40	104	46	44.23
4	50	64	6	9.38

*MS + BA 8.8 μM + NAA 1.64 μM + Sucrose 58.7 mM; 2–4 are male plants.

DIFFERENCES OF DONOR AGE IN PLANTLET REGENERATION. The age of donor trees affected the frequency of plantlet regeneration (Table 4). Many investigators have found that plantlets could be regenerated easily from explants of young trees, but the regeneration frequency was reduced as the donor trees grew older (de Fossard, 1978; Karnosky, 1981; McKeand and Wir, 1984; Scowcroft, 1978). In contrast, our experiments (Table 5) show that the group of younger materials (Number 1) gave lower induction frequencies than the older plants. This may be related to two factors: sex of young trees could not be identified or the inoculated shoot tips were damaged during sterilization because of breaks in bud scales. The third group of explants were taken from a single male tree which showed not only an unexpectedly high frequency of bud regeneration (44.23%) but also fast growth of regenerated plantlets. Consequently, there may be significant differences between trees in regeneration capability.

CHARACTERISTICS OF MORPHOGENESIS AND STABILITY OF CHARACTERISTICS

Shoot tip cultures can be used directly for propagating superior trees. They, therefore, have more practical application than cultures of cotyledons and hypocotyls. At present, the propagation of shoot tips of

Phellodendron amurense in culture depends mainly on the sprouting of axillary buds. In nature, the axillary buds formed during the year are dormant, and they begin to sprout only in the spring of the following year. In vitro culture is different. Barz et al. (1977) found that high concentrations (45–135 μM) of cytokinin would accelerate the sprouting of axillary buds. In shoot tip cultures and subcultures of test tube plantlets, 9 μM cytokinin can effectively promote axillary sprouting. Axillary buds start to develop after 10 days of incubation and grow rapidly. If a shoot is subcultured for 20–30 days, its axillary bud can grow into new shoots 3 to 4 cm long. By this method, a large quantity of materials can be obtained for rapid propagation. Sometimes, the secondary buds were found to grow from one leaf axil during culture, a phenomenon which is rarely seen in nature. The internodes of some test tube plants became gradually larger and developed into multiple buds (Fig. 3).

In recent years, there have been many reports of instability in chromosome number and structure during tissue culture (Bayliss, 1973). Murashige (1974) and D'Amato (1975) found that the regenerated plants were genetically stable if they were derived from leaf axils, rather than calli. As *Phellodendron amurense* callus does not readily differentiate, the problem of variability of the plantlets developed from callus is of no concern, and the genetic stability of regenerated plantlets is ensured. The development of axillary buds was observed by fixing the prepared materials in FAA, making paraffin slices, and staining with iron-hematoxylin. Root tips or shoot tips were collected for chromosome examination under the microscope. The materials were fixed in the solution of alcohol/acetic acid at a ratio of 3:1, hydrolyzed in 1 M HCl at 60°C for 14 min, and stained with iron-hematoxylin.

Chromosomes were counted in 12 regenerated plants subcultured for 15 generations. Their chromosome numbers were $2n = 56$, as in normal trees. No morphological variation has been found in the stems, leaf, etc.,

Figure 3. Multiple shoot buds that formed from a single axillary bud.

of regenerated plants; therefore, the regenerated plants appear to be phenotypically stable.

PROSPECTS

Tissue culture, as a means of asexual propagation, has important application in agriculture and forestry. Jones et al. (1977) obtained 60,000 shoots from one shoot tip explant of an apple tree within 8 months, and Whitehead and Giles (1977) reported that one axillary bud of poplar could produce a million plantlets in 1 year. Two factors must be considered when tissue culture is used for propagation. First, the propagation coefficient should be large. Second, the ploidy and other genetic characteristics of regenerated plants must be stable through generations of subculture. In *Phellodendron amurense*, buds are usually regenerated from shoot tip, cotyledon, and hypocotyl explants in about 25 days. They can develop into plantlets 3–4 cm high in 1 month under favorable environmental conditions. If one plantlet develops four to five pairs of axillary buds in 1 month and one pair of buds forms one plant, more than 8 million plants can thus be produced in 1 year. The plants are genetically stable as they are derived from axillary buds. Therefore, tissue culture is useful for forest production. Great differences exist among cork tree individuals, so explants should be collected from superior individuals with greater regenerating and growing abilities.

In woody plant breeding, at present, there is no method of large-scale propagation of selected F_1 hybrids which have shown superior characteristics in production, except for a few species which can be easily propagated by cutting. Waiting for seed set from F_1 hybrids is time-consuming, and segregation occurs in their offspring. Accelerating the propagation of F_1 hybrids thus remains a problem. However, the success of plant regeneration from tissue culture provides an effective approach to solving this problem. Tissue culture techniques can be used directly to propagate individual trees with superior economic characteristics selected from natural populations. At the same time, the propagation coefficient of tissue culture is great, and a large number of superior plantlets can be produced in a short time.

At present, a common practice in tree breeding is to collect seeds from mother trees, produce seeds in seed plots, and raise seedlings in a nursery; this method cannot guarantee uniformity among individual trees. The establishment and application of tissue culture techniques will alter conventional breeding practices for forest trees. As superior plantlets regenerated via tissue culture are extensively used for afforestation, survival rates and timber production can both be strikingly increased. Tissue culture also has an important role in saving species near extinction and in recovering and developing valuable germ plasm of forest trees.

REFERENCES

Barz, W., Reinhard, E., and Zenk, M. H. 1977. *Plant Tissue Culture and Its Biotechnological Application*. Springer-Verlag, Berlin.

Bayliss, M. W. 1973. Origin of chromosome number variation in cultured plant cells. *Nature* **246**:529–530.

Bonga, J. M. and Durzan, D. J. 1982. *Tissue Culture in Forestry*. pp. 109–149. Martinus Nijhoff, The Hague.

D'Amato, F. 1975. The problem of genetic stability in plant tissue and cell culture. In: *Crop Genetics Resources For Today and Tomorrow* (O. H. Frankel and J. G. Hawkes, eds.), pp. 333–348. Cambridge University Press, Cambridge.

Dong, M., Lin, J., and Huang, Q. 1984. Tissue culture of *Phellodendron amurense. Plant Physiol. Commun*. **4**:36–37 (in Chinese).

de Fossard, R. A. 1978. Tissue culture propagation of *Eucalyptus ficifolia* F. Muell. *Proc. Symp. Plant Tissue Cult.*, Science Press, Beijing, 425–538.

Forest Service, U.S. Department of Agriculture. 1974. Seeds of woody plants in the United States. Washington, D.C., pp. 578–579.

Karnosky, D. F. 1981. Potential for forest tree improvement via tissue culture. *Bioscience* **31**(2):114–120.

Jones, O. P., Hopgood, M. E., and Farrell, D. O. 1977. Propagation in vitro of M. 26 apple rootstocks. *J. Hortic. Sci*. **52**:235–238.

McKeand, S. E. and Weir, R. J. 1984. Tissue culture and forest productivity. *J. Forest*. **4**:212–218.

Murashige, T. 1974. Plant propagation through tissue cultures. *Annu. Rev. Plant Physiol*. **25**:135–166.

Scowcroft, W. R. 1978. A consideration of the main points of the discussion on tissue culture. *Proceedings of Symposium on Plant Tissue Culture*. Science Press, Beijing, 530–531.

Wang, H. 1983. Afforestation and tissue culture. *Sci. Silvae Sin*. **19**(3):292–301 (in Chinese).

Whitehead, H. C. M. and Giles, K. L. 1977. Rapid propagation of poplar by tissue culture methods. *N. Z. J. Forest. Sci*. **7**:40–43.

Yang, N. 1982. List of test tube plants. *Plant Physiol. Commun*. **4**:61–80, **5**:59–71 (in Chinese).

CHAPTER 14
Staghorn Sumac

*Zhiqing Zhu**

INTRODUCTION

Staghorn sumac (*Rhus typhina*) belongs to the family Anacardiaceae. This native North American species is a hardy and tenacious shrub or small tree of 3–10 m in height. It was introduced to the Botanical Garden of the Institute of Botany, Academia Sinica, in 1959, and then distributed to 20 provinces and cities in Northern China. The tree grows well in arid areas and barren hills of this country. It is a salt-tolerant plant surviving in soil having a salt concentration of up to 3%. A fast-growing tree, staghorn sumac grows 1.5–2.0 m/year, so it may be used to produce large amounts of firewood. Although the wood of *R. typhina* is unimportant in industry, it has a wide range of uses in handcraft production because of its good quality and decorative grain. It is also an ornamental tree, with attractive fiery autumn leaves.

POSSIBLE USES OF TISSUE CULTURE TECHNIQUES

Since *R. typhina* has been planted on a large scale in China, ways to improve the tree are being considered. Micropropagation via *R. typhina* tis-

*English translation by Zhiqing Zhu.

sue culture is unnecessary as the species is amenable to both sexual and vegetative propagation. Tissue culture techniques could be useful in the following ways: (1) to speed up proliferation of superior plants selected from natural populations and (2) to isolate more salt-tolerant mutants.

LITERATURE REVIEW

In spite of the difficulties in plant regeneration remarkable progress has been made in tissue culture of perennial crops, especially in broadleaved species. Plants have been regenerated from callus cultures in the following genera of dicotyledon trees: *Acacia* (Skolman and Mapes, 1976), *Acer* (Brown and Sommer, 1982), *Aesculus* (Radojevic, 1978), *Alnus* and *Betula* (Brown and Sommer, 1982), *Broussonetia* (Oka and Ohyama, 1974), *Carya* (Knox and Smith, 1981), *Castanea* (Vieitez and Vieitez, 1980), *Catalpa* (Zhu et al., 1982), *Celastrus* (Wang et al., 1982), *Cinnamomum* (Zhu et al., 1982), *Corylus* (Jarvis et al., 1978), *Eucalyptus* (Cresswell and Nitsch, 1975, Kitahara and Caldas, 1975), *Fagus* (Chalupa, 1979), *Gleditsia* (Brown and Sommer, 1982), *Hevea* (Chen et al., 1978), *Liquidambar* (Brown and Sommer, 1982), *Loropetalum* (Song et al., 1982), *Liriodendron* (Brown and Sommer, 1982), *Parchira* (Cheng and Wang, 1982), *Passiflora* (Cheng et al., 1984), *Paulownia* (Fu, 1978), *Phellodendron* (Dong et al., 1984), *Populus* (Winton, 1971), *Quercus* (Chalupa, 1979), *Robinia* (Brown and Sommer, 1982), *Salix* (Letouze, 1977), *Santalum* (Rao and Rangaswamy, 1971), *Sapium* (Shi and Li, 1982), *Tamarix* (Yang, 1982), *Tectona* (Gupta et al., 1980), and *Ulmus* (Brown and Sommer, 1982). This paper deals with callus induction and plant regeneration of *Rhus typhina*.

CALLUS INDUCTION

Explants

Fully expanded compound leaves of 2-year-old trees were collected for explants. The leaves were surface-sterilized and cut into 1-cm^2 pieces and the leaf petioles cut into 0.5–1.0-cm segments. All explants were inoculated on callus induction medium.

Callus Induction Medium

N_6 medium (Chu, 1978) was used as basal medium. NAA, 2,4-D, and BA are needed for callus initiation, and BA cannot be substituted by KIN. Table 1 shows the callus induction medium used for *R. typhina*.

TABLE 1. The Media of Callus Induction and Regeneration of Houju Tree

Compounds	mM	Compounds	μM
$(NH_4)_2SO_4$	3.5	$MnSO_4 \cdot 4H_2O$	19.7
KNO_3	28.0	$ZnSO_4 \cdot 7H_2O$	5.14
KH_2PO_4	2.96	H_3BO_3	25.8
$MgSO_4 \cdot 7H_2O$	0.75	KI	5.0
$CaCl_2 \cdot 2H_2O$	1.13	Glycine	27
Sucrose	220	Thiamine	3.0
Iron salt: 7.45 g		Nicotinic acid	4.1
Na_2-EDTA and		NAA	2.7
5.57 g $FeSO_4 \cdot$		2,4-D	2.3
$7H_2O$ in 1 l; 5-ml		BA	4.4
solution above in		Casein hydrolysate	500 mg/l
1-l medium (0.2		Agar	8000 mg/l
mM)		pH	5.8

Callus Formation and Proliferation

After culturing about 1 week at 28 ± 2°C under 3000-lux illumination explants of both leaves and leaf petioles gradually turned brown or black, but this browning did not prevent callus formation. After 15–20 days of culture, green callus appeared from the cut edges of the explants, predominantly near veins. When the calli grew to 1 cm in diameter, they could be transferred to newly prepared callus induction medium for subculture or to differentiation medium for shoot induction. Subcultured calli maintained the ability to regenerate plants for at least 1 year.

INDUCTION OF ADVENTITIOUS BUDS FROM CALLUS

Adventitious bud formation in callus depends on BA in the differentiation medium. On an N_6 medium containing 2.2 μM BA and 88 mM sucrose, nearly all calli produced buds and shoots. In general, a callus 1 cm in diameter provided about 30 adventitious buds at the end of 1 month's culture. When both 0.5–1.0 μM NAA and 2.2 μM BA were used, the shoot differentiation frequency was as high as on the medium containing BA only. As NAA concentration exceeded 2.7 μM, shoot formation was depressed. On media without BA, whether auxin and cytokinin were present or not, no bud differentiation was found.

ROOTING AND TRANSPLANTATION

Shoots 1–2 cm tall were excised and transferred to N_6 basal medium containing 88 mM sucrose. In such a medium about 35% of the shoots develop adventitious roots. When the medium was supplemented with 0.49 μM IBA the rooting frequency increased to about 50%, although the same concentration of IAA was ineffective. Intact plants 3–5 cm high could be obtained on rooting medium. The test tube plants were hardened before transplantation by uncapping the tubes or flasks and keeping them in a shady place for 4–6 days. The plants were then removed with forceps, and roots were thoroughly washed in clean running water. They were transplanted to containers with wet, fine sand and covered with beakers or plastic film to maintain humidity. The plants were kept in a culture room at about 20°C with illumination around 3000 lux. When they were large enough, they were replanted into potting soil or the field.

FUTURE PROSPECTS

Staghorn sumac has been extensively planted in arid and high-salt areas of China as a windbreak, so variety improvement is important. Callus culture can be used to produce regenerated plants over a long period, and it may be possible to select somaclonal variants or induced mutants with this system. If the callus is cultured on a specific medium, such as NaCl-containing medium, salt-tolerant somaclones may be selected and some new salt-tolerant varieties may be proliferated via micropropagation techniques. Superior individuals from regenerated plant populations could be screened in the field and the selected lines micropropagated.

REFERENCES

Brown, C. L. and Sommer, H. E. 1982. Vegetative propagation of dicotyledonous trees. In: *Tissue Culture in Forestry* (J. M. Bonga and D. J. Durzan, eds.), pp. 109–149. Martinus Nijhoff, the Hague.

Chalupa, V. 1978. Control of root and shoot formation of sweet gum and production of trees from poplar callus. *Biol. Plant* **16**:316–320.

——— 1979. In vitro propagation of some broad-leaved forest trees. *Commun. Inst. Czech.* **11**:159–170.

Chen, C., Chen, F., Chien, C., Wang, J., Chang, H., Hsu, H., Ou, Y., Ho, T., and Lu, T. 1978. Obtaining pollen plants of *Hevea brasiliensis. Proc. Symp. Plant Tissue Cult*. Science Press, Beijing, 11–12.

Cheng, Z. and Wang, J. 1982. The hypocotyl and young bud culture of *Pachira macrocarpa. Plant Physiol. Commun*. **2**:36 (in Chinese).

———, Wang, J., Liu, D., and Zhao, C. 1984. Tissue culture of *Clerodendranthus spicatus* and *Passiflora edulis. Plant Physiol. Commun*. **5**:39–40 (in Chinese).

Chu, C. C. 1978. The N_6 medium and its applications to anther culture of cereal crops. *Proc. Symp. Plant Tissue Cult.*, Science Press, Beijing, 43–50.

Cresswell, R. and Nitsch, C. 1975. Organ culture of *Eucalyptus grandis* L. *Planta* **125**:87–90.

Dong, M., Lin, J., and Huang, Q. 1984. In vitro culture of *Phellodendron amurense*. *Plant Physiol. Commun.* **4**:36–37 (in Chinese).

Fu, M. I. 1978. Plantlets from *Paulownia* tissue culture. *Fourth Int. Congr. Plant Tissue Cell Cult.*, (T. A. Thorpe ed.), Calgary, Canada, p. 167 (abstract).

Gupta, P. K., Nangir, A. L., Mascarenha, A. F., and Jagannathan, V. 1980. Tissue culture of forest trees: Clonal multiplication of *Tectona grandis* L. by tissue culture. *Plant Sci. Lett.* **17**:259–268.

Jarvis, B. C., Wilson, D. A., and Fowler, M. W. 1978. Growth of isolated embryonic axis from dormant seeds of hazel (*Corylus avellana* L.). *New Phytol.* **80**:117–123.

Kitahara, E. H. and Caldas, L. S. 1975. Shoot and root formation in hypocotyl callus cultures of *Eucalyptus*. *Forest. Sci.* **21**:242–243.

Knox, C. A. and Smith, R. H. 1981. Progress in tissue culture methods for production of 'Riverside' stocks. *Pecan. Q.* **15**:27–31.

Letouze, R. 1977. Croissance du bourgeon axillaire d'une bouture de saule (*Salix babylonica* L.) en culture in vitro. *Physiol. Veg.* **12**:397–412.

Oka, S. and Ohyama, K. 1974. Studies on the in vitro culture of excised buds in mulberry tree. *J. Seric. Sci. Jap.* **43**:230–235.

Radojevic, L. 1978. In vitro induction of androgenetic plantlets in *Aesculus hippocastanum*. *Protoplasma* **96**:369–374.

Rao, P. S. and Rangaswamy, N. S. 1971. Morphogenetic studies in tissue cultures of parasite *Santalum album* L. *Biol. Plant* **13**:200–206.

Shi, Z. and Li, Y. 1982. The plantlet formation from tissue culture of *Sapium sebiferum*. *Plant Physiol. Commun.* **2**:38–39 (in Chinese).

Skolmen, R. G. and Mapes, M. O. 1976. *Acacia kao* Gray plantlets from somatic callus tissue. *J. Hered.* **67**:114–115.

Song, P., Peng, C., and Zhang, X. 1982. Callus formation and organogenesis in tissue culture of *Loropetalum chinesis*. *Plant Physiol. Commun.* **4**:33 (in Chinese).

Vieitez, A. and Vieitez, E. 1980. Plantlet formation from embryonic tissue of chestnut grown in vitro. *Physiol. Plant* **50**:127–130.

Wang, K., Zhang, P., Ni, D., and Bao, C. 1981. Callus formation and organogenesis from tissue cultures of some woody plants. *Acta Bot. Sin.* **23**:97–103 (in Chinese, English abstract).

Winton, L. 1971. Tissue culture propagation of European aspen. *Forest. Sci.* **17**:348–350.

Yang, N. 1982. Organogenesis and clonal propagation of some woody plants in vitro. *Plant Physiol. Commun.* **4**:23–27 (in Chinese).

Zhu, J. 1982. Tissue culture of *Cinnamomum albosericerum*. *Plant Physiol. Commun.* **2**:40–41 (in Chinese).

Zhu, L., Jin, Y., and Guo, Y. 1982. Tissue culture of axillary bud of *Catalpa bungei*. *Plant Physiol. Commun.* **2**:38 (in Chinese).

PART C
Fruit Trees

CHAPTER 15
Apple: Anther Culture

Guangrong Xue and *Jianzhe Niu**

INTRODUCTION

Apple is one of the principal high-yielding fall fruit tree species. Apples are available for fresh fruit, cider juice, dried apples, jam, and preserves as well as for canning. Apple trees are long-lived and can be cultivated on hillsides, riversides, and light alkali-saline soil because of their adaptability.

The apple was the first tree cultivated in history. It originated in middle and southeastern Europe, central Asia, and the Xingjiang Province of China. Most of the varieties of so-called Big Apple in China were introduced from abroad.

In China, the provinces of Liaoning, Hebei, Shandong, and Henan are the major apple growing regions. The growing regions for Small Apple varieties in China are the cool and arid district of the Loess Plateau, North West Yellow Highland, including the provinces of Shanxi, Shaanxi, Gansu, and Ningxia. New regions in southwest China are the provinces of Yunnan, Guizhou, and Sichuan as well as Heilongjiang, Jilin, and Inner Mongolia District (Institute of Pomology, Chinese Acad-

*English translation by Xiaxian Zhou.

emy of Agricultural Sciences, 1959; Sun, 1983; Hebei Agricultural University, 1979).

The apple tree *Malus* Mill. belongs to the family Rosaceae. There are about 35 species in the world, 23 of which originated in China. The following species of *Malus* are cultivated:

1. *Malus pumila* Mill.
2. *M. pumila* var. *medzwetzkyana* Dieck
3. *M. asiatica* Nakai
4. *M. prunifolia* Borkh.
5. *M. micromalus* Makino

(Institute of Pomology, Chinese Academy of Agricultural Sciences, 1959; Sun, 1983; Hebei Agricultural University, 1979; Zhejiang Agricultural University, 1980).

Malus pumila Mill. is the most important species in the genus. Most of the apple varieties cultivated in the world are derived from it or its hybrids. There are over 8000 apple varieties in the world, but fewer than 100 are involved in extensive production. To date, there are only 20 varieties suitable for use in China: American Summer Pearmain, Jonathan, Golden Delicious, Delicious, Starking, Richard Delicious, White Waite Pearmain, Indo, Rainier, and Ralls. The varieties Starkrimson, Gold Spur, and Fuji, which were introduced in recent years, are being tested in the field and released for production. Newly bred varieties such as Sheng Li and Qin Guan are cultivated on a large scale in some regions. Varieties now used in production in China are chiefly old ones introduced from abroad. Most of the early seasonal varieties are not of high quality; late-maturing varieties such as Ralls, though productive and good for storage, do not have superior hue, scent, and flavor; the most popular varieties—Golden Delicious, Starking, and Jonathan—do not store well. The breeding of new, improved varieties is an urgent task at present.

BRIEF REVIEW ON ANTHER CULTURE OF APPLE

Apple trees are self-incompatible and perennial. It is very difficult to obtain completely homozygous clones. In the last decade, research on obtaining homozygous apple clones has been carried out by scientists in China and other countries.

Lespinasse et al. (1983), working at the fruit trees research station of the French Academy of Agriculture, selected haploids from hybrid seedlings by using irradiated anthers and delayed pollination. A homozygous diploid line was obtained after chromosome doubling. In 1976–1977, a haploid was isolated from 9969 seedlings in a Topred and *M. pumila* var. *medzwetzkyana* Dieck. cross. They were the first to develop a process for selecting apple haploids and a screening procedure for obtaining pure lines in haploid apple breeding. However, their method is tedious, time-consuming, and inefficient. Lespinasse assumed that anther culture was necessary in order to obtain more haploids. Anther culture may be the most desirable means for producing homozygous apple strains.

The apple anthers were cultured 3 to 5 days before flowering (Nakayama et al., 1971, 1972) in Miller medium supplemented with 2,4-D, IAA, and KIN. The callus induction frequency was highest in the Ralls variety, and lowest in the varieties Jonathan and Fuji. Roots and a few budlike organs, but no buds, were differentiated in 12 differentiation media tested. Kubicki et al. (1975) cultured uninucleate anthers of Jonathan on modified MS medium. After 5 weeks of culture, embryogenic clusters of 32 to 64 cells, but no plantlets, were formed.

Studies of apple anther culture started in China during the 1970s. In 1978, embryos were induced at Shangdong Agricultural College, but no plantlets were differentiated. Embryos were induced in the Big Apple variety Delicious in the Institute of Pomology, Chinese Academy of Agricultural Sciences, in 1979–1980, and plantlets were differentiated, with chromosome number $2n = 17$ (Fei and Xue, 1981). Intact plantlets were also induced in the crab apple variety Huang Tai Ping in the Northeast Agricultural College of China. Most plantlet cells had a chromosome number of $2n = 17$, indicating that they were haploids (Wu, 1981).

APPLE ANTHER CULTURE TECHNIQUE

Optimum Stage for Anther Inoculation

Choosing the optimum stage of anther for inoculation is critical for successful anther culture of apple (Xue and Niu, 1984). In a 20-year-old Delicious tree, the relationship between external morphological development of flower buds and the developmental stage of the microspores was studied continuously for 2 years. Acetocarmine squash observations are shown in Table 1.

Table 1 shows that the beginning and duration of different developmental phases varied in different years, but that the external morphological characteristics of the flower bud and the anther developmental stage were correlated. Embryos could be formed from the anther when microspores were at the tetrad and uninucleate stages, but the latter was preferable. The uninucleate pollen stage was characterized by fully open flower buds, four or five fully spread leaves, a red central alabastrum, enlargement of the surrounding alabastrums, and an inflorescence. These

TABLE 1. Relationships between Flower Bud External Morphology and Pollen Developmental Stage in the Apple Variety Delicious

Observation Date		External Morphological Traits of Flower Bud	Developmental Stage of Microspore
1980	1981		
Apr. 28	Apr. 19–20	Length 1.5 cm, width 0.7 cm, half open, with 1 or 2 coiled young leaflets spreading out and top alabastrum 0.9 cm in length, 0.35 cm in width	Pollen mother cell
Apr. 30	Apr. 21–22	Length 1.5 cm, width 0.9 cm, with top alabastrum slightly enlarged and 4 coiled young leaflets opening	Early tetrad
May 4	Apr. 22–23	Length 2 cm, width 1.2 cm, with coiled leaflets stretching out and well opened	Late tetrad
May 6	Apr. 25	Fully open, with 4 or 5 leaves fully expanded; central alabastrum became red with the surrounding alabastrums markedly enlarged; inflorescence did not open	Uninucleate microspore

external morphological characteristics are useful for identification of developmental stages of anther not only in Delicious but also in the varieties Golden Delicious, Ralls, Rainier, American Summer Pearmain, and Jonathan. Anthers inoculated at this stage readily formed embryos and were easily sterilized. If anthers were taken for inoculation prior to the uninucleate stage, they were easily broken during inoculation because of their delicate texture. If anthers were taken after the uninucleate stage they were difficult to sterilize. Too long a sterilization period damaged the anther.

Culture Procedure and Media

The induction of anther-derived plantlets consisted of three steps:

1. Embryo induction: Some embryos emerged at the break points of browned anther after 70 days of culture; most of them formed 90 to 120 days after inoculation. An induction medium for Delicious was MS basal medium supplemented with 1.8 μM 2, 4-D, 0.93 μM KIN, 22 μM IAA, 88–235 mM sucrose, and 0.6–0.75% agar, pH adjusted to 5.8.
2. Rootless plantlet induction: Embryos were transferred onto bud dif-

ferentiation medium to form secondary embryos or clustered buds, which developed further into rootless plantlets. This step took 40 to 100 days or longer. Medium was MS basal medium with 0.29 μM GA, 2.5 μM IBA, 4.4 μM BA, 88–235 mM sucrose, and 0.6–0.75% agar, pH 5.8.

3. Root system induction: Rootless plantlets were transferred to root induction medium to form intact plantlets. This took 15 to 20 days. Medium was ½ MS medium with 8.6 μM IAA, 44–58.7 mM sucrose, 0.6–0.75% agar, pH 5.8.

Sterilization of Materials and Culture Conditions

Sudden changes in temperature in the springtime could sometimes make collection of anthers at the optimum stage for inoculation difficult. To prevent this problem anther development was delayed by keeping flower buds in water in a 1–3°C refrigerator. Flower buds inoculated after 40 days of low-temperature storage could still successfully form embryos.

STERILIZATION METHOD FOR MATERIALS. Flower buds were picked, the young leaves were removed, and flower buds were carefully separated. The materials were immersed in 70% ethanol for about 30 sec, surface-sterilized with the supernatant of 10% bleaching powder solution for about 15 to 20 min, then washed three to five times with sterilized water. The materials were put into sterilized culture dishes, in which there were one or two layers of sterilized filter paper. In each flask 35 to 50 anthers were inoculated. Solutions of 0.1% mercuric chloride and 10% sodium hypochlorite were also tested for sterilization for 10 to 15 min and 15 to 20 min, respectively. Both gave satisfactory results.

The temperature of the culture room was maintained at 25–30°C, the light intensity at 1500–2000 lux for 10 hr/day. Embryos could be induced in either light or darkness.

Problems to Be Noted during Anther Culture

MEDIUM AND SUPPLEMENTS. MS medium was used as basal medium. Calli but not embryos were formed when N_6 medium was used instead of MS medium. For embryo induction, the concentrations of plant hormones were kept at low levels: 0.93 μM KIN, 22 μM IAA, and 1.8 μM 2,4-D. The Delicious experiments showed that embryo induction frequency increased up to 1.2% in the presence of 148 μM adenine sulfate and 2.9 μM GA, from a control level of 0.6%. The experiments with Delicious also indicated that only calli were formed in the medium with 58.7 mM sucrose. Neither callus nor embryo was produced on 293 mM sucrose; embryos were induced on 88–235 mM sucrose. Anthers inoculated in the optimal medium were yellowish green at first, eventually turning brown as they enlarged. Naked eye observation indicated that the anther walls shriveled. Small amounts of callus gradually appeared at break points on the anther wall after 30 days' culture. Most of the calli were a semitransparent milky white color with a bright granular surface. After 90 to 120 days in culture, embryos were produced at some of the anther break points. The embryos were milky white, 1 to 2 mm in size, firm and compact, and had two cotyledons. Abnormal embryos with mono- or multicotyledons were occasionally formed. Some anthers gave only one embryo; others gave two to four embryos at the same break point.

VARIATION IN MATERIALS. Induction frequencies were calculated on the basis of data from experiments (1981–1983) on anther culture with different varieties. Table 2 shows that embryos could be induced from all varieties studied, with the exception of Ralls. The induction frequencies were not very high. Delicious gave the highest induction frequency (1.9–2.5%), and embryos were induced repeatedly each year in culture from 1979 to 1984. Rainier gave the lowest induction frequency (0.3–1.2%). Embryo induction frequencies in different years varied to a great extent. Some varieties, such as Ralls in 1983, failed to induce embryos at all in certain years. Embryo induction frequencies were probably related to the internal physiology of the tree from which the material was taken: its fruit production, nutrient availability, endogenous hormone levels, or changes in photoperiods. Variation in induction frequency was also related to genotypic differences reflected in different requirements for plant hormones and nutrition. Such variability could be controlled by adjusting supplements in the medium. For instance, when Delicious anthers were inoculated in the medium with 1.8 μM 2,4-D, 0.93 μM KIN, and 22 μM IAA, embryos could be obtained each year, although no embryos were obtained in other varieties using the same medium. At the same time, media that induced embryos from other apple varieties did not work with Delicious. It was noted that induction of embryo in different varieties had a specific medium requirement that included a specific range of exogenous hormones. Some varieties did not have specific media requirements; for example, embryos were obtained from Delicious even if the medium supplements were slightly changed, e.g., 2,4-D removed or IAA decreased. Embryos could be obtained from Ralls, Rainier, Jonathan, and American Summer Pearmain when all the supplements from the medium for Delicious were removed and GA and BA were added.

PROLIFERATION OF CLUSTERED ROOTLESS BUDS. Usually each embryo could develop into only one intact plantlet. Many plantlets would

TABLE 2. Embryo Induction Frequencies in Anther Culture of Different Apple Varieties

Variety	1981			1982			1983		
	No. of Anthers Inoculated	No. of Embryos Obtained	Induction Frequency (%)	No. of Anthers Inoculated	No. of Embryos Obtained	Induction Frequency (%)	No. of Anthers Inoculated	No. of Embryos Obtained	Induction Frequency (%)
Delicious	462	9	1.9	150	3	2.0	366	9	2.5
Rainier	323	5	1.5	390	1	0.3	697	4	0.6
Ralls	511	6	1.2	240	1	0.4	322	0	0
Golden Delicious	87	3	3.4	300	2	0.7	915	2	0.2
American Summer Pearmain	—	—	—	171	3	1.8	613	5	0.8
Jonathan	—	—	—	60	1	1.7	447	1	0.2

Figure 1. Pollen-derived apple plant.

die early after subculture, chromosome identification, and transplantation. To prevent the loss of materials, embryos were transferred into proliferation medium: MS medium with 0.29 μM GA, 2.5 μM IBA, and 4.4 μM BA. (For other supplements, see Culture Procedure and Media.) The embryos produced many secondary embryos or differentiated buds directly (Figure 1). Some proliferated embryos or buds turned green rapidly and rootless plantlets differentiated. The time required for differentiation of rootless plantlets varied among embryos: the shortest time was 40 days; the largest, a few months or even 1 to 2 years. Some embryos differentiated budlike tissue after turning green. The budlike tissue cut from the embryos differentiated into rootless plantlets or into clusters of leaflike tissue which later differentiated rootless plantlets. No plantlets differentiated from anther-derived callus during the differentiation experiments.

Pollen-derived embryos of different apple varieties differentiated into plantlets in the same basic medium. However, it was necessary to adjust the plant hormone concentrations to improve growth of some of the plantlets. Delicious pollen-derived plantlets were the first produced, followed by Ralls, Rainier, Golden Delicious, and American Summer Pearmain. Their differentiation frequencies (differentiated plantlets per embryo) were Delicious 33.3%, Rainier 40.0%, Golden Delicious 33.3%, and Ralls 50%. Plantlets could be subcultured in differentiation medium so as to proliferate more rootless plantlets continuously; all plantlets derived from one original embryo had the same genotype. The proliferation of rootless plantlets varied between different lines. For example, in the Delicious line 80-3, 300 plantlets readily proliferated within 2 months, whereas in the Delicious line 79-1 proliferation was very difficult. Pollen-derived plantlets in test tubes grew differently from stem-tip-derived plantlets (diploids). Pollen-derived plantlets of different lines of the same variety also grew differently. Some of them proliferated readily, some only with difficulty.

ROOTING OF PLANTLETS. The proliferated plantlets 3 cm high were planted for rooting into half strength MS medium supplemented with 8.6 μM IAA, sucrose concentration decreased to 44–58.7 mM, 0.6–0.75% agar, pH 5.8. Roots were induced after 15 to 20 days. When the rooting medium consisted of half strength MS plus IBA, the quality of rooting was inferior, since callus often formed at the root-stem transition zone. The plantlets thus rooted did not often survive after transplantation. Rooting frequencies of pollen plantlets of different varieties and those of different lines of the same variety varied remarkably. The majority of pollen plantlets from Delicious rooted readily; rooting of the pollen plantlets of other varieties was difficult. It had been expected that rooting would take place easily in juvenile apple trees and rooting ability would be lost in adults, but this was not the case for the apple pollen plantlets. The ability to root was more dependant on varietal differences. Stem-tip-derived plantlets from adult trees of variety Golden Delicious developed healthy roots very easily, whereas pollen plantlets developed from embryos of the same variety were very difficult to root. In contrast, stem-tip-derived plantlets of Delicious were difficult to root, although most of its pollen plantlets rooted quite readily.

Transplantation of Plantlets and Scion Grafting

TRANSPLANTATION OF PLANTLETS. Rooted plantlets were planted in pots in spring and autumn with sandy soil as substrate. Survival rates reached 20–40%. Temperature, humidity, and season of transplantation were the main factors affecting the survival rate. The optimum temperature for pot planting is 15–20°C. Transplanted plants should be covered with glass or plastic for 15 to 20 days after planting to reduce evaporation. In spring and autumn, the transplanted plants generally survived well in the field, but in summer and winter, they were difficult to maintain even when carefully managed. When plantlets transplanted in autumn survived and grew in greenhouses throughout the winter, they grew well in the field next spring.

YOUNG SCION GRAFTING. Graft testing was carried out in the field with tube-grown plantlets to increase plantlet proliferation and to observe the plantlets' botanical characteristics as well as to preserve them when they failed to root. The grafting was started in spring when the flower buds started to grow, the sap began to flow, and bark could be easily removed. Grafting could be continued up to flowering. A bark grafting method was used, giving a relatively high survival rate when 2- or 3-year-old branches of fruit-bearing trees were taken as stock and the young shoots of cultured plantlets as scions. High survival rates also resulted when cut grafting methods were used during the shoots' growing season from late May to early June, and new branches were used as stocks and test tube shoots as scions. However, the survival rate was extremely low when grafting was done in the flowering period, July or August in China. This failure probably resulted from lack of water in the bark since the nutrients and water were directed to the developing flowers. In summer, scions probably died as a result of mildew, high temperature, and humidity. Cut grafting could also be done in September, but the branch could not become mature and lignified and thus could not survive winter. A more useful modification was to graft onto seedlings, which were transferred to the greenhouse for further growth.

Technical proficiency is essential for graft survival: the cut should be smooth, the manipulation skillful and quick, the seam tightly bound up and covered with a two-layer bag (the inner, of plastic; the outer, of paper), so as to reduce evaporation and sunlight exposure. The grafts were examined after 15 to 20 days. The scion survived as long as it was emerald green and grew straight upward. The plastic bag was scissored when the scion survived to promote aeration and then taken off when the scion growth was established. The paper bag could be removed whenever the young branch and the leaves grew large. The young branches grew rapidly as soon as they survived top budding. They could grow up to 30–50 cm in the first year.

Many factors affect survival after grafting, including the weather at the time of grafting (generally it is better to graft on cloudy days or early in the morning or late in the evening) and the quality of the scion. Survival was also tied to the grafting compatibility of the scion. Pollen plantlets of clones from the varieties Delicious, Ralls, and Rainier survived field planting and top budding. The pollen plantlets from Delicious were most successful.

Observation of Chromosomes

Chromosomes of root tip cells of Delicious plantlets were observed after staining with iron haematoxylin. The root tips were treated with 0.1% colchicine solution for 2 to 3 hr or treated at 3°C for 48 hr, fixed by Carnoy's fluid for 48 hr, put into a maceration solution (95% alcohol; condensed hydrogen chloride at 1:1 v/v) for 30 to 50 min, in 4% mordant solution of iron alum for 30 to 40 min, then stained with 0.5% haematoxylin solution for 3 to 3.5 hr. They were softened with 45% acetic acid for 1 hr, heated, and squashed. Chromosome numbers were counted under microscope and photographed. The chromosome number of pollen plantlet root tip cells was $2n = 17$, demonstrating that they were haploids. For the pollen plantlets obtained each year from the varieties Delicious, Ralls, Rainier, and Golden Delicious chromosome numbers in stem tips or root tips of test tube plantlets were determined by iron haematoxylin staining. Most cells of the differentiated plantlets were haploids ($2n = 17$). However, one pollen plantlet from variety Golden Delicious did not survive planting and grafting and was thus subcultured in vitro for about 3 years. Tetraploid cells ($2n = 68$) were found during the chromosome number examination. This phenomenon has also been found in pollen plantlets of other tree species (Chen et al., 1977; Fu and Tang, 1983; Yang and Wei, 1981).

PROSPECTS

In the last 5 years, pollen-derived plants have been successfully obtained by anther culture in the important apple varieties Delicious, Golden Delicious, Ralls, Rainier, and the middle seasonal variety American Summer Pearmain (Xue and Niu, 1984). These are major apple cultivars which are often used as parent materials in conventional hybrid breedings. If pure lines of these varieties were used as parental materials for crosses, the heterosis between pure lines would be even stronger than at present. Obtaining pure lines would also promote the study of apple genetics. It can be expected that the embryo induction rate and plantlet differentiation frequency can be improved further. It is necessary to increase the cytological study of chromosome variation and the induction of mutations in pollen-derived plants of apple.

KEY REFERENCES

Fei, K. and Xue, G. 1981. Induction of haploid plantlets by anther culture in vitro in apple cv. Delicious. *Sci. Agric. Sin.* **4**:41–44 (in Chinese, English abstract).

Kubicki, B., Telezynska, J., and Milewskapawliczuk, E. 1975. Induction of embryo development from apple pollen grains. *Acta Soc. Bot. Pol.* **44**(4):631–635.

Lespinasse, Y., Godicheau, M., and Duron, M. 1983. Potential value and method of producing haploids in the apple tree (*Malus pumila* Mill.) *Acta Hortic. In Vitro Cult.* **131**:223–230.

Wu, J. 1981. Haploid plantlet obtained by anther culture of apple. *Acta Hortic. Sin.* **8**(4):36 (in Chinese, English abstract).

Xue, G. and Niu, J. 1984. A study on the induction of apple pollen plants. *Acta Hortic. Sin.* **11**(3):161–164. (in Chinese, English abstract).

REFERENCES

Chen, C., Chen, F., Chien, C., Wang, C., Chang, S., Hsu, H., Ou, H., Ho, Y., and Lu, T. 1978. Induction of pollen plants of *Hevea brasiliensis* Muell. Arg. *Proc. Symp. Anther Cult*. Science Press, Beijing, 3–8 (in Chinese, English abstract).

Fu, L. and Tang, D. 1983. A study on induction of pollen plants of litchi. *Acta Genet. Sin*. **10**(5):369–374 (in Chinese, English abstract).

Hebei Agricultural University (ed.). 1979. Apples. In: *Cultivation of Fruit Trees*. Vol. 1, pp. 1–9. Agriculture Press, Beijing (in Chinese).

Institute of Pomology, Chinese Academy of Agricultural Sciences. 1959. Apples. In: *Cultivation Science of Fruit Trees in China*. Vol. II. pp. 311–328. Agriculture Press, Beijing (in Chinese).

Nakayama, R., Saito, K., and Yamamoto, R. 1971. Studies on hybridization in apple breeding. *Bull. Fac. Agric. Hirosaki Univ*. **17**:12–19; **19**:1–9.

Sun, Y., Du, S., and Yao, K. 1983. Fruit Trees Growing Center in China. In: *History of Fruit Trees and Fruit Tree Resources in China*. pp. 6–7 Shanghai Science and Technology Press, Shanghai (in Chinese).

Yang, Y. and Wei, W. 1984. Induction of longan haploid plantlets from pollen cultured in certain proper media. *Acta Genet. Sin*. **11**:288–293 (in Chinese, English abstract).

Zhejiang Agricultural University. 1980. Apple breeding. In: *Guoshu Yuzhongxue*. pp. 219–257. Shanghai Science and Technology Press, Shanghai (in Chinese).

CHAPTER 16

Apple: Shoot Tip and Embryo Culture

*Jixuan Wang**

SIGNIFICANCE

The apple is one of the most widely cultivated fruit tree species. It is grown in almost all temperate regions of the world, and in many countries its cultivated area and yield have increased in recent years. According to the United Nations FAO, the 1979 world apple crop was 36.2 million tons, comprising 50% of the total yield of deciduous fruit trees.

China contains vast areas of mountainous sandy soil in which it has been difficult to get good yields from most crops. However, peasants and pomiculture scientists have succeeded in cultivating apples in these marginal regions. The apple growing area has expanded rapidly in various regions of China, causing a shortage of nursery stocks for apple propagation. In fact, the demand for nursery stock of dwarf apple has greatly exceeded supply, because of the low rate of propagation by conventional methods. The application of shoot tip culture in vitro can increase apple propagation rates and should provide additional benefits in more effec-

*English translation by Jifang Huang.

tive usage of marginal areas. The method of shoot tip culture for raising nursery stocks is advantageous because of the savings in time and space and lack of seasonal limitations on work. It is also relatively easy to obtain virus-free propagation in apple through shoot tip culture, because the cells of the apical meristem are usually free of virus.

Young embryo culture involves the in vitro culturing of embryos which are removed shortly after fertilization. This method may provide a solution to the problem of embryo abortion resulting from remote hybridization. It may also enable numerous plantlets to be produced from one seed and thus permit many plants to be regenerated from mutagen-treated seeds, resulting in a higher probability of selecting useful variations.

LITERATURE REVIEW

In addition to sexual propagation, the methods of layering, cutting, and grafting have been used for propagating apple stocks for many years. Jones (1967) first reported the effects of BA on culture of apple tissues and succeeded in differentiating plantlets. Later he suggested that xylem supplements should be added to the culture medium to promote growth and rooting of new shoots. Subsequently other reports indicated the existence of GA_4 and GA_7 in xylem. Xylem also contains phlorigen and its derivatives, which cause the regeneration of adventitious roots of tube-cultured plantlets (Jones and Hatfield, 1976). Later it was reported that a large number of rooted plantlets were obtained from shoot tip cultures of apple stock M_{26}. A few plantlets that were transplanted in soil survived. The authors proposed that 60,000 green shoots could be theoretically obtained from a single cultured shoot tip segment within 8 months (Jones et al., 1977).

In England it was reported that the rate of root regeneration of apple stock Mq could be increased by transferring the cultures onto root regeneration medium. The regeneration rate was raised to over 60%, and the quantity of roots on each plant increased fourfold (James and Thurbon, 1979). In the United States, Werner and Boe (1980) reported the propagation of apple stock M_7 through in vitro culture.

In China successful culture of shoot tips from apple stocks M_7 and M_9 was first achieved in 1979 by scientists of the Institute of Botany at the Chinese Academy of Sciences (Differentiation Group, The Sixth Laboratory Institute of Botany). Since then Chinese scientists have obtained plantlets from in vitro culture of axillary bud meristems of the stocks M_4, M_7, M_9, and M_{26}. In addition, they have achieved differentiation of plantlets from apple graft stocks of *Malus asiatica* and *Malus prunifolia* (Chinese flowering crabapple Lao Shan) (Wang and Sui, 1978).

Meristems of M_4 which were inoculated in April 1978 yielded 6734 transplanted plantlets in April 1982 (Wang, 1983). This is an efficiency 100 times greater than that of conventional techniques. Scions of cultivars Fuji and Gold Spur were grafted on these plantlets, growing at seven experimental plots for studying growth. Some of them have grown to bloom and bear fruit.

Studies of propagation of apple cultivars by plantlets obtained through tissue culture started in the early 1970s, but transplantation and growth of plantlets in the field have only recently been successful. Scientists in the United States have inoculated axillary buds of apple seedlings of the cultivar Northern Spy in MS medium and obtained differentiated shoots, which had a rooting rate of only 1% (Dutcher and Powell, 1972). In Canada researchers found that over 50% of the shoots of the apple regenerated roots on LS medium (Walkey, 1972). English scientists have reported inoculating shoot tips of the cultivar Ju Pin onto MS medium supplemented with KIN. The cut bases of the shoots were dipped into an IBA (490 μM) solution, and shoots were transferred onto a hormone-free medium. Over 50% of the cultured shoots could regenerate roots, and a few plantlets which survived transplantation were obtained (Abbott et al., 1976). Other work has demonstrated production of plantlets for transplantation from culture of stem segments; this method was successful in five apple varieties, including Golden Delicious and Ju Pin (Jones et al., 1979). Chinese scientists reported that roots were regenerated from shoot tip cultures derived from an adult tree of the variety Golden Delicious (Chen and Yang, 1980). We have previously obtained plantlets for transplantation in over 10 apple cultivars and lines, including Fuji, Tian Huang Kui, and Golden Spur (Wang and Sui, 1981). Those plantlets were used as self-rooting nursery stocks in trials of adaptation and fertility response at several experimental locations in Liaoning Province and Beijing.

Investigation of shoot micrografting for virus-free propagation of apple has begun. In the United States apical meristems of new shoots from Griffith apples were used as scions for grafting onto seedlings of Golden Delicious derived from cultured embryos; intact plants were obtained (Huang and Millikan, 1980). We have obtained green plantlets by means of micrografting, with survival rates of 39.1–82.1% for various apple species, cultivars, and lines. Plantlets derived from shoot tip cultures of cultivars M_4, M_7, and M_{26}, as well as *Malus baccata* and *Malus spectabilis* (round leaf), were used as graft stock; shoots derived from shoot cultures of Fuji, Golden Delicious, and Tian Huang Kui apples were used as scions. These grafted plantlets have been planted in the field (Wang et al., 1985).

The problem of abortion of seeds from interspecific and intergeneric hybridization might be solved by in vitro culture of the initially developing hybrid embryo. Chinese researchers cultured young embryos 40 to 50 days after fertilization of the varieties American Summer Pearmain, Golden Delicious, and Jonathan and obtained rooting plantlets (Wu and Liu, 1977). We produced a large number of plantlets through culture of embryos of mature or nearly mature seeds in eight cultivars, including Fuji and Ralls (Wang et al., 1980). This technique will enable us to obtain more mutant plants from mutagen-treated seeds.

CULTURE TECHNOLOGY

Culture Procedures and Media

PREPARATION OF MATERIALS FOR IN VITRO CULTURE. Shoot tips and stem segments were used for apple propagation. The optimum size of shoot tips varied according to the purpose of the study. To obtain virus-free plants from material which may be infected, the excised shoot tip should be about 0.1–0.2 mm. From virus-free branches larger shoot tips or even a segment of stem might be used as culture materials. Apical points of axillary buds from both stock and scions in our experiments were 0.1–0.3 mm in length. We hope to use apical points of this size to obtain virus-free nursery stocks. We have also found that the larger the shoot tip in culture, the higher the rate and speed of differentiation.

Sampling experiments showed that explants from apple tree stocks and scions could be taken for inoculation in any season of the year. All the materials taken in different seasons could differentiate, but with different frequencies and different growth rates after differentiation. Materials cut near the time of axillary bud sprouting showed the highest frequency of differentiation after inoculation, and the differentiated shoots showed faster growth rate than other materials.

Young embryos were excised for culture 30–50 days after fertilization. Embryos at early stages of development were easy to culture successfully. Mature embryos were excised from seeds which were mature or nearly mature in fruit. Comparison of young and mature embryo cultures showed that the embryos from fully mature seeds with brownish yellow coats had greater differentiation than those from seeds with pale yellow or piebald coats.

SHOOT DIFFERENTIATION. The basic medium used in shoot tip cultures was mainly MS medium; however, LS medium was also used. The exact composition was as follows: MS medium supplemented with BA (2.2–8.8 μM), CH (300–500 mg/l) or LH, sucrose (88 mM) and agar (0.7–1%). The medium for embryo culture was modified accordingly: ½

MS medium supplemented with BA (1.1 μM), IAA (2.9 μM), sucrose (88 μM), and agar (0.7–1%).

Screening media for apple graft stock differentiation showed the beneficial effects of a medium supplemented with BA (2.2 μM). The number of buds increased, but there were fewer differentiated shoots. When the concentration of BA was higher than 8.8 μM, the number of buds of stock M_{26} developing in culture was very large. All of the buds had leaves, but they were clustered closely together and did not develop new shoots. Similar results have been reported from laboratories outside China. When the concentration of BA was 4.4 μM, stem elongation and leaf formation were promoted, whereas at concentrations over 44 μM the growth of young stems was inhibited (Jones, 1967). MS medium supplemented with BA (8.8 μM) facilitated differentiation of roots from the shoots.

SUBCULTURE AND PROLIFERATION. MS medium supplemented with BA (2.2–4.4 μM), CH, or LH (300 mg/l) was used for proliferation of shoot tip culture from apple stocks and cultivars. The meristematic tissue for subculture turned green when it was illuminated prior to dark culture, which increased the propagation coefficient. Successive subcultures in darkness were done at intervals of 30–50 days. Shoots more than 1 cm in height were cut into segments with two to three nodes; clusters of shoots less than 1 cm were separated into bunches of five to seven shoots each. After two to three subcultures in the dark, the cultures were grown in the light to encourage healthy growth. Propagation was encouraged by separating the clustered shoots. Shoots taller than 2 cm were transferred to medium for root regeneration. The remaining shoots were separated into lots of five to seven shoots each and transferred onto freshly prepared medium for subculture. Shoots cultured for 20 days after separation are depicted in Fig. 1.

Figure 1. Shoots cultured for 20 days after separating.

The medium and culture procedures used in embryo culture for extensive propagation were the same as those used in stem tip culture.

It was observed that the removal of callus from the cultured shoots during the propagation period promoted the regeneration and growth of new shoots.

ROOTING OF SHOOTS. Half-strength MS media or other media with low salt content such as White's or Nitsch's medium were quite effective for root regeneration. The MS (½) supplemented with IAA (5.7–8.6 μM) promoted rooting in green shoots from various graft stocks and from apple embryos, with rooting rates of 71.4–83.3% and 61.1–75.0%, respectively. This rooting medium was different from that used for green shoots derived from apple cultivars. The composition of the latter medium was as follows: MS (½) medium supplemented with IAA (5.7 μM), IBA (0.98 μM), GA_3 (8.7 μM). The medium was less effective at promoting root regeneration since the rooting rate was 42.9–72.7%.

Concentrations of IBA or NAA higher than 4.9 or 5.4 μM, respectively, caused callus proliferation on the cut plane of most of shoots. In such cases roots differentiated on the callus. The callus dried as the plantlets were taken out of the container and transplanted; thus a layer formed between shoot and roots and decreased the survival rate. In contrast, a medium supplemented with IAA (5.7–8.6 μM) and/or a small quantity of IBA caused roots to regenerate directly from the shoot base. Such plantlets survived readily and had high survival rates after transplantation.

Rooting was accomplished in shoots of apple stock by dipping cut ends in a high concentration of auxin and then transferring shoots to rooting medium which contained little or no auxin. This procedure promoted rooting. Green shoots from the apple graft stock M_9, which were difficult to root, were inoculated onto LS medium containing phloroglucinol (20 μM) for 4–7 days and then transferred onto a hormone-free medium. The roots regenerated rapidly and with high frequency (James and Thurbon, 1979).

GRAFTING OF PLANTLET IN CULTURE. If a plantlet derived from a good cultivar was grafted as scion onto a rooted shoot derived from a dwarf stock, the resulting tree maintained the superior dwarf characteristics and remained virus-free. Some researchers have grafted meristems from newly developing branch apices of the cultivar Griffith onto self-rooting shoots derived from in vitro culture of the cultivar Golden Delicious, and obtained virus-free plantlets (Huang and Millikan, 1980).

We used green plantlets derived from cultures of shoot tip of M_4 and M_7 as stocks and plantlets from shoot tip cultures of Fuji and other varieties as scions. We used the methods of approach grafting, cut grafting, and insert grafting. The plant hormone BA, GA, or IAA was applied to the joint between the stock and the scion. Insert grafts treated with ZEA (460 μM) or GA (290 μM) had high graft survival rates of 40.7–82.1% and 39.1–54.3%, respectively.

For insert grafting, a longitudinal cut was made in the middle part of the rooting shoot. The shoots used were over 2 cm in height. Into such a shoot a green shoot derived from elite cultivars with its base part formed into a wedge was inserted as scion. Onto the joint of the graft two to three drops of ZEA (460 μM) or GA (290 μM) was applied. The grafted plantlets were inoculated onto a newly prepared rooting medium and cultured for around 2–3 weeks and then transplanted in soil. The survival rate of the plantlets was 26.7–56.9%.

Essentials of Culture Techniques

STERILIZATION AND INOCULATION. Inoculation materials for in vitro culture must be viable and aseptic. The following procedures were used for preparation of material.

1. Pretreatment: Thorough sterilization usually involves pretreating the materials with surfactants. Germicides can permeate the inoculation materials readily after pretreatment. Various kinds of surfactants can be used, including 70–75% ethanol for ½–1 min, which not only permeates the living organism well but also acts as a disinfectant.
2. Sterilization: The sterilization is done as soon as the material has been moistened. Various conventional germicides have been used in apple culture. For our experiments, the inoculation materials were soaked in 10% sodium hypochlorite solution for 10–15 min, the supernatant of saturated calcium hypochlorite (bleaching powder) solution for 15–30 min, 10–12% peroxide solution for 10–15 min, 5–10% Neogermicide solution for 10–15 min, or 0.1% mercuric chloride solution for 3–7 min. The first four germicides are especially safe for explants. They do not cause death to plant tissues. Because of the exposure time needed for sterilization, germicides often injure the plants by oxidating outer tissue layers. Mercuric chloride is an effective means of sterilization, and it causes less injury to dormant materials; it is the most commonly applied germicide.
3. Rinsing: The materials must be fully rinsed with sterilized water as soon as the sterilization has been performed. In general, the rinse is performed three to five times to remove excess disinfectant.
4. Inoculation: The size of shoot tips inoculated for producing virus-free plantlets is often very small. The shoot tip meristems must be

Figure 2. Grafting of plantlets in culture tube.

carefully peeled and isolated for inoculation. Solid culture medium is used and the shoot tip should be shallowly embedded into the surface layer of the medium.

CULTURE CONDITIONS. During the period from inoculation to formation of rooted shoots, cultures were kept at a temperature of 20–30°C. We found that in apple shoot tip culture, explants cultured under illumination of 1500–7000 lux for 7–10 days became swollen and green. When they were transferred to the dark, differentiation, as well as growth of green plantlets, was accelerated. The cultures usually underwent two to three subcultures in darkness for proliferation before being exposed to light. This procedure produced strong plantlets with healthy growth. For embryo culture the entire procedure from inoculation to formation of rooted shoots was carried out under an illumination of 1500–2000 lux.

For grafting plantlets (Fig. 2) differentiation of shoot tips took place in the dark. After two to three rounds of subculture the cultures were illuminated until grafting was performed. Subsequent cultures were illuminated for about 3 weeks and then transplanted into sand.

PLANTLET TRANSPLANTATION

Selecting Plantlets

The selection of strong, well-developed rooted shoots was critical to the successful transplantation of plantlets derived from tissue culture. With apple stocks, plantlets which regenerated well-formed adventitious roots directly from the basal part and produced more than 15 dark green leaves with thick leaf blade survived well during soil cultivation after they were transplanted. In contrast, weak plantlets, especially those whose adventitious roots regenerated from callus of the basal part of stem and which produced fewer leaves, survived with difficulty.

Sand Culture as a Transitional Measure for Soil Cultivation

We confirmed the utility of sand culture as a transitional measure for soil cultivation compared to other transplantation methods. It clearly enhanced the survival rate after transplanting. Sand culture involved taking plantlets from culture and initially transplanting them into clean sterilized river sand for approximately 2–3 weeks. The plantlets were covered with either a beaker or a plastic cover for the first week of sand culture to maintain high humidity and constant temperature. The relative humidity was gradually lowered to ambient levels by exposing the plantlets to air. The plantlets were then transplanted into pots, vases, or wooden boxes containing soil.

Regulation of Illumination, Temperature, and Humidity

Intensive illumination, low ambient temperature, proper soil water content, and high relative humidity favored survival of transplanted plantlets of apple. In general, the plantlets required a higher intensity of illumination than did proliferation of rooting cultures: 2000 lux and higher. The air temperature during transplantation should be lower than in previous stages and was kept at 15–20°C. In sand culture the water content was kept at 7–10%. The initial relative humidity was maintained at over 80% and then gradually lowered to 50–60%. After transplantation the soil water content was about 20% and the relative humidity was 50–70%. During the period of sand culture and soil cultivation, fresh water was used for irrigation instead of nutrient solution.

PROSPECTS

One of the crucial factors today for developing apple production is shortening the breeding time for new varieties. The use of shoot tip culture in vitro effectively accelerates the propagation rate of new cultivars, thereby guaranteeing the production of nursery stock of new cultivars.

The use of stem dwarfing and close planting is an important trend in various apple-growing countries. In France, Holland, and Hungary, dwarfing and close planting cultivation are generally used; in other major producing countries, such as England, Italy, the Federal Republic of Germany, and Spain, this type of cultivation has been adopted in many areas. Chinese fruit tree growers have recently begun to favor stem dwarfing and close planting, but the development of the method is still very limited because of the low propagation coefficient of dwarfing stocks. We have applied shoot tip culture to this problem, and 6734 plantlets survived after transplantation from one cultured axillary bud meristem of line M_4. This is 100 times larger than the propagation coefficient of conventional branch layering. Shoot tip culture is clearly a promising approach to the development of stem dwarfing and close planting.

Long-term vegetative propagation with apple stocks and cultivars can lead to severe virus infection. According to reports from the United States, by 1976 over 20 kinds of virus had been identified as apple pathogens. Some of these viruses affect the normal branch and leaf growth; others injure the flowers and fruit or even cause death of the tree. Many apple orchards where seedling stocks are used have also suffered from viral diseases. Shoot tip culture may permit the virus-free propagation of apple stocks and control these viral diseases.

KEY REFERENCES

Differentiation Group, Sixth Laboratory, Institute of Botany, Chinese Academy of Sciences. 1977. Branch apex culture of dwarfing stocks in apple. *Acta Bot. Sin.* **19**(3):244–245 (in Chinese, English abstract).

Wang, J. and Sui, C. 1978. Preliminary study on application of in vitro culture of growing point of bud for obtaining stock plantlets of apple. *Liaoning Pomicult.* **4**:23–26 (in Chinese).

——— and Sui, C. 1981. Study on application of tissue culture for propagation of apple stock. *China Agric. Sci.* **6**:33–37 (in Chinese, English abstract).

———, Sui, C., Ren, E., Li, B., and Li, S. 1980. Embryo culture of apple cultivar and several fruit tree stocks. *Liaoning Pomicult.* **3**:30–33 (in Chinese).

——— 1981. Preliminary study on application of tissue culture to obtain root regenerating plantlet of apple cultivar. *Liaoning Pomicult.* **3**:1–6 (in Chinese).

REFERENCES

Abbott, A. J. and Whiteley, E. 1976. Culture of *Malus* tissues in vitro: I. Multiplication of apple plants from isolated shoot apices. *Sci. Hortic.* **4**:183–189.

Chen, W. and Yang, S. 1980. Branch apex culture of adult apple (cultivar Golden Crown). *Acta Bot. Sin.* **22**(1):93–95 (in Chinese, English abstract).

Dutcher, R. D. and Powell, L. E. 1972. Culture of apple shoot from buds in vitro. *J. Am. Soc. Hortic. Sci.* **97**:511–514.

Huang, S. and Millikan, D. F. 1980. In vitro micrografting of apple shoot-tip shoots. *Hortic. Sci.* **15**(6):741–743.

James, D. J. and Thurbon, I. J. 1979. Rapid in vitro rooting of the apple rootstock M_9. *J. Hortic. Sci.* **54**(4):309–311.

Jones, O. P. 1967. Effect of benzyladenine on isolated apple shoots. *Nature* **215**:1514–1515.

Jones, O. P. and Hatfield, G. S. 1976. Root initiation in apple shoots culture in vitro with auxins and phenolic compounds. *J. Hortic. Sci.* **51**:495–499.

Jones, O. P., Hopgood, M. E., and O'Farrell, D. 1977. Propagation in vitro of M_{26} apple rootstocks. *J. Hortic. Sci.* **52**:235–238.

Jones, O. P., Pontikis, C. A., and Hopgood, M. E. 1979. Propagation in vitro of five apple scion cultivars. *J. Hortic. Sci.* **54**(2):155–158.

Walkay, D. G. 1972. Production of apple plantlets from axillary-bud meristem. *Can. J. Plant Sci.* **52**:1085–1087.

Wang, J., Li, A., Zheng, S., Sui, C., and Li, B. 1985. Studies on grafting of tissue culture of plantlets in apple. *Liaoning Agric. Sci.* **1**:1–3 (in Chinese).

Werner, E. and Boe, A. A. 1980. In vitro propagation of Malling 7 apple rootstock. *Hortic. Sci.* **15**:509–510.

Wu, X. and Liu, S. 1977. Mutagenesis of hypocotyl and seedling above cotyledon. *Acta Genet. Sin.* **4**(2):140–145 (in Chinese, English abstract).

CHAPTER 17

Crabapple (*Malus prunifolia*): Anther Culture

*Jiangyun Wu**

INTRODUCTION

Problems in Breeding

The crabapple *Malus prunifolia* (Wild) Borkh is cold-resistant and is grown in Heilongjiang Province, Jilin Province, and Inner Mongolia Autonomous Region of China. Its fruit is of low quality: small, sour, and intolerant of storage. These defects frequently appear in offspring derived from sexual hybridization, so hybrids of *Malus prunifolia* and apple rarely give an ideal cultivar. However, *M. prunifolia* can still be used as a parent plant in breeding programs for cold resistance. But in such cases repetitive backcrossing for many generations must be employed for establishment of superior apple characteristics in the hybrid. If haploid breeding techniques were used, then it would be possible to combine the desirable characteristics of apples in one generation and accelerate improvement of the genotype.

*English translation by Jifang Huang.

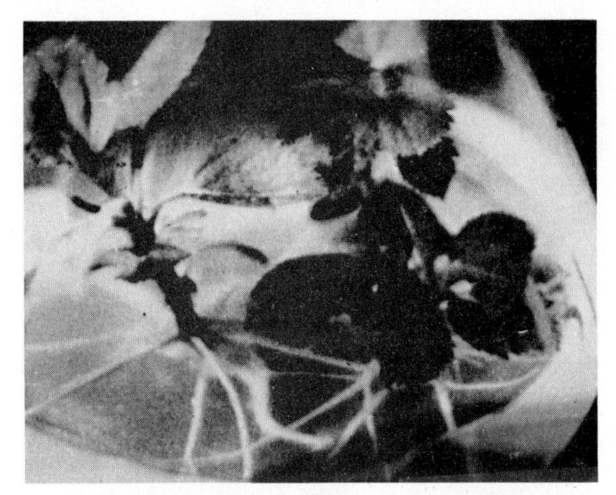

Figure 1. Pollen plantlet with root system.

Literature Review

In recent years, plant haploid breeding techniques have been developed in fruit trees. Haploids have been obtained in *Citrus* (Chen et al., 1980), *Vitis vinifera* (Zhou and Li, 1981), *Poncirus trifoliata* (Hidaka et al., 1979), *Euphoria longana* (Yang and Wei, 1984), and *Litchi chinensis* (Fu and Tang, 1983). Chinese scientists have also successfully induced intact plantlets in the apple cultivar Delicious (Fei and Xue, 1981).

Our studies of anther culture in the crabapple *M. prunifolia* began in May 1978. The first induction of callus from pollen, subsequent differentiation of shoots and roots, and formation of intact plantlets were accomplished with the cultivars Huang Tai Pin and Jin Hong in October 1979 (Wu, 1980, 1981a, 1981b) (Fig. 1).

The oldest pollen-derived trees are now 6 years old. Microscopic observations of root tip cells indicate a chromosome number $n = 17$ (Fig. 2).

Grafting was done in 1986 on 500 plants in suburban orchards near Shengyang City, Liaoning Province. One tree flowered and bore fruit that year (Fig. 3). In 1985, over 40 plants of *M. prunifolia* flowered and produced fruit.

ANTHER CULTURE TECHNIQUES

Culture Procedures and Media

SELECTION OF CULTURE MATERIALS. The cultivars Huang Tai Pin, Da Qiu, Jin Hong, and Long Quang were used as experimental ma-

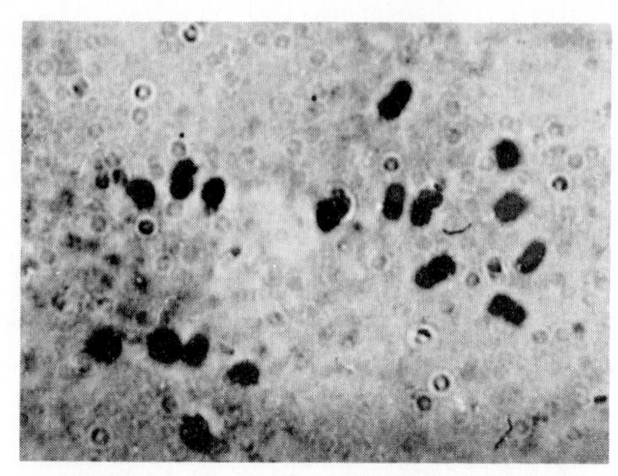

Figure 2. Chromosomes of root tip cells of pollen plantlet, $n = 17$.

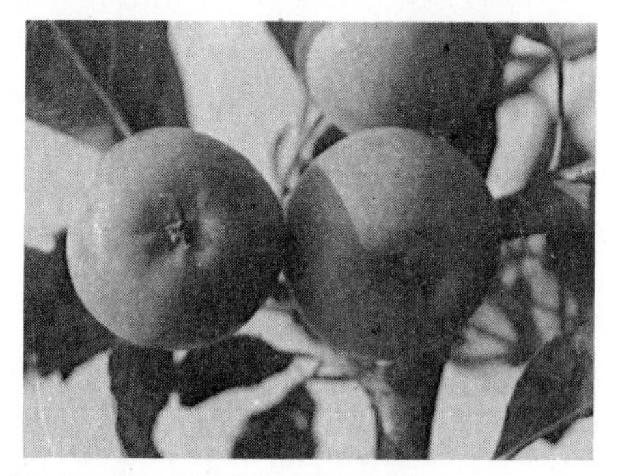

Figure 3. Top grafted pollen plant bearing fruit.

terials. Pollen in different developmental stages differed greatly in its callus forming capability. Pollen at tetrad, middle mononuclear, late mononuclear, and binuclear stages was inoculated in culture. The experiments with Huang Tai Pin indicated that the anthers in the middle and late mononuclear stages had the highest callus induction frequency. Almost no callus was formed from the anthers in the tetrad-spore stage. Very few anthers in the binuclear stage formed callus, and most of them turned dark brown within a week after inoculation. Only the anthers in the late mononuclear stage were suitable for inoculation in culture.

When the flower bud opened but the color did not change, most of the pollen in the bud was at the late mononuclear stage except a few that were binuclear. This stage provided the best material for subsequent inoculation into culture.

CALLUS INDUCTION. The anthers were inoculated onto MS medium supplemented with KIN (9.3 μM), 2,4-D (4.5–9.0 μM), LH 100 mg/l, sucrose 88 mM, agar 0.7%. The pH of the medium was 5.8.

TABLE 1. Effect of Different Media on Callus Induction in Four Crabapple Varieties

	Huang Tai Pin			Da Qiu		
Medium*	No. of anthers	No. of calli	Induction frequency	No. of anthers	No. of calli	Induction frequency
MS	150	70	46.7	90	26	28.9
White's	150	31	20.7	151	18	11.9
B_5	150	12	8.0	150	14	9.3
	Jing Hong			**Long Guang**		
Medium*	**No. of anthers**	**No. of calli**	**Induction frequency**	**No. of anthers**	**No. of calli**	**Induction frequency**
MS	150	120	80.0	150	94	62.7
White's	150	42	28.0	150	31	20.6
B_5	150	19	12.6	150	24	16.0

*Supplemented with 9.3 μM KIN and 9.0 μM 2,4-D.

Cultures on hormone-free MS, B_5, or White's medium did not differentiate any callus; cultures on MS medium supplemented with KIN (9.3 μM), and 2,4-D (9 μM), gave the best results (Table 1).

In another experiment, the effects of three hormones, 2,4-D, NAA, and IAA, at concentrations of 4.5–11 μM were studied. A month after inoculation, the cultivars Huang Tai Pin, Jin Hong, and Long Guang showed the highest frequency of callus induction on the medium supplemented with 4.5 or 9 μM of 2,4-D; for cultivar Da Qiu, the medium with 5.4 or 11 μM of NAA was better, with a callus induction frequency of 36–43%. IAA gave an induction frequency of only 3%.

Cultured anthers of Huang Tai Pin and Jin Hong began to form callus after 10 days in culture; those of Da Qiu and Long Guang formed less callus and only after 15–20 days culture. The calli fell into three types: (1) callus with hard, compact texture and white surface, which sometimes turned green then brown and died; (2) callus with loose texture, watery surface and rapid growth but no differentiation; and (3) callus with compact texture, intermediate hardness, yellow surface, and slower growth rate. Some of these calli turned green and subsequently differentiated into shoots after transfer onto a differentiation medium.

INDUCTION OF CALLUS DIFFERENTIATION. In order to promote differentiation, the calli were transferred to MS medium supplemented with BA 2.2 μM, NAA 0.54–2.7 μM, biotin 20.5 μM, and insulin 8 mg/l, or onto White's medium supplemented with BA 8.8 μM, NAA 0.54–27 μM, biotin 20.5 μM, insulin 8 mg/l, and CH 500 mg/l. Calli were transferred when they became the size of a rice grain. The cultures differentiated on both media. Cultures on these two media without biotin and insulin did not differentiate any shoots. When these media were supplemented with BA, the calli differentiated multiple shoots which were compacted into difficult-to-separate masses and did not grow further. When BA supplement was replaced by KIN, the number of shoots decreased markedly and most of the shoots were strong and well developed.

INDUCTION OF HEALTHY SHOOTS. The shoots were separated and transferred onto MS medium without hormones but containing CH. When the young shoot grew to a height of 1.5–2.0 cm in 20 days or so, it was transferred to root medium for root regeneration.

ROOT INDUCTION. It was difficult to promote root regeneration in *M. prunifolia* when shoots were used for cuttings. An experiment on the effects of seven different auxin combinations was performed using MS (½) macroelements as basal medium. The combination of IAA and IBA showed good effects, especially in medium supplemented with IBA 0.98 μM and IAA 1.1 μM. After transferring young shoots over 1 cm high onto such medium, white young roots began to grow in 15 days, and a good root system formed within 15–20 days. There were three to five roots on the basal part of a shoot with no callus formation. The frequency of induction reached 90% and more.

The base of most of the young shoots grew callus when ½ MS medium was supplemented with NAA 2.7 μM or IAA 2.9–5.7 μM + NAA 2.7–5.4 μM. Some other hormone combinations could initiate and induce roots, but the roots were not connected to the shoot through the vascular bundle and rarely survived.

Inoculation Techniques and Culture Conditions

Calli were initially induced in the dark at ambient temperature (22 ± 2°C). Higher temperatures produced loose calli which differentiated with difficulty. Calli were transferred onto differentiation medium when 3–4 mm in diameter. Differentiation was done at 22–25°C, since higher temperatures often resulted in the cessation of differentiation except in proliferating calli. The illumination was provided by fluorescent tubes for 10–12 hr/day.

Transplantation of Plantlets

Plantlets of *M. prunifolia* derived from anther culture rarely survived when they were transferred from the culture medium to soil. One problem was that they were very susceptible to damping-off infection, and leaves often wilted 5–6 days after transplantation. Use of a transitional sand culture was often an effective solution. The cotton stop or seal of the flask was removed to expose the plantlets to the environment for 3 days. The plantlets were removed from the flask, rinsed with water to clean the medium remaining on the roots, transplanted into sand culture boxes, and incubated at 20+°C. After 15–20 days root hairs developed and the plantlets were transplanted into soil and sprayed with water two or three times a day. The survival rate was more than 90%.

ORIGIN OF PLANTLETS

Observations on Androgenesis

Androgenesis of perennial crops has been reported in papers describing anther culture of the rubber tree (Chen et al., 1978). Our observations in *M. prunifolia* indicate that three types of androgenesis occur: (1) The mononuclear pollen cell forms two daughter cells after the first division, which are identical to each other in size and do not undergo morphological differentiation. These two daughter cells probably divide continuously, forming multicellular pollen, then pollen embryos or callus. (2) The mononuclear pollen first divides into two isolated nuclei, which divide successively to form pollen with multiple isolated nuclei and form further cell aggregates through cell wall formation. These aggregates undergo successive cell divisions and proliferate as a callus. (3) A vegetative nucleus and a germ nucleus form after the first division in the mononuclear pollen cell and continue to divide to form multiple-nuclear pollen. The second type of androgenesis was the most frequent in *M. prunifolia*.

Observation on Chromosome Number in Plantlet Root Tip Cells*

METHODS. The haematoxylin staining and carbolic acid–fuchsin staining methods gave good results for observation of chromosome number in plantlet root and shoot tip cells. The methods were as follows:

1. The white newly emerged young roots were removed from the cultures. Root tip segments (0.3–0.5 cm) were cut from the root tip.
2. The root tips were cleaned and soaked in a saturated aqueous solu-

*Zhu, 1982

tion of para-dichlorobenzene (0.25% w/v) for 2 hr. The specimens were incubated at 60°C for 4 hr and then chilled.

3. Fixation: The root tips were fixed in Carnoy's solution for 1–2 hr (the fixation time depended on the size of the root tips).
4. Sample preservation: The root tips were preserved in alcohol (70%).
5. Dissociation: The root tips were dissociated by 1 N HCl for 12–15 min at 60°C.
6. Staining: The dissociated samples were stained with carbolic acid–fuchsin or ferric alum hematoxylin for 1 hr and cleaned with 45% acetic acid for 45 min.

RESULTS OF OBSERVATION. The chromosome number of *M. prunifolia* is $2n = 34$, and the haploids have the chromosome number $n = 17$. Observations of root tip cells indicated that there were a large number of haploid cells and a small number of aneuploid cells in most of plantlets. In isolated cases, the cultured plantlet had more aneuploid cells and fewer haploid cells. No diploid cells were observed. These observations are similar to those reported for *Hevea brasiliensis* (Chen et al., 1978) and for maize (Gu et al., 1978). Chromosome number observations showed that in *M. prunifolia* plantlets the haploid cells were a majority, proving that the plantlets originated from pollen.

In potted plantlets shoot tips from the first two- to three-leaf stage were used for determining chromosome number. The leaf primordia around the shoot apex were used for smearing, because they had more dividing cells than the apex itself. The smear technique used for shoot tips was the same as that for root tips. The chromosome number of shoot tip cells from potted plantlets varied with the length of time in the pot. This seemed to reflect a natural tendency for the chromosome number to double such that most of the plantlets were mixoploids as reported in poplar (Wu et al., 1980).

PROSPECTS

A large number of crabapple plantlets derived from pollen have been transplanted, survived, and borne fruit. These plants exhibit much variation in botanical characteristics: erect or spreading growth, large or small leaves, alternate or opposite phyllotaxy, dark brown or yellowish brown branches, and various densities of lenticel distribution on the bark. These characteristics of pollen plantlets were distinctly different from those of the parent trees. It is clear that plantlets derived from pollen fully reveal the plants' genotype and will be very useful in breeding work. To select further superior pollen plantlets from anther cultures, crabapple growers still need a great quantity of pollen plantlets from various lines. Therefore, it is necessary to extend the spectrum of apple cultivars in anther culture, to enhance the induction frequency of pollen plantlets, and to obtain homozygous diploid plants of various genotypes for apple breeding.

KEY REFERENCES

Fei, K. and Xue, G. 1981. Brief report on plant formation by using anther culture in cultivars of Delicious apple. *China Agric. Sci.* **10**(5):41 (in Chinese, English abstract).

Wu, J. 1980. Preliminary results on induction of pollen plants in *Malus prunifolia. Proc. Northeast Coll. Agric.* **2**:69 (in Chinese).

———. 1981a. Studies on obtaining haploid plants through anther culture in crabapple. *Proc. Northeast Coll. Agric.* **2**:105–108 (in Chinese).

———. 1981b. Haploid plant obtained through anther culture in cultivars of crabapple Huang Tai Pin. *Sci. Bull.* **21**:1344 (in Chinese).

Zhu, C. 1982. *Plant Chromosome and Its Stain Technology*. Science Press, Beijing (in Chinese).

REFERENCES

Chen, C., Chen, F., Chien, C., Wang, C., Chang, S., Hsu, H., Ou, H., Ho, Y., and Lu, T. 1978. Induction of pollen plant in *Hevea brasiliensis. Proc. Symp. Anther Cult.* Science Press, Beijing, 3–8 (in Chinese, English summary).

Chen, Z., Wang, M., and Liao, H. 1980. Induction of pollen plant in Citrus. *Acta Genet. Sin.* **7**(2):189–191 (in Chinese, English abstract).

Fu, L. and Tang, D. 1983. Induction of pollen plant in *Litchi chinensis. Acta Genet. Sin.* **10**(5):369–375 (in Chinese, English abstract).

Gu, M., Zheng, W., Guo, L., Guan, Y., An, X., and Huang, J. 1978. Studies on induction of pollen plant in maize. *Proc. Symp. Anther Cult.* Science Press, Beijing, 25 (in Chinese).

Hidaka, T., Yamada, Y., and Shichijo, T. 1979. In vitro differentiation of haploid plants by anther culture in *Poncirus trifoliata* (L.) Raf. *Jap. J. Breed.* **29**(3):248–254.

Wu, K., Xu, M., and Xie, Q. 1980. Studies on the natural doubling of chromosome number of pollen plant and the artificial doubling technique in poplar. *Forest. Sci. Technol.* **2**:6–10 (in Chinese).

Yang, Y. and Wei, W. 1984. Induction of pollen plant in *Euphoria longan. Acta Genet. Sin.* **11**(4):288–293 (in Chinese, English abstract).

Zhou, C. and Li, P. 1981. Induction of pollen plant in *Vitis vinifera* L. *Acta Bot. Sin.* **23**(1):79–81 (in Chinese, English abstract).

CHAPTER 18
Pear

Huixiang Zhao and *Nailiang Gu**

INTRODUCTION

There are two groups of pear in *Pyrus*: The "oriental pears" and *P. cammunis* L.. The "oriental pears" are the pears originally found in China, so they are also called "Chinese pears." The *P. cammunis* are the pears originally found in Europe and Asia. Both types are widely cultivated throughout the world; they are the most important deciduous fruit tree after apple in temperate zones.

Economic Importance

The pear fruit is of great value for human nutrition. In addition to the 80% water content, the sugar content is 8% and may be as high as 20%. The fruit also contains free acids, pectin, protein, fat, calcium, iron, phosphorus, ash, and various vitamins. The nutritional composition per 100 g of edible parts of fresh pear fruit is 0.1 g protein, 0.1 g fat, 12 g carbohydrates, 5 mg calcium, 6 mg phosphorus, 0.2 mg iron, 0.01 g carotene, 0.01 g thiamine, 0.01 g riboflavin, 0.2 mg nicotinic acid, and 3 mg ascorbic acid. It is one of the traditional fruits in China because of its abundant juice, sweet smell, and crisp or tender texture. The late-

*English translation by Jifang Huang.

maturing cultivar White Pear, originally Chinese, is tolerant of storage and transportation like the apple, so, although it is harvested seasonally, it can be brought to market year round. Pears can be eaten raw and be processed into dried pears, preserved pears, canned pears, pear jam, juice, and paste as well as wine and vinegar.

In China pears have a high value in traditional medicine. According to some Chinese medicinal references, pears promote digestion, relieve cough, allay fever, relieve inflammation, and prevent alcoholism. Pear paste is currently used as a drug for relieving cough and dispelling phlegm.

The pear tree is very adaptable. It can grow on high mountainous land, hilly land, sandy wasteland, and low-lying land and on saline, alkaline, and clay soil. It tolerates various kinds of weather and environment. This is the main reason pear trees are widely distributed in China. The pear tree is a high-yielding fruit tree. In China, pear trees 200 to 300 years old remain and bear fruit in various regions of both north and south China. Pear trees hundreds of years old often have yields as high as 1000 kg per plant, making them an important resource in the rural economy.

Geographical Distribution and Classification

The oriental pear has a long history of cultivation in China and is traditionally named "ancestor of fruit." It is distributed all over China from the Ussuri River of the north to the Pearl River Delta of the south; from the Tibet Plateau of the west to the eastern provinces along the Chinese coast. The *P. cammunis* L. originated in middle and southeastern Europe as well as Asia Minor and the Middle East. It has been cultivated in Europe for 2000 years since Roman times. Many good cultivars have been bred since the 16th century. Wild *P. cammunis* trees are mainly distributed in southern and middle Europe, Caucasia, and Asia Minor and even grow in tracts of forest in some regions. The main countries of pear production in Asia are China and Japan; in Europe, Italy, Spain, France, the Federal Republic of Germany, Switzerland, Greece, and the Union of Soviet Socialist Republics; in America, the United States and Argentina; in Africa, South Africa; in Oceania, Australia (Institute of Fruit Trees, Chinese Academy of Agriculture Sciences, 1963).

The pear belongs to the family Rosaceae. There are 30 species in the genus *Pyrus*, including 13 species that originated in China. The cultivars of pear in China are mainly those of *Pyrus bretschneideri* redh., *P. pyrifolia* (Nakai Redh.), and *P. ussuriensis* L. Chromosome number observations show that $2n = 3x = 51$ in *Pyrus ussuriensis* maxim (cultivar Qiu Zhi pear) and $2n = 4x = 68$ in *P. bretschneideri* Redh (cultivar White Pear). The chromosome number $2n = 2x = 34$ is common to most cultivars, such as *P. ussuriensis* var. *ovoidea* (cultivar Xiao Xiang Sui) (Gu, 1956). The Chinese cultivar Chuan pear (*Pyrus pashia*, D. Don) is distributed in India, Burma, Bhutan, and Nepal.

Production Goals

1. To breed new high-yielding cultivars resistant to adverse circumstances, to renew cultivars used presently, and to supply the market with pears earlier in the season.
2. To propagate nursery stocks rapidly as well as to obtain virus-free plants for use as seedlings.

Literature Review

The embryos of several fruit trees sometime fail to develop on the plant. Tukey successfully carried out embryo culture during the 1930s on apple, pear, peach, and plum trees (Laboratory of Cytology, Shanghai Institute of Plant Physiology, 1978). Lammerts (1942) attempted to shorten the breeding cycle of deciduous fruit trees (*Prunus salicina, Pyrus persica* var. *nectarina*, and *Prunus persica*) and to increase the germination rate of hybrid seeds by using the techniques of embryo culture. In 1975, pear embryos were successfully cultured by A. E. Zdruikovskaya-Richter and A. X. Horolikov using the cultivar Green Magdalina. Sixty-six plantlets were obtained, and among them 5 plantlets had qualities superior to those of their mother plants. One of the plantlets which mature in June was called Ultra-early Pear. Another success occurred in 1972, when hybrid embryos from a remote hybridization of pear and quince (*Cydonia oblonga*) were cultured in vitro and from them "plant twins" were developed. Since 1979, we have been working on embryo culture of hybrids between two early cultivars, Early 208 × Early Crisp; superior plants have been obtained, which mature even earlier than the parent plants. Our experiments in embryo culture have investigated media, culturing conditions, transplantation techniques, and orchard growth of mature fruit-bearing trees.

We have carried out in vitro culture of hypocotyl, transverse segments of cotyledon, and young embryos. One-half of MS medium was used as the basic medium, to which was added 2.2 μM of BA. The hypocotyl cultures regenerated adventitious buds. Callus formed on transverse cotyledon segments, and adventitious shoots differentiated from the callus. From culture of young embryos, slender and weak plants were obtained.

Studies on the origin of adventitious shoots and their induction by hormones in embryo culture were carried out in 1983 by Shi Yinpin (Shi, 1983).

Investigations on endosperm culture in vitro have been carried out in over 30 kinds of plants, of which about a dozen are perennial crops (Laboratory of Cytology, Shanghai Institute of Plant Physiology, 1978). Endosperm plantlets and/or plants have only been obtained in *Malus* (Wu, 1980), *Citrus grandis*, and *Actinida chinensis* (Gui et al., 1982). In other species, only callus, embryos, or differentiated roots were obtained. Successful formation of plantlets from endosperm culture in pear has not previously been reported. We have now induced intact plantlets from endosperm culture in the pear cultivar Jin Feng and successfully transplanted and potted the cultured plantlets, which have grown to young trees (Zhao, 1983).

Shoot tip culture has been investigated in *Malus* grafting stock (Abbott and Whiteley, 1976; Sixth Laboratory of Beijing Institute of Botany, Chinese Academy of Sciences, 1977; Jones and Ferrell, 1977; Liu et al., 1978; Zhao, 1981). In *Pyrus*, plantlets have been obtained from cultures (David, 1979). In the same year, we grew intact plantlets from in vitro culture of shoot tips which were taken from a test tube–cultured pear seedling. The cultured plantlets were subsequently transplanted in soil and survived (Zhao, 1981a, 1981b, 1982).

EMBRYO CULTURE IN VITRO

Culture Methods and Media

PREPARATION OF MATERIALS. Nearly mature hybrid fruit from the cross of Early 208 × Early Crisp cultivars was used as explant. It was stored in the refrigerator (0–3°C) for 3 months. The fruit was then surface-sterilized with 70% alcohol. The pear fruit was cut aseptically into transverse sections. The seeds were taken out and placed in a sterilized petri dish, and the seed coats were removed. The embryo was taken out and inoculated onto the medium. The temperature of the culture room was 26° ± 2°C. Illumination was 10hr/day, 3000 lux.

MEDIUM. The medium used for embryo culture was V_1 medium without hormone. The embryos expanded their two cotyledons 3 days after culturing, grew into shoots with small leaves within 10 days, and regenerated roots 5 days later. It could be seen that V_1 medium was suitable for embryo culture of hybrid Early 208 × Early Crisp. V_1 medium was also suitable for seed germination of the apple cultivars Ralls, India, and Qiu Jin and for embryo culture of the pear cultivar "Jin Feng" into plantlets. As compared to Tukey medium, V_1 medium did not precipitate and was less expensive because a smaller amount was required for culture (Zhao and Ding, 1984).

Transplantation of Plantlets

TRANSPLANTATION OF PLANTLETS. The survival rate of transplanted plantlets could reach 80–100% if they had been cultured for 10 days and developed white main roots 1.5–2.0 cm in length without any fibrous roots. Plantlets 20 days old with yellowish main roots 4–5 cm in length and fibrous roots had a survival rate of over 70%. Thirty-day-old plantlets grew main roots with a grayish yellow color, and most of them had large fibrous roots. They survived at a low rate of 40% after transplantation. Therefore, the best time for transplantation of plantlets was the 10th day of culture, when they grew vigorously but had not developed fibrous roots. These plantlets had a good survival rate because the root system held up well during transplantation. The root system of 30-day-old plantlets grew excessively, resulting in weakness of the main root and in fibrous roots which broke easily when transplanted. The plantlets 20 days old showed intermediate results.

METHOD OF TRANSPLANTATION. The cultured plantlets were placed in pots containing ordinary soil. Watering was performed by immersing the pot in water until it permeated the pot. The pot was covered with a glass plate until new leaves developed. The appearance of newly developed leaves indicated the recovery of physiological functions in the root system. The time required for recovery varied, depending on the season when the transplantation was carried out. Recovery took about a month in late December or early January. Such plantlets died from lack of water if the covering glass plate was removed before the recovery of the root system. When transplantation was carried out in April, the root system recovered more rapidly, and the plantlets brought forth new leaves 10 days after transplanting. The potted plantlets were placed in a cultivation room, where the ambient temperature was kept in the range 15–20°C. Temperatures over 20°C raised transpiration so high that the roots failed to regenerate and the cultivated plantlets died. Temperatures below 15°C also caused high plantlet death rates.

A relative humidity of 80% was suitable for potting plantlets. Because the plantlet stem was slender and crisp, water spraying was applied with care. When the soil in the pot dried, it had to be scarified and the next irrigation was then carried out. No fertilizer or nutritional solution was applied to the soil because excessively high ion concentrations could result in damage to the plantlets.

FIELD SETTING OF PLANTLETS. If the plantlets were transplanted into pots in late December, they should be planted in the field the following May. The plantlets should be transplanted with soil around the root system to ensure optimal results. When test tube–cultured plantlets were transferred from pots in April, a 10- to 20-day-potting period was sufficient before the plantlets were planted in the field. When setting in the field, it is necessary to embed the root system deeply, so that the soil around the roots is below the field soil surface, to prevent frost damage in the winter. Spacings of 2 × 4 m within and between the rows are suitable. The orchard was maintained without pruning the plantlets.

Comparison of Embryo Cultured Plantlets and Seedlings from Matured Seeds

HIGHER SURVIVAL RATE IN VITRO. Two hundred forty hybrid seed embryos were inoculated on medium in culture tubes. All of them germinated and developed into plantlets, except one that was injured mechanically. Among 1956 hybrid seeds from the same hybridization, only 37 seeds germinated through conventional methods, a rate of 1.8%.

SHIFT OF FRUIT BEARING AND MATURING TO EARLIER DATES. Another superior characteristic of embryo culture is that fruit bearing begins earlier, shortening the years in each breeding cycle. None of the 3-year-old offspring from 1956 hybrid seeds of Early 208 × Early Crisp bore fruit, whereas 5% of the plants obtained from 240 cultured hybrid embryos set fruit within 3 years. Moreover, the time to reach maturity in most of these trees was less than that of their parents (Fig. 1). Thus, it could be seen that there was good opportunity for obtaining early cultivars through embryo culture.

BREEDING OF EARLY CULTIVARS THROUGH EMBRYO CULTURE. The problem of embryo abortion or low rates of seed germination in hybrids may be overcome with embryo culture. This technique permits the expression of various genotypes in offspring derived from segregation in hybrid progeny. Low germination rate for seeds of early cultivars is a persistent problem in conventional breeding. This difficulty may be overcome by using in vitro embryo culture.

ENDOSPERM CULTURE IN VITRO

Culture Methods and Media

SELECTION OF MATERIALS. The cultivars Jin Feng and Early Crisp bred by the Institute of Pomology were used for endosperm culture. Ob-

Figure 1. Fruit-bearing tree aged 3 years.

servation of the stages of seed development showed that the endosperm tissue was formed in the last third of June. At that time the endosperm was excised and inoculated onto medium.

INDUCTION AND DIFFERENTIATION OF CALLUS. Endosperm tissue was inoculated on Medium No. 1 with the composition MS medium + 2.3 μM 2,4-D + 6.6 μM BA + 2.9 μM IAA with 88 mM sucrose. After 4 weeks of culture, callus was induced in a few cultures, and subsequently bud primordia gradually differentiated.

ACCELERATION OF DIFFERENTIATION. After 4 weeks' culture, callus with bud primordia was transferred onto Medium No. 2: MS medium + 29 μM GA + 8.8 μM BA with 88 mM sucrose. Differentiation accelerated on this medium.

FURTHER DEVELOPMENT INTO SHOOTS. After 34 days on Medium No. 2, callus was transferred to Medium No. 3 for shoot differen-

TABLE 1. Composition of Media

Chemicals	V_1 Medium	AS Medium	ASH Medium
	(mM)		
$CaCl_2 \cdot 2H_2O$	0.75	1.5	1.5
NH_4NO_3	5.1	10.3	10.3
KNO_3	4.7	9.4	9.4
$MgSO_4 \cdot 7H_2O$	0.38	0.75	0.75
KH_2PO_4	0.31	0.63	0.63
Na_2-EDTA + $FeSO_4 \cdot 7H_2O$	0.025	0.05	0.05
	(μM)		
KI	0.5	0.5	0.5
$Na_2MoO_4 \cdot 2H_2O$	0.1	0.1	0.1
$CuSO_4 \cdot 5H_2O$	0.01	0.01	0.01
$CoSO_4 \cdot 6H_2O$	0.01	0.01	0.01
$MnSO_4 \cdot 4H_2O$	10	10	10
$ZnSO_4 \cdot 7H_2O$	3.6	3.6	3.6
H_3BO_3	10	10	10
Glycine	2.7	2.7	2.7
Thiamine-HCl	0.12	0.12	0.12
Pyridoxin-HCl	0.24	0.24	0.24
Nicotinic acid	0.41	0.41	0.41
Inositol	55.5	55.5	55.5
Sucrose	44 mM	88 mM	44 mM
Agar	7.5g/l	7.5g/l	7.5g/l
LH	0	250 mg/l	0
pH	5.8	5.8	5.8

tiation. The composition of Medium No. 3 was AS medium + 29 μM GA + 4.4 μM BA + 250 mg/l LH with 88% sucrose (Table 1).

INDUCTION OF ROOT DIFFERENTIATION. After culturing on Medium No. 3 for 5–6 weeks, the rootless shoots were transferred onto the root regeneration medium: ASH medium + 8.6 μM IAA with 44 mM sucrose (Table 1).

The agar content of all the media mentioned was 0.6–0.65% and the pH was 5.8.

Other Critical Variables

STERILIZATION AND INOCULATION. The surface of young fruit was wiped with 75% alcohol, the fruit was cut into transverse sections, and the uninjured seeds were removed and placed in a petri dish with forceps to peel off the seed coat and cut away the whole embryo. A portion of the endosperm tissue was taken and inoculated onto the medium.

Figure 2. Endosperm-derived plantlets after transplantation.

CULTURE CONDITIONS. The ambient temperature during the day was 26 ± 2°C and in the range of 20–25°C during the night. Illumination was 10 hr/day with intensity of 3000 lux.

TRANSPLANTATION. When the roots grew to 1–2 cm in length, the endosperm plantlets were transplanted using the same procedure as for embryo plantlets. The potted plantlets grew well, developing broad, thick, dark green leaves (Fig. 2).

Morphogenesis

The endosperm from seeds of the cultivar Jin Feng pear were cultured on Medium No. 1 for 28 days; callus appeared on the 11th day. Only 2% of the calli differentiated. The center of the callus developed a depression, and a light green outgrowth appeared from the center and gradually turned dark green, forming bud primordia. After 17 days of culture, calli with green bud primordia were transferred onto Medium No. 3 for 34 days, until the bud primordia developed into small leaflike structures. The basal part of the callus continued to differentiate new, small bud primordia. The cultures grew rapidly after transfer to Medium No. 3. The small leaflike structures developed, and the basal bud primordia grew into clustered shoots (Fig. 3). These young shoots were cultured for 23 days and were separated into individuals which were inoculated onto the same medium for subculture. The young shoots grew into clustered shoots in 1 month. These clustered shoots were separated once again and subcultured on fresh prepared Medium No. 3. The shoots were transferred to rooting medium when they were 2–3 cm in height. Fifteen days later, root primordia developed at the base of the shoots. The develop-

Figure 3. Plantlet from endosperm culture.

ment of an intact tube-cultured plantlet from the initial inoculation of endosperm took about 180 days.

Cytological Observation

A staining technique was used to determine the chromosome number of young leaf and stem cells of endosperm plantlets. The procedure was as follows:

1. Pretreatment: The young shoots and leaves were collected from 6:30 to 10:30 A.M. from test tube plantlets and were immersed in 2 mM 8-hydroxyquinoline in 0.02% colchicine for 2 hr. The materials were cut into pieces and immersed in 75 mM KCl for half an hour.
2. Fixation: The KCl was discarded and a 3:1 mixture of methanol and glacial acetic acid was added for half an hour to fix the materials.
3. Enzyme treatment: The fixation solution was discarded; the materials were rinsed two to three times with distilled water and were then treated with a mixture of 2.5% cellulase and 2.5% pectinase solution at 26°C for 1.5–4 hr.
4. Wash: The enzyme solution was discarded, the materials were rinsed once with distilled water, and distilled water was added for 10–15 min.
5. Smearing: The materials were taken from the vessel and excessive water was blotted away. A small amount of fixation solution was added to the materials. Two to three drops of fixation solution were dropped on a glass slide (previously cleaned and stored in the freezer). The materials were then evenly smeared on the slide.

6. Drying: The slide smeared with materials was dried over a alcohol burner.
7. Staining: The materials on the slide were stained with Giemsa solution for 3–4 h. The stained slide was rinsed with tap water and air-dried. This method digests the cell wall with cellulase and pectinase and swells the nucleus under low osmotic pressure prior to smearing and staining. This procedure dispenses the chromosomes and facilitates counting (Chen et al., 1979).

Mixoploidy in Endosperm Plantlets

The cytological observations demonstrated very unstable ploidy in endosperm plantlet cells. The plantlets were mixoploids composed of cells with different numbers of chromosomes: $2n = 3x = 51$, $2n = 2x = 34$, $2n = 27$, $2n = 32$, and $2n = 30$.

Wu investigated variation of ploidy in callus cells of apple endosperm. It was found that most cells were polyploid or aneuploid; triploid cells were very few (Wu and Liu, 1977; Sun and Zhu, 1981). These mixoploid materials have advantages for breeding work since they can be used to obtain cell lines with different numbers of chromosomes.

SHOOT TIP CULTURE

Culture Methods and Media

SELECTION OF MATERIALS. Since 1978, shoot tips from the cultivars Jin Feng, Early Crisp, Jin Xiang, Ya Pear, Qiu Bai Pear, Ba Pear, Red Ba Pear, Conference, Beurre Precoce Mor Ettiri, Jin Bai, Qiu Zhi, Wu Jiu Xiang, etc., have been cultured in vitro. The materials for culture were taken from trees of different ages and included scion buds 7 days after grafting to mature trees 20 years old. Sampling took place from the middle of April to June 20. Culture of shoot tips from different-age trees on AS medium showed that the younger the age of the tree, the faster the explant growth. The growth rate of the scion bud 7 days after grafting was five times that of buds from 3-year-old trees (measured at a month after inoculation).

PRETREATMENT WITH GIBBERELLINS. Shoots about 15 cm in length were soaked in 100 ml of 145–580 μM GA and placed in an incubator at 26 ± 2°C. Distilled water was added to maintain the original volume of solution. After 7 days' soaking in GA, the culture materials were transferred into distilled water for another 11-day soaking.

INDUCTION OF SHOOT FORMATION. Segments of shoot tip meristem 0.5 mm in length were inoculated individually onto medium with the composition AS medium + 4.4 μM BA + 29 μM GA with 3% sucrose.

The materials which were inoculated on AS medium differentiated into roots and shoots and intact plantlets without transfer to another medium. The differentiation rate of cultured shoots (the ratio of the number of differentiating shoots to the number of shoots inoculated) increased with increasing concentration of GA. Highest rates of differentiation (45%) were found with 580 μM GA. Differentiation rates on MS medium supplemented with the same hormones as in AS medium (4.4 μM BA and 29 μM GA) decreased with increased concentration of GA.

AS medium was more suitable for shoot tip differentiation. Different cultivars responded differently to the same culture medium. For instance, on AS medium, the highest shoot tip multiplication rate from Early Crisp pear was higher than that of Jin Feng pear: 456%, compared to 300%.

ROOT INDUCTION. The rootless shoots which grew 2–3 cm in height were immersed in 100 ml IBA solution for 25 hr and transferred onto ASH medium containing no supplemental hormones. The roots were induced rapidly in about 3 weeks. The lignification level of the rootless shoots greatly affected root differentiation. Both high and low levels of plantlet lignification were disadvantageous to root differentiation.

Summary of Culture Techniques

1. Method of Inoculation: The buds for culture were sterilized in 70% alcohol for 15–30 sec (for newly growing shoot tip, the sterilization took 15 sec, and for dormant buds, 30 sec) and then rinsed three times with sterilized water.
2. The materials were transferred into newly prepared supernatant of filtered bleaching powder (10%) or into 20% of commercial bleach for 10 min.
3. The bud apical cone was taken out under the microscope, then the apical point less than 0.5 mm in diameter was cut off and inoculated onto the medium.
4. Conditions of culture: The ambient temperature in the culture room was kept at 20–28°C and 20°C at night. The illumination was 10hr/day with an intensity of 1000–3000 lux.

Transplantation and Field Setting of Plantlets

SMALL-SCALE TRANSPLANTATION OF PLANTLETS. The transplantation procedures for endosperm plantlets was the same as for embryo plantlets. Transplantation was carried out when the plantlets grew to 1.5 cm in height. Low survival rates occurred when fibrous roots were present on the plantlet prior to transplantation. The plantlets were transplanted into earthen pots filled with soil from the pear orchard. A glass plate was placed over the pot, and the temperature was maintained in a range of 15–20°C. The glass plate was removed as the plantlets grew new leaves. The potted plantlets were set in fields in the first 10 days of May.

KEY REFERENCES

Zhao, H. 1982. In vitro culture of apex in pear. *Acta Bot. Sin.* **24**:392–394 (in Chinese, English abstract).

———. 1983. Induction of endosperm plant and its ploidy in pear cultivar Jin Feng. *Bull. Bot.* **1**:38–39 (in Chinese).

——— and Ding, A. 1984. Study on embryo culture in vitro in pear. *Shanxi Fruit Tree* **2**:8–11 (in Chinese).

REFERENCES

Abbott, A. T. and Whiteley, E. 1976. Culture of *Malus* tissue in vitro: I. Multiplication of apple plants from isolated shoot apices. *Sci. Hortic.* **4**:183–189.

Chen, R., Song, W., and Li, X. 1979. New method in sample preparation of chromosome preparation during mitosis in plant. *Acta Bot. Sin.* **2**:297–298 (in Chinese and English abstract).

Cytology Labolatory, Shanghai Institute of Plant Physiology, Chinese Academy of Sciences. 1978. Organ Culture; Embryo Culture. In: *Plant Tissue Culture and Cell Culture*, pp. 190–230. Shanghai Science and Technology Press, Shanghai (in Chinese).

David, L. W. 1979. Regeneration of pear plants from shoot meristem-tips. *Plant Sci. Lett.* **16**:337–342.

Differentiation Group, Sixth Laboratory, Beijing Institute of Botany, Chinese Academy of Sciences. 1977. Apex culture of dwarfing stock in apple. *Acta Bot. Sin.* **19**:244–245 (in Chinese, English abstract).

Gui, Y., Wi, X., and Xu, T. 1982. Study on morphological differentiation in endosperm plant of Kiwi fruit (*Actinidia chinensis*). *Acta Bot. Sin.* **24**:216–221 (in Chinese, English abstract).

Institute of Fruit Trees, Chinese Academy of Agriculture Sciences. 1963. *Introduction in Flora of Chinese Fruit Trees*. Vol. III, Shanghai Science and Technology Press, Shanghai (in Chinese).

Jones, O. P. and Ferrell, O. D. 1977. Propagation in vitro of M. 26 apple rootstocks. *J. Hortic. Sci.* **52**:235–238.

———, Pontikis, C. A., and Hopgood, M. E. 1979. Propagation in vitro of five apple scion cultivars. *J. Hortic. Sci.* **54**:155–158.

Liu, S., Chen, W., Wang, H., and Yang, S. 1978. In vitro culture of apices of dwarfing stock and seedling in apple. *Acta Bot. Sin.* **22**:93–94 (in Chinese, English abstract).

Shi, Y., Jia, Y., Li, X., Wu, M., Hu, Z., and Wang, Q. 1983. Studies on test tube mutation breeding of fruit trees: I. Origin and induction of adventitious shoots and branches in pear embryo culture. *Shandong Agric. Coll.* 1–8 (in Chinese).

Sun, J. and Zhu, Z. 1981. Vegetative reproduction and chromosomal ploidy of endosperm plant of barley. *Acta Bot. Sin.* **8**(3):11–14 (in Chinese, English abstract).

Wu, J. 1980. Induction of callus from endosperm and plant differentiation in crabapple. *Proc. Northeast Agric. Coll.* **2**:97–98 (in Chinese).

Wu, X. and Liu, S. 1979. Studies on occurrence of callus from embryo and variation of ploidy in apple. *Acta Bot. Sin.* **21**:309–319 (in Chinese, English abstract).

———, Liu, S., Zhou, Y., Qian, N., Chang, P., Xie, H., Zhang, F., and Yan, Z. 1977. Induction of callus from endosperm and differentiation of plant. *Sci. Sin.* **4**:355–359 (in Chinese).

Yu, D. 1956. Pome In: *Taxonomy of Fruit Trees in China*, pp. 87–147. Agriculture Press, Beijing (in Chinese).

Zhao, H. 1981a. Embryo culture in pear. *Liaoning Agric. Sci.* **3**:31–32 (in Chinese).

——— 1981b. Shoot tip culture of Zhu Mei crabapple stock of apple. *Zhongguo Guoshu* **3**:41–42 (in Chinese).

——— 1981c. Pear plantlet obtained by culture of shoot tip of test tube seedling. *Acta Bot. Sin.* **24**:216–221 (in Chinese, English abstract).

CHAPTER 19
Peach

Niyun Hu, Zenghai Yang, and *Guangming Lu*

INTRODUCTION

History of Cultivation and Economic Importance

The peach (*Prunus* sp.) originated in the upper reaches of the Yellow River Valley, Shaanxi and Gansu provinces, in highlands elevated 1000–2000 m above sea level. The wild peach can be found in the regions of Henan and Yunnan provinces, Tibet Zhuang Autonomous Region, and the watersheds of the Yangtze River and the Yellow River. It has been cultivated for more than 3000 years.

The peach fruit has a bright color, sweet smell, delicious taste, juiciness, and nutritional value. In addition to its fresh market use, peach fruit can be processed into concentrates, canned peaches, tinned peaches, dried peaches, and peach jam. Of the processed peaches, canned peach varieties with yellow mesocarp (fruit pulp) have had good sales internationally. The shell of peach pits can be processed into activated carbon. The roots, leaves, flowers, and seed are also used in Chinese herbal medicines.

The peach has the advantage that it bears fruit and reaches high yields early in its life, realizing profits a short time after planting. The peach has strong viability and is well adapted to different climates and soils; it is cultivated in southern and northern mountainous regions and plains in China. The peach tree has a small crown, so that it is easy to care for. The peach fruit is rather delicate and juicy and not able to endure transportation and storage, so for fresh market produce it is most appropriately planted in suburbs. The peach holds an important position in Chinese fruit production. It is also widely distributed in Asia, Africa, Europe, America, and Australia. There are more than 80 peach growing countries in the world, with a total yield of 72 million tons/year (1979). The European crop is the largest in the world; those in America and Asia rank second. These three continents represent 72.3% of the world total. The African and Australian peach crops are relatively small.

Germ Plasm Resources

BOTANICAL TAXONOMY. The peach belongs to the family Rosaceae and the subfamily Pronoideae. Some species important for peach improvement are as follows.

***Prunus persica* Batsch, the peach.** It is also named floky (hairy) peach or common peach. Its fruit is spherical with fine soft hair. Its leaf is oval and lance shaped, and its lateral veins form reticulate veination before reaching the blade edge. The peach pit is long, flat, and oblate with furrows on the surface. This species includes most peach cultivars, is most widely distributed, and includes five varieties:

1. Var. *compressa*, the flat peach: Its fruit is oblate in shape.
2. Var. *nectarina*, such as oil peach: It has no hair on the surface of the fruit coat.
3. Var. *densa*, the Longevity God peach: A dwarf tree with shorter branches and internodes.
4. Var. *pendula*, the branch-bowing peach: Its branches are soft, bowing downward.
5. Var. *duplex*, jasper peach: Its flowers are multicolored.

***Prunus davidiana* Franch, the mountain peach.** This species originates from the mountainous regions in north and northwest China. It is a dwarf tree with smooth bark and slender branches. The fruit is spherical and cracks and dries out when matured. The fruit is not usually edible, but it is cold-resistant and drought-tolerant. There are three types in this species: red-flowered, white-flowered, and smooth-leaved. This is one of the main species used as rootstock in north China and is also often used as a breeding parent for its resistance characteristics.

***P. mira* Koehne, the smooth pit peach.** This is a wild type of peach which is found scattered in the highlands of Sichuan Province and Tibet Zhuang Autonomous Region. The fruits are small, spherical, and covered by bristles, with a flat, smooth, furrowless pit. It is highly cold-resistant, bearing edible fruit which can be dried.

***P. fergarensis* Kostet Riel., the Xinjiang peach.** It is widely cultivated in Kashi District, Xinjiang Uygur Autonomous Region. The surface of the pit is furrowed. This species includes most sweet-kerneled peaches and many cultivated varieties highly resistant to drought.

***P. communis* Fritsch., the almond.** It originated in Persia (Iran) and is widely cultivated in Xinjiang Uygur Autonomous Region, China. The fruit is shaped as a flattened sphere with a tough and tensile pulp that is not suitable for eating. After maturing, the fruit pulp splits and exposes the pit. The seed inside the pit is large, fleshy, edible, and good for medicinal uses. The almond is characterized by high resistance to cold and drought. There are two varieties in this species. *P. communis* var. *typica* Schneider has a hard fruit pit and is usually used as root stock, giving a slight dwarfing effect. *P. communis* var. *fragilis* Schneider. bears a thin coated fragile pit. The sweet-seeded type of this variety is suitable for eating; the bitter-seeded type is usually used for medicine. Both are used as parent plants for crossing in breeding work.

***P. tenella* Batsch., the dwarf almond.** It is also called the Mongolian dwarf almond. It is a shrub native to the cold regions of the U.S.S.R. The fruit is dry, hard, and inedible. The surface of the pit is smooth. This variety is highly resistant to cold so that it is often used as a parent in breeding for cold resistance.

ECOLOGICAL CLASSIFICATION. Peach cultivars can be divided into three ecological groups based on their climatic adaptation.

North Chinese Cultivar Group. The cultivars in this group are mainly distributed in the Yellow River Valley, north China and northwest China, where the temperature in summer is high with a large diurnal range and the temperature in winter is low. The annual precipitation ranges from 400 to 600 mm. The cultivars in this group are characterized by high resistance to cold and drought, but they cannot bear in high temperature and high humidity. They can be subdivided into types as follows.

1. Honey peach type: This type of peach has large fruit. Its pulp is fine and solid and of good flavor. The fruit is of high quality and is able to endure transportation and storage. The cultivars Feicheng Peach, Shenzhou Peach, and Autumn Honey belong to this type.
2. Tough peach type: There is a distinct tip on the top of the fruit. Early in the ripening period, the fruit pulp is stiff and crisp with little juice. After full maturation, the pulp becomes soft. The cultivars May Delicacy, June White, and Eagle Beak belong to this type.
3. Yellow pulp peach type: It is mainly distributed in Shaanxi and

Gansu provinces and Xinjiang Uygur Autonomous Region. This type has a large number of cultivars and lines with various characteristics. Important cultivars include Pit-Adhered Yellow Peach; Separated Pit Yellow Peach; Yellow Sweet Peaches from Lingwu, Gansu Province, and Wugong, Shaanxi Province; and Yellow Pulp Eagle Beak Peach.

South Chinese Cultivar Group. They are mainly distributed in the Yangtze River Valley of east, middle, and southwest China. In these regions, the temperature in summer is high, the diurnal temperature range is narrow, and it is warm in winter. The annual rainfall in these regions is from 1000 to 1200 mm. The main types in this group are the following:

1. Honey peach type: The fruit pulp is soft and juicy, with a nice balance of sweet and sour tastes. The fruit top is a flattened sphere. The cultivars Shanghai Honey, Hundred Flowers, Jade Dew, and White Phoenix belong to this type, as do most Japanese cultivars.
2. Flat peach type: This type is characterized by a flat oblate-shaped fruit, concave on both ends; the fruit pit is also oblate. The pulp is delicate, soft, and juicy with a sweet smell. The cultivars Bai Yun Flat Peach, Chen Pu Flat Peach, and Sha Hua Hong Flat Peach belong to this type.
3. Stiff pulp peach type: The fruit pulp is fine and firm in texture, stiff, and crisp with little juice. It endures transportation and storage well. When overmatured, the fruit pulp becomes floury. The cultivars Pin Bei Zhi, Hanging-down White, Yunnan Chen Gong Early Peach, and Luling Peach belong to this type.

European Cultivar Group. The cultivars in the United States and the countries of southern Europe belong to this group. This group was formed when the peach was transferred from China to Europe in ancient times. These peaches are mainly distributed in East Asia and the coastal countries of the Mediterranean, where the climate is dry in the summer and sunshine is intense and winter temperatures are low. The group includes cultivars with yellow pulp and white pulp and oily peaches. The cultivars Amidon June, Early Clawform, Phillips, Tuscan, and Triumph have been introduced into China.

Problems for Breeding

PROBLEM OF AVAILABILITY OF FRESH PEACHES. Peaches of most cultivars mature during July and August. Therefore, it is necessary to breed early-maturing cultivars to extend the period of availability of fresh fruit. The harvest date for early-maturing cultivars should be in June or even earlier. Most early-maturing cultivars are characterized by fruit of small size and low quality, so a major goal of breeding is production of cultivars with large high-quality fruit (Morphology Group, Beijing Institute of Botany, 1974). The harvest dates for late-maturing cultivars should be from late September to the middle of October. There are germ plasm resources of winter peach in China, and they should be considered as breeding materials.

PROBLEM OF BREEDING CULTIVARS FOR CANNING. Canned peaches can be supplied all year and are in high demand in world markets. A great deal of work has been done in China, and several good cultivars for canned fruit production have been bred and selected. However, there are few cultivars and the cultivation areas are still insufficient to meet demand. In northwest China there are rich yellow peach plants. They should be further investigated for breeding superior cultivars.

PROBLEM OF BREEDING VIRUS-FREE PLANTLETS. Virus-free plantlets are of great importance for increasing peach yields. In the northeastern part of the United States, the yellow virus disease of peach is devastating. Green spot virus infection of leaves not only reduces the survival rate of grafts but also aggravates other diseases on the peach tree. Thus, growing virus-free plantlets via shoot tip culture in vitro is an important task for peach breeders.

BREEDING OF DWARFING CULTIVARS AND GRAFTING STOCK. Using a cultivation pattern in which closely planted small trees are substituted for thinly planted large trees is an effective way to obtain higher fruit yields earlier in the season.

BREEDING PROGRAM FOR MULTIPLE TRAITS. Breeding peach cultivars resistant to cold, high temperature, high humidity, insect pests, diseases, and waterlogging will extend the area that can be cultivated.

LITERATURE REVIEW

Tukey (1933) investigated the relationship between peach embryo and fruit growth and the maturity time and suggested that normal fruit development could be divided into three periods. The first period lasted 48 days in both early- and late-maturing cultivars. The duration of the second period varied from 5 to 24 days, depending upon the cultivars. The duration of the third period also varied with different cultivars. Tukey found that the second period was very short in early-maturing cultivars, and embryo growth extended into the third period. For this reason, the embryo was incompletely developed when the fruit was mature enough for harvest, and these seeds failed to germinate. Tukey achieved successful embryo culture in many kinds of fruit trees, such as apple, pear, peach, and plum. Lammerts (1942) showed that embryo culture could be used to shorten the breeding cycle in deciduous fruit trees and to increase germination rate for hybrid seeds. Shen Dexu et al. (1974) studied the development of peach fruit and embryo as well as embryo culture. They found that embryos which were taken from fruits at 84–90 days old were able to develop into seedlings. Embryos from fruit at day 100 gave a higher rate of plant regeneration, 48%. The Beijing Institute of Botany, the Chinese Academy of Sciences, and the Institute of Pomology and Forestry at Beijing Academy of Agricultural Sciences carried out cooperative investigations from 1963 to 1974 in which 80–90-day-old embryos were cultured in vitro. The plantlet induction rate was 50–60%. Plantlets that formed were transplanted and have borne fruit for several years. The early-maturing cultivar Amidon June was used by Lu Guangming et al. (1977) as a female parent to produce hybrid seeds. Embryos were taken at 70–75 days. Plantlets were successfully induced with a frequency of 31.5%. After several years of selection and cultivation, a single plant line of extra-early-maturing peach was chosen for propagation. This line of peach matures at the end of May or the first of June in Yangling, Shaanxi Province, i.e., 15 days earlier than the female parent and early-maturing cultivars in production in general and 26 days earlier than the male parent. The fruit of this cultivar is brightly colored and of high quality. Yang Zenghai et al. (1983, 1984a, 1984b) reported that embryos of five cultivars taken from fruits grown for less than 100 days were cultured, and the plantlet induction rate was higher than 70%. Zhuang Enji et al. (1981) successively bred early- and extra-early-maturing peaches via embryo culture. They obtained an extra-early-maturing juicy peach, in which fruit developed in only 56 days. Regular peach cultivars with development to maturity of less than 60 days have not been reported. Embryo culture has been used to grow the peach cultivar Shanghai 005, maturing at the end of May and early June in Shanghai. Zhu Jijun et al. (1984) have carried out embryo culture using naturally pollinated seed from cultivars including Yu Hua Dew, Rosy Dawn, Gangshan Early, Shazi Early, and Shanghai Honey Early. Six cultivars in which the pit had grown for 70–82 days produced green embryo plants at a rate of 42.4–95.0%, and five cultivars in which the fruit had grown for 80–88 days yielded green plants at a rate of 86.6–96.6%. Hesse and Kester (1955) suggested that an embryo development index (PF_1 = embryo length/seed length) of approximately 0.7 is a lower limit for successful embryo culture. Yang Zenghai et al. (1983) used the cultivar Amidon June, which matures in mid-June, for embryo culture. Fruit development took only 65 days, giving a rather small embryo size with PF_1 of 0.42. Nonetheless, they achieved a seedling induction rate of 85.7%. Zhu Jihun et al. (1984) cultured very young peach embryos with a PF_1 index lower than 0.4. Using a modified medium, they succeeded in culturing young embryos with PF_1 of 0.2–0.4 and had a plantlet induction rate of 68.7%.

As in other plants, work on peach shoot tip cultures has progressed swiftly since the 1970s (Yang Zenghai et al., 1984). Tabachnik et al. (1977) developed shoot apex cultures in almond and in hybrids from almond and peach. Hammerschlag (1980, 1982) investigated methods for rapid propagation from bud culture in vitro. Skirvin et al. (1978a, 1978b) pointed out that rapid propagation of peach through tissue culture techniques might cause a revolution in nursery stock production, but they did not obtain intact plantlets. In 1981 they reported a root regeneration medium suitable for culturing shoots of the peach cultivar Herbrite and obtained intact plants (Skirvin, 1978b, 1981). Dong Yihu et al. (1984a) successfully used shoot tip culture in Nanking cherry (*Prunus tomentosa* or *Cerasus tomentosa*), a rootstock for peach grafting. Since 1979, we have performed shoot tip cultures using tissue-cultured peach seedlings as well as shoot tips from trees of peach, cultivars, and the rootstocks Nanking cherry and David peach (*Prunus davidiana* or *Amygdalus davidiana*). These experiments provided the basis for peach micropropagation through shoot tip culture in China (Yang Zenghai, et al. 1984, Dong Yihu et al. 1984b).

Utility of Embryo and Shoot Tip Cultures

PROMOTION OF SEED GERMINATION IN BREEDING OF EARLY- AND EXTRA-EARLY-MATURING CULTIVARS. Early- and extra-early-maturing peach cultivars are usually bred by hybridizing an intermediate-maturing cultivar as the female parent and an early-maturing one as the male parent. However, the time of fruit maturation in hybrids usually is intermediate between the female and male parents and is often closer to the later date. It is thus difficult to breed new cultivars or lines that mature earlier than their parents successfully. It is clearly necessary to perform crosses between two early-maturing parent plants. Since the second period of fruit development is quite short in early-maturing plants, the embryo taken from such hybrid fruits will not develop. Seeds formed will germinate and sprout infrequently or not at all. In vitro embryo culture of immature seeds may permit seed germination. Embryo culture in peach is important to breeding early-maturing cultivars or lines and is an indispensable tool for selecting and breeding cultivars maturing earlier than their parent plants.

TO OVERCOME EMBRYO ABORTION RESULTING FROM DISTANT HYBRIDIZATION. In peach breeding, distant crossing often results in seeds which are not viable because embryos develop poorly or development is interrupted. Culture of the immature embryos gives good results. Embryo culture can also shorten the breeding cycle for deciduous fruit trees, such as peach, apricot, and plum, and is useful for investigations of the metabolic and physiological changes during embryo development.

RAPID PROPAGATION OF SUPERIOR CULTURES. Seedling propagation via shoot tip culture has many advantages, such as rapid multiplication rate, lack of restrictions due to seasonal and environmental conditions, and shortening of the reproduction cycle. The combination of peach breeding with shoot tip culture would accelerate the breeding process and make it possible for a new cultivar to be planted widely 2 years after its release. Shoot tip culture would also provide genetically identical experimental materials, overcoming the confusing individual differences seen in plants propagated from seeds.

CULTURE OF VIRUS-FREE PLANTLETS. Virus-free plantlets possess the advantages of high yields, savings in fertilizers, and vigorous uniform growth of plantlets. The technique of shoot tip culture can be effectively applied for propagation and preservation of virus-free cultivars which have been certified by assay or recovered from plants infected by virus. It also can be used for preservation of valuable germ plasm. Apical meristem culture (as the length of the shoot tip is about 0.1 mm with one or two leaf primordia) can be used to obtain plantlets free of ring spot virus and other viruses in peach.

PROTOCOLS: EMBRYO CULTURE

Process and Medium

The development of peach fruit can be divided into three stages:

1. Prehard-mature stage: The fruit has not reached its maximum volume, the surface of the fruit coat is green, and the fruit is bitter to the taste and unsuitable for eating.
2. Hard-mature stage: The peach fruit almost reaches its maximum volume, and the surface color changes to yellow or yellowish. The fruit pulp is hard and crisp and acidity is reduced.
3. Soft-mature stage: The fruit has reached its maximum volume, and its characteristic flavor develops. The fruit pulp becomes soft with a bright colored coat. The embryos used in our experiments were taken from fruit in the hard-mature stage.

In vitro culture of embryos from the peach cultivar Da Ju Bao (fruit development period of 100 days) was carried out in the Department of Horticulture, Zhejiang University of Agriculture. Embryos taken from fruit in the hard-mature stage gave a much higher rate of seedling formation (80%) than those from fruit in the soft-mature stage (48%). Because fruits at different stages of maturity were picked from the same tree on the same day, the difference in seedling formation rate was probably due to differences in embryo maturity. Yang Zenghai et al. (1983) used five cultivars as explants for embryo culture; embryos were taken from fruits at three different stages of maturity from the same tree on the same day and stored at 1–5°C for 80 days. Embryos from all three stages could be induced to develop into seedling plantlets 1 cm in height and bearing three secondary leaves or more. The embryos from fruit at the hard-mature stage gave the highest rate of plant regeneration. Table 1 confirms that the optimum stage for embryo culture is the hard-mature stage of fruit.

Low-Temperature Treatment. Like the completely matured seeds, developing peach embryos require a low-temperature dormancy period for normal growth. Lesley et al. (1952) kept peach fruit at 2°C for 7 weeks. Brook et al. (1958) placed cultured embryos at 4°C for 2–4 weeks after inoculation. Embryos were inoculated immediately after the fruits were picked in their hard-mature stage and then cultures were kept at 1–5°C for a low-temperature pretreatment. The optimum duration of the low-temperature pretreatment varied, depending upon cultivar. For the cultivars Early Fragment Jade and Dong Wang Mu, in which the embryo developmental period was less than 82 days, pretreatment for 80–110 days at 1–5°C was suitable. After pretreatment, the rate of seedling formation reached 62.5–70.8%. The cultivar Triumph had an embryo development stage of 88 days, and its rate of seedling formation was 74.3–88.3% with pretreatment for 40–80 days. In Orange Early, which requires 90 days of development, 40 days was enough for pretreatment, to give a seedling formation rate of 70%. However, for most cultivars, the pretreatment period should not be less than 70 days. In general, the

TABLE 1. Effect of Different Embryo Maturities on the Rate of Plantlet Formation*

	Plantlet Formation (%) from Embryos of Different Maturities		
Cultivars	Early-Hard Mature Stage	Hard-Mature Stage	Soft-Mature Stage
1–3	66.7	86.2	71.4
Dong Wang Mu	77.1	82.9	50.0
Triumph	67.6	86.7	60.6
Orange Early	82.9	91.2	63.6
4–1	74.3	100.0	85.7

*In each treatment 30 to 35 embryos were inoculated.

shorter the developmental period, the longer the low-temperature pretreatment should be. Cultured peach embryos which were not pretreated at low temperature developed only very short embryonic roots (0.3–1.0 cm in length). The epicotyl was elongated, but the internodes were very short. The leaves at the shoot tip formed rosettes. With treatment at 1–5°C, the rosettes were reduced or even disappeared (Janick et al., 1975).

Tukey's medium was used as basal medium, but its iron salt was replaced by that of Heller's, ie., 0.1 mM Na_2-EDTA and $FeSO_4 \cdot 7H_2O$. The microelements of MS medium and 0.5% activated charcoal were added to 1 l of Tukey's medium to yield superior results. After 25 days of culture, the embryos were transferred to Norstog's culture medium, where most germinated in 2–3 weeks. Hough's culture medium was also effective. The components of the media are listed in Table 2. Yang Zenghai et al. (1983) cultured young embryos which were 65 days old

TABLE 2. Components of Media Suitable for Embryo Culture in Peach*

	Medium		
Components	Tukey (1934)	Norstog (1963)	Hough (1980)
Macroelement			
KNO_3	1.35 mM	1.6 mM	—
$Ca(NO_3)_2 \cdot 4H_2O$	—	1.2 mM	4.92 mM
KH_2PO_4	—	—	1 mM
$NaH_2PO_4 \cdot 4H_2O$	—	1 mM	—
$Ca_3(PO_4)_2$	0.61 mM	—	—
Na_2SO_4	—	1.4 mM	—
$MgSO_4 \cdot 7H_2O$	—	3.0 mM	2.0 mM
$MgSO_4$	1.41 mM	—	—
$CaSO_4$	1.25 mM	—	2.5 mM
$(NH_4)_2SO_4$	—	—	1 mM
KCl	9.4 mM	1.9 mM	—
NaCl	—	—	1 mM
$FePO_4 \cdot 7H_2O$	—	—	9 μM
$FePO_4 \cdot 2H_2O$	0.91 mM	—	—
$FeC_6H_5O_7$ (1%)	—	29.8 μM	—
Microelements			
H_3BO_3	—	8.3 μM	9.5 μM
$CuSO_4 \cdot 5H_2O$	—	1 μM	0.16 μM
$MnSO_4 \cdot 4H_2O$	—	13.4 μM	3.45 μM
$ZnSO_4 \cdot 7H_2O$	—	1.7 μM	1.5 μM
$Na_2MoO_4 \cdot 2H_2O$	—	1 μM	—
MoO_3 or	—	—	0.1 μM
$H_2MoO_4 \cdot H_2O$	—	—	0.14 μM
$CoCl_2 \cdot 6H_2O$	—	1 μM	—
EDTA	—	—	11.4 μM
Sucrose	29.2 mM	29.2 mM	58.4 mM

*For solidifying the three media 0.6–1% of agar was used.

with an embryo development index (embryo length/seed length) of 0.42. After 25 days on Tukey's medium, the young embryos were transferred to Norstog's medium. After 2–3 weeks, their germination rate reached 87.5% and the survival rate of seedlings planted in soil was 85.7% of the germination rate. Zhu Jijun et al. (1984) cultured immature embryos of the cultivar Rain Flower Dew on Tukey's medium and found that supplements of activated charcoal or MS microelements promoted germination of the embryos and formation of green seedlings. The medium supplemented with both activated charcoal and microelements of MS medium gave the best results. According to Hough and his associates, increasing sucrose concentration was the key to successful culture of peach embryos. Hu Niyun et al. added 0.15–0.3 M of sucrose to the medium for culturing young peach embryos with good results. As the embryo developed, the sucrose concentration had to be reduced gradually. In peach embryo culture, sucrose is usually used, but glucose and fructose can also be used. The concentration range of BA used in peach embryo culture is usually 0.88–2.2 μM to a maximum of 4.4 μM. If the concentration of BA is higher than 4.4 μM, embryo growth will be inhibited. For young embryos in early stages of development, supplements of vitamins, such as thiamine, pyridoxin, and nicotinic acid, were applied with positive effects. The pH of the medium also had remarkable effects on embryo growth. In general, the pH should be 5.6–5.8, but for immature young embryos, 6.0–6.5 is suitable.

Key Steps

The peach seed was carefully removed from the cracked pit. The seed was sterilized with 70% ethanol for 1 min and immediately rinsed three to four times with sterilized water. It was then immersed in a saturated bleaching powder solution and stirred for 10–20 min. The bleaching powder solution was discarded and the seed was again rinsed three or four times with sterilized water. The seed coat was removed with forceps and scalpel, and the embryo was picked out and placed onto the medium. The cultured embryo should be inserted upright in the medium and buried to about one-third of its height.

After transfer to culture medium, the embryo had to be exposed to low temperatures (0–5°C) for more than 70 days. The culture tubes containing the embryos were placed in an incubator. The culture conditions were as follows: illumination by fluorescent lamp (40 W), 16 hr/day, with an intensity of 1500–3000 lux.

Transplanting plantlets into soil is the critical event for cell-cultured seedlings, because their initial conditions are very different from the normal growing conditions of seedlings. The only feasible way to ensure that the cell-cultured plantlets survive after transplantation is to reduce the differences between these two environments as much as possible: the plantlets then adapt themselves gradually to normal growing conditions. For this reason, strict soil disinfection and intermittent water spraying must be emphasized (Janick et al., 1975). Many researchers (Zhuang Enji et al., 1981; Yang Zenhai et al., 1983) have tried to harden the cell culture–derived seedlings before transplanting, exposing them to natural conditions by removing the stoppers of the culture tubes and exposing them to sunshine. Afterward, transplantation was more successful even when the soil was not disinfected. Zhuang Enji et al. (1981) also reported that maintaining diurnal temperature differences could increase the survival rate. The temperature in daytime should be no higher than 30°C, and at night, no lower than 15°C. The average temperature should be under 25°C and the diurnal difference should be greater than 8°C. Yang Zenghai et al. (1983) reported that, in addition to the measures mentioned, they had undertaken the following procedures for increasing seedling survival rates after transplantation.

1. Preparation before transplanting: When the seedlings had grown to a height of 4–6 cm and developed four to six leaves, they were placed in a greenhouse for hardening under natural illumination for 3–5 days. The stoppers of culture tubes were then removed and the cultures exposed to air for 2–3 days. One to two drops of distilled water was dispensed into the tube every day, for reducing not only evaporation of medium but also concentration of the inorganic salts. The medium left on the roots had to be cleaned off carefully by water rinsing before transplanting to control microbial growth.
2. Zineb treatment of the root: The cleaned roots were immersed in 0.1% Zineb solution for 3–5 min. This procedure resulted in an obvious increase of survival rate to 75–100% even in the weakest seedlings. In the cultivar Amidon June, the survival rate after transplanting into soil could be as high as 85.7%, whereas the survival rate for plantlets which had been immersed in water for 20 min was only 63.3%. Activated charcoal also promoted root development and growth of stout plantlets.
3. Transplanting: Three parts of soil and one part of sand were mixed in a plastic pot. The seedlings that had gone through the procedure described earlier were planted in pots immediately after removal from the culture flasks. In order to reduce transpiration from the seedlings, it was necessary to water spray intermittently or to cover the pot with a glass cover. After transplantation, the potted seedlings were placed at a lower temperature of 16–20°C, 70% relative humidity, and illumination of 1500 lux. Two weeks later, the temperature

was gradually raised to 25°C, intensity of illumination was changed to 3000 lux, and nutrient solution was added when watering.

PROTOCOLS: STEM APEX CULTURE

The media used for shoot apex culture in peach and its rootstocks can be divided into three groups:

1. Differentiation media for axillary and adventitious buds
2. Subculture media
3. Root regeneration induction media

In the initial culture period, the Media G, L, and MS were suitable for culture of axillary and adventitous buds (Table 3). In the subculture period, the Media MS-R, G, and Half Macro-G (containing half the macroelements of Medium G) were used. Half MS medium and half Medium G were used for root regeneration. None of these media contained cytokinin; whether the media were supplemented with auxin or not depended upon the need for root regeneration. Some researchers have sug-

TABLE 3. Media for Shoot Tip Culture in Peach and Its Grafting Stock

Components	Medium		
	G	L	MS-R
KH_2PO_4	3.68 mM	1.99 mM	1.25 mM
$MgSO_4 \cdot 7H_2O$	1.78 mM	1.46 mM	1.5 mM
NH_4NO_3	—	5 mM	5.14 mM
$(NH_4)_2SO_4$	1.19 mM	—	—
KNO_3	29.7 mM	17.8 mM	4.7 mM
$CaCl_2 \cdot 2H_2O$	1.5 mM	—	2.73 mM
$Ca(NO_3)_2 \cdot 4H_2O$	—	5 mM	—
$ZnSO_4 \cdot 7H_2O$	30 μM	30 μM	30 μM
$MnSO_4 \cdot 4H_2O$	100 μM	4.5 μM	100 μM
H_3BO_3	100 μM	100 μM	150 μM
KI	5 μM	0.5 μM	5 μM
$Na_2MoO_4 \cdot 2H_2O$	1 μM	1 μM	1 μM
$CuSO_4 \cdot 5H_2O$	0.1 μM	0.1 μM	0.1 μM
$CoCl_2 \cdot 6H_2O$	0.1 μM	0.1 μM	0.1 μM
$FeSO_4 \cdot 7H_2O$*	100 μM	100 μM	200 μM
Glycine	26.6 μM	26.6 μM	26.6 μM
Vitamin B_1	1.19 μM	1.19 μM	1.19 μM
Vitamin B_6	2.96 μM	2.96 μM	2.96 μM
Nicotinic acid	4.06 μM	4.06 μM	4.06 μM
Inositol	555 μM	555 μM	555 μM

*Fe-Na-EDTA is substituted for $FeSO_4 \cdot 7H_2O$.

gested that supplementing the media with activated charcoal enhances the regeneration rate.

During peach apex culture, the explants developed poorly after transfer to freshly prepared MS medium for subculture, although it was the same as used in the initial culture. The MS medium was, therefore, modified and named MS-R. MS-R medium was more suitable for subculture and growth of healthy shoot apices. Another medium, the G Medium, was also designed for apex culture in peach and was even better than MS medium. Its distinct characteristics are that the nitrate nitrogen content, 30 mM KNO_3, is greatly increased, and the ammonia nitrogen content is reduced: $(NH_4)_2SO_4$ (8.66 mM) was substituted for NH_4NO_3. In addition, the content of $MgSO_4 \cdot 7H_2O$ (1.78 mM) and K_2HPO_4 (3.68 mM) was increased and the content of $CaCl_2 \cdot 2H_2O$ (1.5 mM) was reduced.

The multiplication rate is an economic criterion of propagation in quantities through tissue culture; hence, it needs successive subcultures to produce the maximum number of multiplication units, i.e., newly developed apices. A large number of newly developed apices grew up to 1 cm in length as the cultures were carried for 4–6 weeks. At that time, the newly developed shoots could be cut into sections bearing apical buds and axillary buds, and the latter were transferred onto freshly prepared medium for subculture. Thirty days later they could be cut into segments and transferred again for other subcultures, resulting in a great number of new shoots. Newly developed shoots over 1 cm in height were taken from the culture tube and transferred for root induction in one of the following ways:

1. The basal part of the shoot was immersed in IBA solution for several hours, then transferred into hormone-free root regeneration medium.
2. The shoot was cultured on medium containing auxins for a week, then transferred into hormone-free root regeneration medium.
3. The new shoot was transferred directly onto a root regeneration medium containing auxins.

Although new shoots were able to regenerate roots by all three methods, the first two were more suitable for root regeneration. Auxins encouraged formation of root primodia, but after they had formed, a high auxin concentration was unfavorable to young root development.

Media for Apex Culture

Media in which the average hormone levels were 2.2 μM BA, 9.8 μM IBA, and 5.8 μM GA, with 500 mg/l LH and 146 mM sucrose, were used.

MS medium, G medium, and modified MS medium (in which the contents of KNO_3 and NH_4NO_3 were reduced to one-fourth and the iron salt was doubled) were compared. G medium was most effective for differentiation of new shoots. Sixteen shoot apices were inoculated on each medium. After culture, 42 shoot apices were obtained from G medium, and 31 and 24 shoot apices were obtained from MS medium and modified MS medium, respectively. The shoot explants on G medium were healthier and much greener with large leaves and longer internodes than on MS medium and modified medium. Variance analysis indicated that there were significant differences between G and MS media as well as between G and modified MS media. Screening of different growth regulator concentrations showed that 2.2 μM BA plus 9.8 μM IBA was the best combination. Auxins were screened by testing the effects of three auxins, each at three concentrations: IAA (2.9, 11, 22 μM); IBA (2.5, 9.8, 19.6 μM); and NAA (2.7, 11, 22 μM). Four explants were placed in each culture flask with 10 flasks for each treatment. The basic medium was supplemented with 2.2 μM BA, and the medium without auxin was used as the control. The treatments with 2.5 μM or 9.8 μM IBA were the most beneficial for explant growth and differentiation. The number of viable buds was obviously higher than that of the control, although their differentiation rates (65.0% and 67.5%, respectively) only approached that of the control (70%). The combination of 9.8 μM IBA and 2.2 μM BA increased the number of viable buds. This combination was better than that of IAA and NAA. Cytokinins were screened by testing the following concentrations—BA, 2.2, 4.4, 11 μM; KIN, 2.3, 4.6, 11.5 μM; and ZEA, 2.3, 4.6, 11.5 μM—in experiments with four explants (shoot apices) for each culture flask, four flasks per treatment. The basic medium was supplemented with 9.8 μM IBA, and the medium without cytokinin was the control. BA and KIN was beneficial for new shoot growth and differentiation. The differentiation rate for the former was as high as 85%; for the latter, it reached 87.5%. However, in treatments in which the KIN was added to the medium, the new shoots no longer grew after they reached 1 cm in height, only expanded their leaves. In the treatment in which BA was used, the new shoots, after culture for a month, grew vigorously and could reach 4–5 cm in height. The optimum concentration of BA was 2.2 μM.

In cultures of the shoot apex and axillary buds, the shoot segments containing nodes were cultured on MS medium supplemented with 2.2 μM BA, 9.8 μM IBA, and 500 mg/l LH. They formed many calli and the explants became obviously swollen. The new shoots grew well and rapidly into seedlings which bore many viable buds, regenerating one to two lateral branches from the basal area. Inhibition occurred when LH was raised to 1000 mg/l, but the production of viable buds was improved when 500–1000 mg/l of LH was added. The differentiation rate for new shoots decreased as the concentration of LH increased. The effect of CH was less than that of LH: the differentiation rate decreased with increasing concentration, but there was no change in the rate of formation of viable shoots.

On MS medium supplemented with 2.2 μM BA, 9.8 μM IBA, and 500 mg/l LH, differences in growth of the cultured explants were detected. Some shoots grew quickly and healthily, and others grew slowly and weakly. This was probably due to differences in the explants' axillary buds prior to culturing.

Two experiments were conducted to compare culture media. In the first experiment, 4.6 μM ZEA and 4.9 μM IBA were added to the three basic media: L medium, G medium, and MS medium. In the second, basic media, L and G, were supplemented with 4.4 μM BA and 2.5 μM IBA. In each treatment, 33–42 buds were inoculated to initiate cultures. Apices of Nanking cherry were able to grow on all three media, with no obvious difference between them. The average bud multiplication rate was 1.1. The rate of production of viable shoots on L medium in the two experiments reached 56.3% and 30.6%, respectively, rates much higher than on G medium (22.2% and 14%, respectively). On G medium some cultured shoots and leaves turned yellowish; however, on MS medium, the leaf color was normal but their growth was poor. Therefore, among the three media, L medium was the most suitable for shoot apex multiplication in Nanking peach. IBA was the best among tested auxins; BA was the best among tested cytokinins. ZEA also promoted the growth of cultures. In order to find out the optimum combination of hormones for shoot apex culture in Nanking cherry, experiments on different combinations of ZEA, BA, and IBA were completed.

When the supplements were 4.9 μM IBA and 4.6 μM ZEA, the multiplication rate was 1.5 times that of the control's rate. The percentage for formation of newly developed shoots approached 33.3%. In the medium with 9.1 μM ZEA, although many buds multiplied, fewer new shoots developed. BA concentrations of 2.2, 4.4, 8.8, and 13.2 μM were tested with 4.9 μM IBA; 35 explants on average were used for each treatment. All the treatments containing BA gave multiplication rates of 2 or more per bud. Among the concentrations tested, the medium with 4.4 μM BA gave the highest multiplication rate; gave each bud an average of 3.3 new shoots and the most successful culture produced 10 shoots. The medium with 8.8 μM BA was the next best. In media with 8.8 or 13.2 μM BA some cultured explants formed clusters of small buds. However, the highest multiplication rate of viable shoots was in the medium with 2.2 μM BA (26.3%), and the second highest was with 4.4 μM BA (16.7%). With BA concentrations higher than 8.8 μM, almost no new shoots appeared.

Dong Yihu et al. (1984a) reported taking and sterilizing shoot apices of David peach; inoculating them onto three media, MS medium, L medium, and G medium, with 2.2 μM BA, 9.8 μM IBA, and 500 mg/l LH; and culturing them for 4 weeks. After that, the cultures were transferred onto the original medium supplemented with 2.5 μM IBA and 2.2 μM BA and cultured for 4 weeks. Shoot apices grew and multiplied on all three media, but the growth conditions and the bud multiplication number differed among them. Cultures on G medium showed the best results; those on L and MS media were not as good. After transferring and subculturing again, those differences were more distinct. The multiplication rates of the cultures on G medium were two to three times those on L medium and MS medium. On G medium, the number of new shoots from one bud could be as high as 22, the highest multiplication number in the experiment.

The optimal concentrations of hormones for apex shoot culture were investigated. When explants were inoculated on G medium supplemented with 4.4 μM BA and 2.5 μM IBA, they showed higher multiplication rates and more vigorous growth. The average multiplication rate reached 2.3, and the production of viable new shoots was 21.1%. The combination of 2.2 μM BA and 4.9 μM IBA gave the highest rate of viable shoots, but the multiplication rate was not as high as with 4.4 μM BA and 12.3 μM IBA. When the concentration of BA was higher than 4.4 μM, the buds were small, were clustered more closely, and grew poorly. When the concentration of IBA was higher than 2.5 μM, the callus tended to expand. In contrast, explants cultured on the control, hormone-free medium stopped growing after a few leaves appeared.

Media for Root Induction

One-half MS medium was used to induce roots, supplemented with NAA at different concentrations, 0.27, 0.54, 1.1, and 2.7 μM, and the medium without NAA was used as control. After the shoot apices had been inoculated for 5 weeks, the medium with 0.54 μM NAA showed the highest rate of root regeneration (77.4%) with an average root number of 2.4.

The shoot apex cultures with newly developed roots were transferred onto half MS medium containing 0.54 μM NAA for 1 week. The shoots were then transferred onto the same medium without the hormone. Cultures which were retained on the original medium and not transferred were used as controls. They had an average root formation of 80% with 2.8 roots per shoot. The transferred shoot cultures from medium containing 0.54 μM NAA were not significantly different from the controls in root formation rate and root number, but there were rather significant differences in morphology of the roots. The roots from the control shoots were thick and deformed, but the roots from the shoots on the NAA medium grew normally without callus formation.

Shoot apices of David peach were inoculated onto half-L medium supplemented with 0.54–5.4 μM NAA and were induced to regenerate roots. As the NAA concentration increased, the quality of roots decreased, while the root number increased. Treatment with 0.54 μM NAA resulted in the highest root regeneration rate (65.4%) and normal roots. In the treatment with 2.7 μM NAA, callus grew on the basal area of new shoots. In the treatment with 5.4 μM NAA, the roots were thick and short. No roots formed on medium without NAA.

New shoots of David peach were inoculated onto the half-L medium supplemented with 0.54 μM NAA and cultured in the dark for 12 days. After 4 weeks' culture, the dark treatment increased the root regeneration rate and was 20% greater than the control's. When the shoots were cut on one side of their bases and cultured in the dark for 12 days, the root formation rose to 100%. When these shoots were cultured in the dark, they showed roots 2 days earlier than controls.

Half-MS was the most suitable medium for root formation in David peach shoot apex culture, with a root induction frequency 2.7 times that of half-G medium. On half-MS medium, the highest rate of root formation appeared on medium supplemented with 0.54 μM NAA. As the NAA concentration increased, root formation decreased and callus increased. Although their number increased, roots tended to be shorter and thicker, particularly on the medium with 5.4 μM NAA, where many roots became very thick and deformed.

Media for Subculture

When the new shoots differentiated from shoot apex or axillary bud or when they grew more than 1 cm in height, they were transferred to root regeneration medium. The new shoots which were not high enough for root formation were cut into segments and transferred onto multiplication medium. After 30 days, when the new shoots reached 1–2 cm, the larger shoots could be cut for root regeneration, and the smaller ones cut into small segments for multiplication. In this way, the larger shoots could be used for root induction while the apex culture was maintained.

MS medium was unsuitable for shoot apex subcultures. After the shoots were transferred onto MS medium for 2 weeks of culture, there was no distinct elongation of the stems and leaves and the leaves turned yellow. Necrotic tissue appeared on the basal part of the shoot 2 weeks later, resulting in death of the whole explant. With the same concentration of hormones, MS-R medium gave better results than MS medium, with normal growth for newly developed shoots, stems, and leaves as well.

On MS-R medium supplemented with 2.2 μM BA, 2.9 μM IAA, and 500 mg/l CH, the average number of viable new shoots was 4.3 and the rate of viable new shoots reached 75.4% (55 explants were transferred in total), but the new shoots were abnormal, with short internodes and small leaves. On MS-R medium supplemented with 2.2 μM BA, 4.9–9.8 μM IBA, and 300–500 mg/l LH, the internodes of the new shoots elongated, the average number of new shoots reached 52.9–65.8%, and the shoots grew normally. After 4 weeks, the highest shoot was more than 2.5 cm high.

The new shoots from primary cultures of David peach apices were cut into segments of about 5 mm containing an apical or axillary bud, which was then inoculated onto G or half Macro-G media, supplemented with 2.2 μM BA and 4.9 μM IBA. Their multiplication rates ranged from 57.4 to 70.7%. Both G and half Macro-G medium were suitable for subculture of David peach.

With half Macro-G medium, a suitable hormone combination for subculture was 0.44 μM BA + 0.49 μM IBA, giving a multiplication rate and a rate of new shoot production of 2.5 and 71.7%, respectively; the combination of 2.2 μM BA and 2.5 μM IBA was the next best, with a multiplication rate and a rate of viable new shoots of 2.1 and 60.3%, respectively. In four treatments supplemented with 4.9 μM IBA, the new shoots grew weakly and some turned brown and died.

Key Steps

1. Sterilization: Explants were sterilized when materials were taken from the field or from potted plants cultivated for the purpose of developing cultivars.
2. Procedure of sterilization used in apex culture: The leaves were first removed from the shoot and the dormant bud and the bud scales were removed. The shoot apex was rinsed with running water for 2 hr. The explants were then immersed in 70% ethanol for 1 min and in 0.1–0.2% mercuric chloride solution for 10 min. Finally, the materials were rinsed with sterilized water four to five times. Sodium hypochlorite could be substituted for mercuric chloride. Explants which had been treated with ethanol were immersed in a sodium hypochlorite solution or saturated solution of bleaching powder for 15 min, then rinsed with sterilized water. Materials taken from the field were first sterilized by the procedure described previously. The

shoot apices were then inoculated onto a multiplication medium containing no hormones or vitamins and placed in the culture room for 24 hr. After that, the shoot apices were sterilized once again with the disinfectants mentioned, and the shoot apex or axillary buds were cut and transferred onto medium under aseptic conditions.

3. Cutting off the shoot apex: The shoot apices were cut into sections 0.5–1 cm in length. Entire buds could be used instead when the culture was not employed for virus elimination. When the apical meristem was used, it was necessary to remove the young leaves and leaf primordia under a dissection microscope and excise the terminal part of the apex about 0.1–0.2 mm in thickness for use as an explant.
4. Culture conditions: Peach shoot apices grew normally at about 25–30°C. As the temperature was raised from 25°C to 30°C, the growth rate accelerated. Supplementary illumination with an intensity of 2000–3000 lux for 16 hr/day was beneficial. The pH of the medium should be adjusted to 5.8 before autoclaving. Root induction was carried out in a temperature range of 20–25°C. After transplantation to soil, the root-bearing plantlets needed care, especially control of temperature and humidity. In the initial phase of culture, lower temperatures (16–18°C) and higher relative humidity were necessary. The relative humidity around the plantlets was maintained by spraying water intermittently or by covering the plantlets with a glass plate or beaker.
5. Transplantation of plantlets: Plantlets derived from shoot apex culture were transplanted by the same procedure described for plantlets from embryo culture. After being cultured on root regeneration medium for 4 weeks, most shoots developed healthy and strong root systems. Usually plantlets which only grew to a few centimeters in height did not survive direct transplantation into soil. In recent years, most researchers have tried to reduce the difference between the environment inside and that outside the culture tube as well as culturing strong seedlings to promote plantlet survival. During transplantion, the plantlets were taken out of the culture tube or flask and the medium adhering to the roots was washed away with tap water. The cleaned plantlets were placed under low temperatures, 16–18°C, relative humidity of over 70%, and weak illumination for the initial period of growth. When new growth appeared on the plantlets, the intensity of illumination and the air temperature were raised to promote chlorophyll accumulation and enhance survival rate. Transplanting the plantlets when root primordia were initiated (but roots were not yet growing) could also raise the survival rate.

FUTURE PROSPECTS

Embryo culture has unique value in breeding early and extra-early peach cultivars. When the components and proportions of standard culture media are adjusted, the conditions for embryo culture in peach should be improved. Embryo development, culture condition, and plantlet formation from young embryos less than 60 days old should also be studied. The result will be the breeding of cultivars with superior quality and earlier maturity.

Immature embryos, especially those taken very early in development, are very difficult to culture successfully in vitro. During culture, young embryos undergo precocious germination, which often results in malformation and weakness of shoots or plantlets, and may lead to the death of plantlets after transplantation. This is a problem which must be further studied.

The application of peach tissue culture for rapid propagation has a promising future. With continuing efforts, tissue culture for rapid propagation of seedling nursery stocks will move from the laboratory to commercial production. At present, much effort is needed to increase the propagation rate. Investigations for obtaining virus-free seedlings of peach and their cultivation are of great importance as well. Breeding and cultivation studies on virus-free cultivars are needed, especially to accelerate propagation of virus-free nursery stocks via tissue culture.

KEY REFERENCES

Dong, Y., Lu G., Hu N., and Yang Z. 1984a. Shoot apex culture in *Prunus davidiana* Franch. *Plant Physiol. Commun.* **3**:43 (in Chinese).

———, Yang, Z., Hu N., and Lu G. 1984b. In vitro culture in *Prunus tomentosa* (Nanking cherry). *Plant Physiol. Commun.* **4**:36 (in Chinese).

Lu, G. and Hu, N. 1977. The experiment of artificial culture of embryo of early-maturing peach. *J. Northwest Agric. Coll.* New Series 39–44 (in Chinese).

Yang, Z., Hu, N., and Lu, G. 1983. Study on the embryo culture technique of early maturing peach. *J. Northwest Agric. Coll.* **1**:15–25 (in Chinese).

———, Hu, N., and Lu, G. 1984a. Study on shoot tip culture in tube cultured seedling. *J. Hortic.* **1**:7–13 (in Chinese, English abstract).

REFERENCES

Bowen, H. H. 1980. Earli Grandi peach. *Hortic. Sci.* **15**:207–208.

Brooks, H. J. and Hough, L. F. 1958. Vernalization studies with peach embryos. *Proc. Am. Hortic. Sci.* **71**:95–102.

Hammerschlag, F. 1980. Peach micropropagation. *Proc. Conf. Nursery Prod. Fruit Plants Tissue Cult. Appl. and Feasibility*, 48–52.

———. 1982. Factors affecting establishment and growth of peach shoot in vitro. *Hortic. Sci.* **17**:85–86.

Hesse, C. O. and Kester, D. E. 1955. Germination of embryos of *Prunus* related to degree of embryo development and method of handling. *Proc. Am. Soc. Hortic. Sci.*, p. 65.

Hu Shiyi. 1983. *Angiosperm Embryology*, pp. 196–203. Higher Education Press, Beijing (in Chinese).

Janick, J. and Moore, J. N. 1975. *Advances of Fruit Breeding*. Purdue University Press, West Lafayette, Ind.

Lammerts, W. E. 1942. Embryo culture an effective technique for shortening the breeding cycle of deciduous trees and increasing germination of hybrid seeds. *Am. J. Bot.* **29**:166–171.

Lesley, J. W. and Bonner, J. 1952. The development of normal peach seedlings from seeds of early maturing varieties. *Proc. Am. Soc. Hortic. Sci.* :60.

Morphology Group, Fifth Laboratory, Beijing Institute of Botany, Chinese Academy of Sciences, and Group of Pomology, Laboratory of Forestry, Beijing Institute of Agricultural Science. 1974. The breeding of early-maturing peach. *Bot. Mag.* **1**(4):23–25 (in Chinese).

Skirvin, R. M. and Chu, M. C. 1978a. Tissue culture of peach shoot tip. *Hortic. Sci.* **13**:29.

——— and Chu, M. C. 1978b. Tissue culture may revolutionize the production of peach shoots. *Illinois Res.* **19**:18–19.

———, Chu, M. C., and Rukan, H. 1981. An improved medium for the in vitro rooting of Harbrite peach. *Fruit Var. J.* **36**(1):15–17.

Smith, C. A. 1981. The cultivation method of some fruit tree seedlings from immature embryo. *Beijing Agric. Sci. Technol.* **1**:19–27.

Tabachnik, L. and Kester, D. E. 1977. Shoot culture for almond and almond-peach hybrid clones in vitro. *Hortic. Sci.* **12**:545–547.

Tukey, H. B. 1933. Growth of peach embryo in relation to growth of fruit and season of ripening. *Proc. Am. Soc. Hortic. Sci.* **30**:71.

Yang, Z., Wu, N., and Lu, G. M.. 1984b. Peach shoot tip culture in peach. *J. Northwest Agric. Coll.* **1**:13–18 (in Chinese).

Zhuang, E., Xu, Z., Wu, Y., and Cai, X. 1981. The breeding of extra-early peach. *J. Hortic.* **9**(3):1–8 (in Chinese, English abstracts).

Zhu, J., Wu, H., and Wang, Z. 1984. The young embryo culture technique in extra early maturing cultivars of peach. *Jiangsu Agric. Sci. Technol.* **5**: (in Chinese).

CHAPTER 20
Grape: Anther Culture

*Ziyi Cao**

INTRODUCTION

Grape is a perennial vine of the family Vitaceae. There are more than 70 grape species, 20 of which are cultivated. Grape vines are generally divided according to their geographical distributions and ecological traits into three groups as follows: Europe-Asia (vinifera or wine grape), North America (Labrusca), and East Asia.

Economic Significance

Grape vines have been cultured for 5000 to 7000 years and are widely cultivated for fresh fruit, processing, brewing, and canning. They are characterized by early fruit setting and are very adaptable. Grape is also an ornamental plant. Grapes are cultivated on 10 million ha worldwide and yield 65 million tons of fruit per year (Gui, 1984). In both cultivated land and yield, grape holds the first position among fruit trees. In recent years, the area under cultivation has increased about 10% per year in Japan and the United States; the increase is even higher in the developing countries.

*The author thanks Yushu Qi, dean of Department of Horticulture, Gansu Agricultural University, for his advice and assistance. English translation by Xiaxian Zhou.

Requirements for Grape Breeding

Grape has many varieties, now numbering more than 8000. Grapes are vegetatively propagated and highly heterozygous. Grape breeding requires wine varieties with high yield, disease resistance, and good quality; eating varieties should be seedless with large fruit, early maturity, disease resistance, and tolerance to storage and transportation (Wu, 1982; Gui, 1984). Biotechnology has great potential for grape improvement when used in conjunction with traditional methods of conventional breeding. Pure lines can be obtained by self-pollination, but the process requires at least 36 years. However, haploids can be obtained by anther culture, thereby producing pure lines after chromosome doubling. Haploids and diploid pure lines are important for mutation breeding, maximization of heterosis, and genetic studies (Wang, 1980; Hu, 1982; Chen, 1984). To date, no information on natural haploid grapes has been reported.

History of Grape Anther Culture

Mullins (1971) cultured grape anthers in an attempt to obtain embryos and plantlets without success. Gresshoff et al. (1974) obtained haploid callus by anther culture of Europe-Asia species but failed to differentiate plantlets. Hirabayashi et al. (1976) obtained budlike structures from *Vitis thunbergii*, with an unidentified ploidy level. Shinseky Hiroo (1976) subsequently found that the plantlets were diploids and did not survive transplantation (Krul et al., 1977). Zou et al. (1978) inoculated anthers from more than 10 varieties and obtained large amounts of anther-derived callus, but no differentiation was observed. Tajaske et al. (1979) reported that a large quantity of embryos and plantlets were obtained by anther culture of (*V. vinifera* × *V. rupestris*) hybrids, and haploid and diploid callus was observed. However, all the regenerated plantlets were diploids of unknown origin. We induced diploid plantlets from the wine variety Canepabh in 1978 and 1979 and succeeded in transplanting them in the field. The plants have blossomed and produced fruit. Three lines varying from each other in leaf shape, young branchlets, and isozymes were obtained. These were clearly different from the original variety. Further analyses of their origin are in progress (Cao et al., 1980; Qi et al., 1984). In 1979, Zou et al. inoculated a few anthers of the variety Llobema, using our induction and differentiation media, and haploid plantlets were induced (personal communication, 1979). They later inoculated anthers from 112 varieties, lines, and seedling plants, including the European species *Vitis vinifera* L., a hybrid of European and American species (*V. vinifera* × *V. Labrusca*,) and the hybrid *V. vinifera* × *V.*

amurensis. However, only 7, including Llobema, produced embryos and differentiated plantlets. The development of plantlets from anther-derived embryos was studied by Bao et al. (1981). Liu et al. (1982) also obtained large amounts of pollen-derived plantlets from anther culture of the variety Llobema. Rajasekaran et al. (1979, 1983) reported that the anther-derived plantlets they obtained were obviously different morphologically from the original variety. Segregation was evident in seedlings from self-pollinated progeny. It was, therefore, concluded that the anther-derived plantlets originated from somatic cells, not from pollen. Hirabayashi et al. (1982) inoculated anthers of 28 grape varieties, hybrids, and wild species. Embryos and plantlets were induced from 13 of them, and haploid and diploid cells were observed in callus, but all the differentiated plantlets were diploid.

GRAPE ANTHER CULTURE TECHNIQUES

Culture Procedure and Media

INDUCTION OF CALLUSING ANTHERS. B_5 medium was slightly modified for the anther culture of grape. The composition is shown in Table 1. The modified B_5 medium was used with a supplement of 2.3 μM 2,4-D, 8.8 μM BA, 88 mM sucrose, and 0.4–0.6% agar. Large quantities of callus were induced. Chromosome analysis showed that most of them were diploid calli and embryos (Cao et al., 1980). Zou et al. (1981) obtained haploid callus and embryoids from the variety Llobema by using the same induction medium and hormone components. Liu (1982) also

TABLE 1. Components of Modified B_5 Medium

Inorganic Elements		Organic components	
KNO_3	24.73 mM	Inositol	555 μM
$CaCl_2 \cdot 2H_2O$	1 mM	Thiamine-HCl	29.6 μM
$MgSO_4 \cdot 7H_2O$	1 mM	Pyriodoxine-HCl	4.8 μM
$(NH_4)_2SO_4$	1.14 mM	Nicotinic acid	8.2 μM
$NaH_2PO_4 \cdot H_2O$	1.06 mM	2,4-D	2.3 μM
KI	4.52 μM	BA	8.8 μM
H_3BO_3	48.4 μM	Sucrose	88 mM
$MnSO_4 \cdot H_2O$	44.8 μM	Agar	0.4–0.6%
$ZnSO_4 \cdot 7H_2O$	6.9 μM	pH	5.8–6.0
$Na_2MoO_4 \cdot 2H_2O$	1 μM	Distilled water	1000 (ml)
$CuSO_4 \cdot 5H_2O$	0.1 μM		
$CoCl_2 \cdot 6H_2O$	0.1 μM		
Fe salt (same as in H medium)			

obtained haploid callus and embryoid by using the same medium with 600 mg/l LH.

DIFFERENTIATION AND PROLIFERATION OF EMBRYOS. Embryos could be induced by using modified B_5 medium with the addition of 1.1 μM NAA and 2.2–17.6 μM BA, but better results were obtained if 17.6 μM BA was added with 58.7 mM sucrose, 0.6% agar, and pH 5.8 to 6.0 (Cao, 1980; Zou et al., 1981; Liu et al., 1982). B_5 medium used with the two plant hormones mentioned together with 600 mg/l LH was most effective for embryo differentiation.

MEDIUM FOR PLANTLET INDUCTION. B_5 medium supplemented with 2.2 μM BA, 600 mg/l LH, 58.7 mM sucrose, 0.6% agar, pH 5.8–6.0, was used by Cao et al. (1980), Zou et al. (1981), and Liu et al. (1982).

Different conclusions have been reached on the best conditions for grape anther culture by various authors using different methods. Hirabayashi et al. (1976), Cao et al. (1980), and Rajasekaran et al. (1979) all obtained plantlets after culture for 2 to 7 months; Rajasekaran et al. assumed that, just like zygotic embryos which need dormancy, plantlets would develop if the embryos were treated at a low temperature, 4°C. This low-temperature requirement could be replaced by GA_3 treatment. However, embryos were induced in about 1 month after culture by Zou et al. (1981) and by Liu et al. (1982). No requirement for low-temperature treatment was reported in the development of embryos to plantlets (Cao et al., 1980; Zou et al., 1981; Liu et al., 1982; Hirabayashi et al., 1982).

Explant Selection and Sterilization

EXPLANT SELECTION. Anthers at the tetrad to late uninucleate stage were used for culture. To determine the correct pollen development stage, anthers from flower buds located at different parts of the flower head were stained with acetocarmine or I_2-KI solution, then squashed and examined under the microscope. To save time, the optimal pollen development state for inoculation was determined from external morphological criteria: the small buds aggregated at spike had begun to separate; the ratio of crown and floral receptacle length was 2–4:1; and the crown and anther were green or whitish green. For most varieties, anthers collected from flowers with these three morphological characteristics and located on spring branches were suitable for inoculation. Zou et al. (1981) assumed that the induction frequency would be higher if pollen at the tetrad stage were used. However, the flower at that time is small and difficult to manipulate, and it is, therefore, easier to collect anthers at the mid- to late uninucleate stage for inoculation.

STERILIZATION AND INOCULATION. Flower buds were (1) taken from the field, wrapped with clean wet cotton gauze, and kept in a large-mouthed bottle; (2) quickly soaked with 70% ethanol for less than 10 sec; (3) sterilized with 1.0% mercuric chloride or bleaching powder solution for 5 to 8 min; (4) rinsed with aseptic water three times. Then (5) the degenerated crown was removed with small forceps, and anthers were taken with the forceps and inoculated into induction medium with 30 anthers per bottle.

CULTURE CONDITIONS. The temperature for anther dedifferentiation culture was 25–28°C. Calli and embryos were formed in greater numbers and more quickly in the dark than in light. Rajasekaran et al. (1979) demonstrated that the callus induction frequency for anthers treated at low temperature (4°C) and then cultured in the dark was 30.4%; the rate in light cultures was 3.4%. The induction frequency was only 0.3% in light and dark without the low-temperature treatment. Zou (1981) concluded that induction of dedifferentiation in the dark was better than under diffused light. Variation was seen in different genotypes; for instance, in varieties such as Llobema and Muscat Hamburg callus was formed from anthers under diffuse light, but this was not possible in other varieties.

Important Factors in Culture

MEDIA AND SUPPLEMENTS. A suitable medium and its supplements have been described previously. However, the literature contains a divergence of opinion on which medium conditions are best.

1. Concentration of inorganic salts: B_5 or modified medium has been used by most authors. Liu et al. (1982) suggested that callus induction frequency could be increased by increasing the concentration of nitrate. He increased the concentration of KNO_3 to 29.7 mM, decreased $(NH_4)_2$ SO_4 to 1 mM, and obtained better results than with the original B_5 medium. He also found that a better callus induction frequency was obtained when the level of sodium molybdate was reduced from 1 μM to 0.1 μM.
2. Concentration and combination of plant growth regulators: The kinds of plant hormone in dedifferentiation medium have been widely studied. Most people feel that BA in combination with 2,4-D is of prime importance for inducing anther dedifferentiation, and 2,4-D is necessary for embryo formation. Rajasekaran et al. (1979) induced anther-derived embryos and regenerated grape plantlets by using a combination of 1 μM BA and 5 μM 2,4-D. Hirabayashi et al. (1982) induced

Figure 1. Grape pollen plant.

diploid embryos and regenerated plantlets by anther culture from numerous grape species, demonstrating the important effect of 2,4-D on embryo induction by anther culture. We have confirmed that 2,4-D plays a critical role in inducing anther dedifferentiation. Low concentrations of 2,4-D (2.3 μM) and 8.8 μM BA are suitable for dedifferentiation of anthers. Zou et al. (1978) inoculated flowers located on summer branches of nine varieties of this medium. Anthers from the variety Llobema produced numerous embryos, and numerous haploid plantlets were obtained from anthers on this medium (Fig. 1).

The most effective differentiation medium was obtained by adding 1.1 μM NAA and 17.6 μM BA to B_5 medium. It was necessary to reduce or totally remove the plant hormones from the medium for further development of the embryos (Cao et al., 1980; Zou et al., 1981).

Liquid media were considered to have better effects than solid ones by Rajasekaran et al. (1979) and Zou et al. (1981).

EFFECT OF GENOTYPE ON INDUCTION FREQUENCY. Among 23 types of grape inoculated by Gresshoff et al. (1974), only Hei Ke Yin, Ge Lie La Si, and Js 23-46 gave undifferentiated haploid callus. Anthers of different grape varieties were cultivated by Hirabayashi et al. (1976); budlike structures were obtained only from *V. thunbergii*. The rates of callus formation and plantlet regeneration under different conditions from the cultivar Cabernet Sauvignon and hybrids of the ornamental male grape Glory were studied by Rajasekaran et al. (1979). Hermaphro-

TABLE 2. Effect of Genotypes and Sex in Flower on the Formation of Anther Callus

Materials	No. of Anthers Inoculated		No. of Calli Formed	
	Under Light	In Dark	Under Light	In Dark
Camperdown	125	125	10 (8%)	55 (44%)
Thornleigh	250	250	16 (6%)	24 (34%)
Cheltenham	250	250	1 (0.4%)	68 (27%)
Cabernet Sauvignon	100	100	0 (0%)	1 (1%)

Rajasekaran, K. and Mullins, M. G. 1979. Embryos and plantlets from cultured anthers of hybrid grapevines. *J. Exp. Bot.* **30**(116):399–407.

dite Cabernet Sauvignon flowers gave only a small amount of callus with no dedifferentiation and died. Differences in callus formation ability were found in three male hybrids of Glory, with Camperdown giving the highest induction frequency (Table 2). In 1978, anthers from 9 grape varieties were cultured, and large numbers of embryos and regenerated plantlets were induced only from the wine variety Canepabh (Zou et al., 1978). Anthers from another 9 varieties were seeded by Zou et al. in 1979, and callus was obtained only from anthers of the variety Llobema, which differentiated into embryo and haploid plantlets. They inoculated 70 anthers from variety Llobema; 12 calli with differentiation ability were induced with an induction frequency as high as 17.1%. Subsequently, anthers were inoculated from 112 varieties, strains, and seedlings, including Europe-Asia species, Europe-America hybrids, and a hybrid of *V. amurensis* Rupr. Plantlets were differentiated from anthers of 6 varieties: Llobema, Muscat of Alexandria, Gros Colman, Black Hamburg, and Muscat Hamburg (Zou et al., 1981). In 1980, anthers from 5 varieties were cultivated by Liu et al. Haploid anther-derived embryos and plantlets were obtained only from the variety Llobema (Liu et al., 1982). Anthers from some grape varieties and hybrids were cultivated by Rajasekaran et al. (1979), and embryos were obtained only from *V. vinifera* varieties. It was assumed that anther callus and embryo would be more easily produced from male flower plants than from female or hermaphrodite flower plants. Kim et al. (1981) inoculated anthers from 8 grape varieties; anther callus was induced but did not differentiate. Hirabayashi et al. (1982) inoculated anthers from 28 grape varieties, hybrids, and wild species, including three male hybrids (*V. thunbergii* and two intervarietal hybrids) and eight interspecific male hybrids (hybrids of rootstock and wild grape). Diploid embryos and plantlets were induced from 13 materials: Kosha Sanujak, Neo Muscat, Kaiji, Asker, Gray Reisling, Black Hamburg, Muscat Hamburg, Delaware, D x K 151, and three hybrids of *V. thunbergii*; reciprocal crosses of *V. thunbergii* ×

Muscat Bailey A and Delaware × *V. thunbergii*, as well as the wild species *V. thunbergii*. Grape varieties, strains, or hybrids for which embryo induction and plantlet regeneration have been reported are Llobema, Muscat Hamburg, *V. thunbergii*, Canepabh, Black Hamburg, Muscat of Alexandria, Gros Colman, Delaware Gray Reisling, Kosha Sanjaka, Neo Muscat, Kaiji, Asker, D x K 151, Glory male hybrid, and a hybrid from *V. thunbergii*. The pedigree of Muscat Alexandria is depicted in Table 3. This variety was found in the genotypes of Nos. 2 to 6; in the genotypes of Nos. 7 to 10, the pedigree of *V. thunbergii* was found. Inducibility is probably heritable. It may be possible to increase the utility of anther culture by culturing progeny of inducible genotypes crossed with noninducible genotypes with desirable traits. Different genotypes require different dedifferentiation and differentiation media, types and concentrations of growth regulators, as well as different culture conditions.

TRANSPLANTATION OF PLANTLETS AND CYTOLOGICAL INVESTIGATION

Transplantation of Anther- or Pollen-Derived Plantlets

Budlike structures were obtained from anther culture of *V. thunbergii* by Hirabayashi et al., and plantlets were formed but failed to survive planting (Krul and Worley, 1978). The death rate after transplantation of the regenerated plantlets induced by Rajasekaran (1979) was 75%. Later it was replaced by healthy shoots. Shoots with two nodes were dissected, transplanted to a mixture of 1 part peat to 3 parts sand, and survived. The procedure we used was the following:

1. When young plantlets developed three or four healthy leaves, the cotton plug was withdrawn, the culture vessel was covered with a flask under sterile conditions, and light intensity was increased to promote plant growth. The flask was removed when the young plantlets grew to the top of the bottle. They were then kept under light for hardening.
2. After 3 to 4 days, the young plantlets were taken out of the bottle, the medium adhering to the root was washed off, and the plantlets were kept in clean water for 3 to 5 days.
3. The young plants were transplanted into soil containing one third sand and kept under natural diffuse light.
4. The light intensity was gradually increased when new leaves appeared. The plants were transferred into the greenhouse after 7 days of hardening. Thirty-four out of the 43 plants transferred survived:

TABLE 3. Pedigree Relations of Inducible Genotypes in Anther Culture of *Vitis*

No.	Variety	Resource	Pedigree Relation	Reference
1.	Black Hamburg	Unknown		Zou Changjie et al., 1981
2.	Muscat of Alexandria	Unknown		Hirabayashi et al., 1982
3.	Muscat Hamburg	Black Hamburg × Muscat of Alexandria	Black Hamburg Muscat of Alexandria	Institute of Grape Brewage, Shandong Province, China, 1981
4.	Neo Muscat	Muscat of Alexandria × Jiachuan Sanda	Muscat of Alexandria	Hirabayashi et al., 1982
5.	Muscat Bailey	Muscat Bailey × Muscat Hamburg	Muscat of Alexandria Black Hamburg	Hirabayashi et al., 1982
6.	Llobema	Charkanski × Muscat Hamburg	Muscat of Alexandria Black Hamburg	Liu Pengchang et al., 1982
7.	*V. thunbergii*	Unknown		Hirabayashi et al., 1976
8.	*V. thunbergii* × Muscat Bailey		Muscat Hamburg *V. thunbergii*	Hirabayashi et al., 1982
9.	Muscat Bailey × *V. thunbergii*		Muscat Hamburg *V. thunbergii*	Hirabayashi et al., 1982
10.	Delaware × *V. thunbergii*		*V. thunbergii*	Hirabayashi et al., 1982
11.	Camperdown	*V. vinifera* × *V. rupestris*	Glory	Rajasekaran and Mullins, 1979
12.	Thornleigh	*V. vinifera* × *V. rupestris*	Glory	Rajasekaran and Mullins, 1979
13.	Cheltenham	*V. vinifera* × *V. rupestris*	Glory	Rajasekaran and Mullins, 1979

the survival rate reached 80.4% (Cao et al., 1980). Zou et al. (1981) reported that the grape pollen plantlets they induced survived just as well as those transplanted by traditional methods. Anther- or pollen-derived plantlets clearly can be transplanted easily.

Chromosome Number in Anther- or Pollen-Derived Plantlets

Anther callus and root tips were examined as follows: the material was fixed with Carnoy's fluid, stained with iron haematoxylin (Cao et al., 1980; Zou et al., 1981) or Schiff's reagent (Rajasekaran and Mullins, 1979), and then squashed. Plantlets induced by Hirabayashi et al. (1976) from *V. thunbergii* were diploids. Only the plantlets obtained through anther culture in grape by Zou et al. (1981) and Liu et al. (1982) from the variety Llobema were haploid. Rajasekaran et al. found that there were haploid and diploid cells in anther callus, although in the regenerated plantlets, all the cells were diploids. They demonstrated through self-pollination that these diploid plantlets were not from pollen, but from the somatic cells of anther wall tissue (Rajasekaran, 1979, 1982). Hirabayashi et al. (1982) seeded 28 specimens and obtained embryos and plantlets from 13 of them. Although haploid and diploid cells were observed, all the plantlets were diploids, and no haploid plantlets were seen.

Zou et al. (1981) examined the root tip cells of 30 previously regenerated plantlets. Although most cells were haploid, others were aneuploid, diploid, and polyploid, so that mixoploidy was prevalent. Liu et al. (1982) examined root tips from 36 plantlets and found that they were mostly haploids, less often diploids, and very rarely aneuploids. The ploidy of anther-derived grape plantlets appears to be unstable. Chromosome number in regenerated plantlets after subculture in Llobema was examined by Bao et al. (1981): 29 of the 48 cells examined at metaphase were haploids (64.4%), 2 cells were diploids, and 17 cells were aneuploids. Most of the aneuploid cells had 23 chromosomes, but a few had 17 chromosomes, indicating mixoploidy. The chromosomes of the haploid anther-derived plantlets obtained by Zou et al. doubled spontaneously.

PROSPECT

Haploid anther-derived grape plantlets have been induced by anther culture, and haploid embryonic cell strains have been obtained. This result is important for grape haploid breeding, selection of cellular mutations, application of heterosis, cell fusion, and genetic investigations. More strains should be examined and the induction frequency enhanced. It is necessary to intensify chromosome studies and investigation of anther-derived plantlets so that the origin of cultured plantlets can be identified. Grape is the most widely cultivated fruit tree, with numerous varieties and rich genetic resources. It is important to promote the induction of haploid materials, the study of androgenesis, and the development of anther-derived embryos. Media should be screened to increase the frequency of callus induction and differentiation and to extend the practical application of anther-derived plantlets.

KEY REFERENCES

Bao, X. Z., Zhang, J. Q., Yu, H. H., Zou, C. J., and Li, P. F. 1981. Study on embryogenesis of grape. *Acta Shandong Univ. (Nat. Sci.)* **3**:101–106 (in Chinese).

Cao, Z. Y., Gui, C. Y., Qi, Y. S., Ao, L. D. Wang, M. X., Hao, J. J., Qin, Y. F. 1980. Anther induced plantlets of grape. *Gansu Nongye Keji* **4**:26–29 (in Chinese).

———, Qi, Y. S., Guo, C. Y. 1979. Rapid propagation of grape in test tube. *Putao Keji* **4**:26–29 (in Chinese).

Liu, P. C., Chen, L. W., Zhao, Y. X. 1982. A study of in vitro anther culture in grape. *Acta Shangdong Teach. Univ.*, 1–5 (in Chinese).

Zou, C. J. and Li, P. F. 1981. Induction of pollen plants of grape (*Vitis vinifera* L.) *Acta Bot. Sin.* **23**(1):79–81 (in Chinese, English abstract).

REFERENCES

Chen, Z. H. 1984. Advances in anther culture of woody plants. *Hereditas* **6**(4):34–37 (in Chinese).

Gui, Q. C. 1984. Suggestions by the advanced countries in brewery production to the development of science and technology in vinous liquor production in China. *Putao Zaipei Yu Niangjiu* **1**:4–16 (in Chinese).

Gresshoff, P. M. and Doy, C. H. 1974. Derivation of a haploid cell lines from *Vitis vinifera* and the importance of the stage of meiotic development of anthers for haploid culture of this and other genera. *Z. Pflanzenphysiol.* **73**(2):132–141.

Hirabayashi, T., Kozaki, I., and Akihama, T. 1976. In vitro differentiation of shoots from anther callus in *Vitis*. *Hortic. Sci.* **11**(5):511–512.

——— and Akihama, T. 1982. In vitro embryogenesis and plant regeneration from the anther derived callus of *Vitis*. *Proc. Fifth Int. Cong. Plant Tissue Cell Cult.*, 547–548.

Hu, H. 1982. Advances in somatic cell genetic study in plants. *Plant Physiol. Commun.* **3**:1–7 (in Chinese).

Kim, S. K. and Paek, K. Y. 1981. Studies on anther culture of grapes: 1. Varietal differences in callus formation. *J. Korean Soc. Hortic. Sci.* **22**(2):86–91.

Rajasekaran, K. and Mullins, M. G. 1979. Embryos and plantlets from cultured anthers of hybrid grapevines. *J. Exp. Bot.* **30**(116):399–407.

——— and Mullins, M. G. 1983. The origin of embryos and plantlets from cultured anthers of hybrid grapevines. *Am. J. Eul. Vitic.* **34**(2):108–113.

Qi, Y. S., Cao, Z. Y., Wang, J. C., Luo, J. C., Zhang, S., and Guo, C. Y. 1983. A grapery established by the method of plant tissue culture—the test tube grapery. *Gansu Nongye Daxue Xiaokan* **6**:1 (in Chinese).

Wang, D. Y. 1980. Tissue culture of fruit trees. *Plant Physiol. Commun.* **5**:1–4 (in Chinese).

Wu, J. J. 1982. *Cultivation of Grape*. pp. 1–12. Liaoning People's Press.

Zou, C. J. and Li, P. F. 1978. Anther culture of grape. In: *Symposium on Anther Culture*. pp. 206–208. Science Press, Beijing (in Chinese).

CHAPTER 21
Grape: Micropropagation

Ziyi Cao

SIGNIFICANCE OF IN VITRO PROPAGATION OF GRAPE

Acceleration of Propagation

In vitro culture is faster than conventional methods for breeding new cultivars, propagating newly selected superior individual plants, and introducing rare cultivars. The propagation rate to increase a selected plant by using conventional methods of vegetative propagation is only 100 to 200 times. Propagation in vitro can achieve rates of several thousand to even 1 million times. Harris et al. (1982) obtained 12,000 plantlets from a grape vine stem segment 3–5 mm in length from the hybrid cultivar Baco within 4 months; 8000 grape plantlets were obtained via apex culture within 4 months by Barlass et al. (1978). Lin Peide et al. (1984) carried out shoot apex culture in vitro and obtained propagation rates of 10-fold in a month, eventually deriving a million plantlets from an initial plantlet. Cao et al. (1984) and Qi et al. (1985) obtained a 2.8–4.1-fold propagation rate in grape stem segment culture. Their calculations indicated that from a single plantlet, 230,000–2,250,000 plantlets might be reproduced in 1 year. Propagation by in vitro culture is not affected by the seasons and climate and results in plants that are free of diseases and insect pests. The method can be greatly shortened by the propagation cycle.

Virus Elimination

Shoot apex culture accompanied by heat treatment can eliminate viruses from cultured plantlets. Recently, the exchange and release of new cultivars have led to an increase in virus infection, leading to a reduction in yields and a decrease in grape quality. These viruses are difficult to control and are a severe problem in production. The viruses are characterized by an uneven distribution in the infected plant, so it is possible to exclude the virus by culturing the shoot apex, which is virus-free, for subsequent propagation of virus-free plantlets. Masahiko et al. (1982) have used shoot apex culture coupled with heat treatment for the grape cultivar Sewoji and obtained 80.4% virus-free plantlets, three times more than from heat treatment alone.

Advantages for Germ Plasm Preservation

There are over 8000 cultivars of grape. Maintaining such a quantity of germ plasm by traditional means requires considerable labor, time, and space. If plantlets are cultured in vitro, it is practical to keep quantities of germ plasm safely in a smaller space. Morel (1975) reported that 800 cultivars of grape could be cultured in a space of 2 m^3 at a temperature of 9°C for as long as 15 years by subculturing at intervals of 1 year. Preserving so many cultivars by conventional means would require at least 1 ha.

Use in Plant Physiology Research

It is possible to produce plantlets year round by in vitro culture and to perform investigations under controlled conditions, excluding disturbances from outside. Srinivasan et al. (1978) treated grape tendrils with BA to induce flowering; this treatment was of great convenience for work on the physiology of flowering.

Use in Pathology Research and Disease Control

Relationships between parasitic pathogens and their hosts have been investigated by Morel et al. (1944, 1947) in grape callus. They founded the "dual culture method," an effective technique for comparing the effects of different fungicides on control of downy mildew in grape and for identifying resistance.

REVIEW OF LITERATURE

Morel began in vitro culture of grape callus in 1944. Fallot (1955) and Pelet (1959) cultured grape vine, forming adventitious roots. In 1960, Morel successfully cultured apical meristems of grape and started investigations on in vitro low-temperature storage of grape. Further investigations on shoot apex culture using various media and culture conditions for virus elimination were carried out by Galzy (1969), Gilford et al. (1961), and Hoeffer et al. (1964). In 1973, Favre succeeded in inducing adventitious buds from grape leaves. In the same year (1973) Ottenwaelter et al. compared survival rates of vine cuttings in the greenhouse and field transplanted shoot apex plantlets. Survival rate of the latter was four times greater than that of the former. In 1976, Bini cultured grape vine apices but obtained very low survival rates except for two of the cultivars. Mullins and Srinivasan (1976) first succeeded in inducing embryogenesis in somatic cells of callus derived from grape nucellus. Srinivasan et al. (1980) then developed methods for high-frequency somatic embryogenesis. Krul and Worley (1977) obtained callus from explants of grape hybrids which differentiated adventitious embryos, resulting in the development of 65 plantlets per cubic centimeter of callus. The first orchard in the world of grape plants obtained through tissue culture was established in Maryland. In China, we also carried out in vitro culture of grape vine segments, and the first tissue cultured grape orchard in China was established in 1979. The transplanted grape plantlets in that orchard began to bear fruit in 1983 (Fig. 1) (Qi et al., 1983, 1985). Jona (1978) initiated shoot apex cultures of Sylvaner Reisling and obtained plantlets in 1978. Barlass and Skene (1978) cultured grape apices in liquid medium. Detailed investigations of the influence of growth regulators and of photoperiodism on the micropropagation of grape apices were carried out by Chee and Pool (1982) and Golodriga et al. (1982). No variation was found in the offspring derived through in vitro propagation, so the researchers suggested that in vitro culture might be used for rapid propagation of newly bred cultivars. Liu and Zhu (1984), Masahiko et al. (1982), and Burzytin et al. (1983) reported the results of research on shoot apex culture and in vitro propagation in grape. Many researchers suggested that the large-scale production of nursery stock of grape might be realized through in vitro culture (Chee and Pool, 1982; Cao et al., 1984; Qi et al., 1985).

Figure 1. The first tissue culture grape orchard.

PROTOCOL OF IN VITRO PROPAGATION

There were two pathways for achieving rapid propagation of grape in vitro: continuous multiplication of shoots and induction of embryos from callus.

Shoot Multiplication

Most research on rapid propagation through in vitro culture has aimed at multiplying apex or axillary buds, and good results have been achieved. This approach has a number of advantages: it is easy to achieve because the medium and the culture process are simple; there is little genotype effect, and most cultivars, hybrids, and lines can be readily propagated; genetic stability is maintained, and few plantlets are malformed; and transplantation is easy. The shortcomings of this approach are that the propagation rate is lower than that via the embryo culture pathway and that with cultured buds improvement at the cellular level is not possible. However, many researchers have favored bud multiplication for breeding superior selected plants and rare cultivars as well as for eliminating viruses.

Stem segments bearing a single bud or axillary buds with a leaf or petiole attached were used as explants. The cultures underwent root regeneration and shoot formation, formed plantlets, and could be repeated indefinitely. This method was first published by Galzy (1969), who named it "nodal cutting." The explant used in this method for in vitro rapid propagation is called an aseptic short stem (Luo, 1981). Farre et al. (1977) suggested that buds with leaves were the best material for obtaining plantlets in culture. Our experiments showed that the method by which single bud stem segments grew directly into plantlets had the following advantages: easy operation, applicability to many genotypes, healthy and strong plantlets which are easily transplanted, genetic stability, and a multiplication rate of 2–5.5 per month. Initially, this method gave a low rate of propagation, but its cycle was short and the number of the shoots accumulated through several culture cycles was equivalent to that of other methods. Twenty-seven cultivars were used as experimental materials, and all formed plantlets except 401 Ju Feng × Himrod (Qi et al., 1985). The axillary buds remaining on the lower part of the segment after shoots had been cut were able to regenerate even if the medium dehydrated. The total number of shoots regenerated from the initial explant in 2 months reached 9–15.

CULTURE PROCESS AND MEDIA. Single-bud stem segments were cultured as follows: Shoots were taken from plants grown in the field and cut into segments each with a single bud. After culturing, some of the shoots developed into intact plantlets and were transplanted into the field. Other shoots were used in successive subcultures for continuous multiplication. The action of disinfectant on the buds taken from the field caused the explants to have a low propagation rate when first cultured; explants taken from subsequent subcultures expressed a much faster propagation rate.

Segments with Single Buds Developed Directly into Plantlets. B_5 medium was used as basic medium to which some modifications were made; it was then named Gs medium. The following compositions were used with various grape lines: For most cultivars Gs-2 medium containing 1.1 μM IAA was used, but 2.9 μM IAA was more suitable for some cultivars, such as Jin Zao Jin. The frequency of plantlet formation and the increase in stem nodes are shown in Table 1. Gs-4 was used for several cultivars and lines; it contained 0.98 μM IBA and 148 μM adenine (Table 1). Gs-5 contained 0.09 μM BA, in addition to IBA (0.98 μM) and ADE (148 μM).

Subculture and Storage of Material. The cultured plantlets were kept in semihermetically sealed glass boxes at room temperature. Diffused illumination was provided. The cultures were transferred to freshly prepared medium at intervals of 5–8 months.

STERILIZATION AND INOCULATION OF EXPLANTS. The techniques of sterilization we described had the advantage that it could be

TABLE 1. Frequency of Plantlet Formation and Multiplication Rates

Medium	Cultivars	No. of Segments Inoculated	Plantlets Formed		Newly Regenerating Stem Nodes	
			No.	%	Total no.	Average No. per Segment ($\bar{x}$±SE)
Gs–2	Jin Zao Jin	148	148	100.0	725	4.90 ± 1.62
	Thompson seedless	132	111	84.1	451	3.42 ± 1.78
	Xnhaothbi	30	27	90.0	246	8.2 ± 2.6
	Thompson seedless (long-fruitery)	31	26	83.9	99	3.19 ± 1.54
	Pannoniariiace	96	86	89.6	288	3.00 ± 1.99
	Ziana	137	107	78.1	353	2.58 ± 0.73
	Black Olimpia	63	61	96.8	221	3.51 ± 1.18
	Benifuji	78	75	96.2	281	3.60 ± 1.12
	Pioneer	158	157	99.4	672	4.25 ± 1.48
	Grenache	30	24	80.0	71	2.37 ± 1.02
	Cabernet sauvignon	35	30	85.7	147	4.20 ± 0.41
	Riesling	40	34	85.0	176	4.40 ± 1.22
	Nohuk	35	28	80.0	135	3.86 ± 1.43
	401 Ju Feng × Himrod	8	0	0	4	0.50 ± 0.80
Gs–4	Monukka	78	75	96.2	257	3.29 ± 0.88
	Undefined germ plasm	60	59	98.3	222	3.70 ± 0.94
Gs–5	Kichmich noir	29	22	75.9	81	2.79 ± 1.54
	V_{70} anther plant line	75	74	98.6	283	3.77 ± 1.10
	Chouche	36	32	88.8	148	4.11 ± 1.20

used to obtain aseptic explants even from rare material. The procedures were as follows:

1. Vine cuttings were inserted into sandy beds, which were placed in a clean culture room for sprouting. Growing shoots were taken when they developed three or more nodes.
2. Young vines were used for explant material. The leaf blades were cut away, leaving the petiole. Then the prepared materials were placed in previously sterilized flasks.
3. Sterilized water was poured into the flask to soak and rinse the culture materials. The materials absorbed enough water to reduce the detrimental effect of disinfectants. This procedure was performed under sterile conditions.
4. The water was discarded and 0.1% mercuric chloride was poured on the explants (2% ethanol was added to the mercuric chloride solution if the tissue had trichomes). The flasks were vigorously shaken several times for 5–8 min.
5. The mercuric chloride solution was discarded, and the materials were rinsed three times with sterilized water.
6. The old cut surface of the material was cut away, and the remaining stem was cut into segments, each bearing a single bud. One segment was inserted upright into the medium in each culture tube or flask.
7. The culture was checked 7–10 days later, and contaminated materials were discarded.

For collecting explants from field-grown vines, young shoots were taken from the upper part of vines, were wrapped in wet gauze, and were taken back to the laboratory immediately. Explants were surface-sterilized by treatment with disinfectants in accordance with the procedures outlined previously. The materials which were taken from grapes grown at other places were sterilized locally and returned to the laboratory as soon as possible, then put into culture.

Since young vines were susceptible to ethanol, 70% ethanol treatment caused injury in 32–100% of explants. We therefore eliminated any use of alcohol in our experiments. Dormant vines were severely contaminated even after two sterilizations. The procedure described had only a 20.9% contamination rate for 301 cultures of various inoculated materials in the first 6 months of 1984. We have now obtained aseptic plantlets of rare cultivars, such as Pioneer, Red Fuji, and Black Olimpia, which were derived from just a few young vines or even from a single bud. Aseptic material could be obtained more easily without injury because sprouts growing from dormant vines in a sterile culture room contained fewer fungi and bacteria than plants from the field. The explants were treated with a highly effective disinfectant and then were repeatedly rinsed with sterilized water to buffer and reduce the effect of the disinfectant on the interior tissue of the explant. The cut plane of the segment was removed to ease the injury to the remaining segments. The inoculation of single buds into separate tubes prevented cross-infection of explants. We have also tried the sterilization method for dormant buds developed by Burzuin (1983) but have not obtained good results. The water culture method applied by Masahiko (1982) to shoot apex cultures to produce aseptic shoots was similar to the sand culture method described.

Bacterial and fungal contamination always occurred at the leaf axil or at the surface of explants in contact with the medium when sterilization was incomplete. In such cases, transferring the cultures or taking the upper buds from recently developed shoots and reculturing could save some contaminated materials. Contamination caused by *Bacillus*, however, continuously proliferates in successive subcultures, so the contaminated materials must be discarded as soon as possible. The sterilization method described was ineffective for materials with viruses within the tissues. These could be eliminated only by shoot apex culture combined with heat treatment.

For inoculation, the in vitro plantlets and aseptic shoots grown from sterilized materials were cut into segments with a single bud and leaf. The segments were inserted upright into the medium. Usually, three to five segments were inoculated per culture tube. The smaller apices with one leaf were cultured under incandescent lamps of 2000–3000 lux and kept at 25–29°C. They required 7–10 days for regenerating roots, 10–15 days for sprouting shoots, and 1 month for growing into plantlets with four to five leaves. The plantlets were used either for transplantation or for subsequent multiplication. This method was suitable for culturing rare materials and those in an early stage of propagation. To speed propagation, as soon as sufficient shoots had formed, some were taken for subculture and the remaining stem segments with one or two nodes were left in the culture tube. After another 15–20 days, a plantlet with one to three branches regenerated, and once again segments could be cut for the next culture until the medium dried out. We could perform three to four successive stem cuttings from a single cultured plantlet, raising the propagation rate significantly.

CRITICAL VARIABLES. *Genotypic Differences.* Different propagation coefficients were achieved for different cultivars, lines, and hybrids cultured on the same medium and under the same culture conditions. Some cultivars were adapted to a wide range of media and growth regulator concentrations, such as Jin Zao Jin. Other cultivars possessed little adaptability to medium and supplements, such as Red Fuji. Our 26 cultivars and lines gave good response on Gs medium with different growth regulator combinations, leading to increases in propagation of 2.4–4.5. Only the cultivar 401 Ju Fent had difficulty in developing into plantlets, so that its rate of propagation was less than 1 (Table 1). More than 20 media, including the hormone-free media, MS and B_5, were used with this cultivar, but they often caused callus formation. The cultures of the other 26 cultivars and lines developed shoots and roots simultaneously when cultured on suitable medium. They generally regenerated two to eight roots and four to five leaves in a month. Root regeneration and shoot growth were inhibited by callus formation on the base of the segment, which occurred if the medium was unsuitable. In some instances both the root and shoot failed to grow normally and even turned brown and died. Table 1 indicates that the cultivars which were easily propagated were adapted to a wide range of media. Cultivars with vigorous germination and branching capacity, such as Jin Zao Jin, propagated rapidly; cultivars with low rates of germination and weak branching capability, such as Seedless Black and Seedless Red, propagated at low rates.

Combination of Hormones. BA plays an important role in anther culture and adventitious embryogenesis (see Chapter 20), but cultured stem segments with single buds usually did not need BA or needed only very low concentrations (0.09 μM). A BA concentration of over 0.44 μM inhibited root regeneration, and when the concentration reached 2.2 μM, the cultures failed to regenerate roots (Table 2). Gs-3 medium was used as the basic medium and was supplemented with 0.09 μM BA. When the concentration of IBA was increased in this medium, the average number of regenerated roots per cultured stem segment increased. The nodal propagation rate and the plantlet regeneration rate were highest at 1 μM IBA; other concentrations, either lower or higher, decreased the plantlet-forming rate and the nodal

TABLE 2. Effects of BA on Growth of Stem Segments Bearing Single Buds*

		20 ± 1 days		30 ± 1 days	
Concentration of BA (μM)	No. of Stem Segments Inoculated	No. of Roots per Segment ($\bar{x}$ ± SE)	% of Rooting Segments	No. of Stem Nodes Formed ($\bar{x}$ ± SE)	Frequency of Plantlet Formation (%)
0	35	5.55 ± 1.55	83.3	2.83 ± 0.87	69.8
0.09	34	4.54 ± 1.47	100.0	3.75 ± 1.52	98.5
0.44	30	3.64 ± 1.68	95.6	3.54 ± 2.94	69.2
0.88	29	1.20 ± 0.52	83.3	0.86 ± 0.73	66.6
1.32	35	0	0	0.54 ± 0.50	43.0

*Experimental cultivar: V_{70} pollen plant; Gs-1 medium.

propagation rate (Table 3). Another experiment was carried out on the effects of adenine (ADE) at different concentrations, using the basic medium Gs-4 supplemented with 0.09 μM BA and 0.98 μM IBA. The best effects were found in medium supplemented with 74–300 μM ADE. ADE did not give better results than IBA and BA (Table 4) and at high concentrations caused a negative effect. Moreover, the latter two were of lower cost, so they can serve as a substitute for ADE.

Germ Plasm Storage. The length of time the material had been stored impacted the propagation rate (Table 5). As a result of loss of water, oxygen deficiency, and starvation, the stems and leaves of plantlets which were stored for too long often yellowed and dried, and the plantlet formation rate and multiplication rate after transfer were low. The longer the storage time, the greater these effects. Table 5 shows that the average

TABLE 3. Effects of IBA on Growth of Stem Segments Bearing Single Buds*

		20 ± 1 days		30 ± 1 days	
Concentration of IBA (μM)	No. of Stem Segments Inoculated	No. of Roots per Segment ($\bar{x}$ ± SE)	% of Rooting Segments	No. of Stem Nodes Formed ($\bar{x}$ ± SE)	Frequency of Plantlet Formation (%)
0	35	2.29 ± 0.90	79.2	1.91 ± 0.81	70.8
0.10	34	2.52 ± 1.20	100.0	2.52 ± 1.43	85.7
0.49	32	2.66 ± 1.02	83.3	2.02 ± 0.96	75.0
0.98	35	4.52 ± 1.13	100.0	3.75 ± 1.24	98.5
2.5	34	—	—	2.80 ± 1.43	66.6
4.9	29	5.75 ± 1.48	61.9	0.80 ± 0.82	37.3

*Experimental cultivar: V_{70} anther plant; Gs-3 medium.

TABLE 4. Effect of Adenine on Growth of Stem Segments Bearing Single Buds*

		20 ± 1 days		30 ± 1 days	
Adenine Concentration (μM)	No. of Stem Segments Inoculated	No. of Roots per Segment ($\bar{x}$ ± SE)	% of Rooting Segments	No. of Stem Nodes Formed ($\bar{x}$ ± SE)	Frequency of Plantlet Formation (%)
0	35	1.20 ± 0.98	66.6	1.22 ± 0.94	55.5
74	35	2.00 ± 1.20	100.0	2.33 ± 1.34	100.0
148	34	4.50 ± 1.84	100.0	3.75 ± 1.78	98.5
296	36	2.00 ± 1.56	90.0	2.80 ± 1.11	80.5
592	33	2.70 ± 1.22	80.0	0.80 ± 0.88	53.3

*Experimental cultivar: V_{70} pollen plant; Gs-4 medium.

TABLE 5. Effects of Preservation Time on Propagation Rate

Cultivars	Preserved for 5 Months			Preserved for 13 Months			Cultured Directly after Transfer		
	Total No. Observed	% of Plantlet Formation	No. of Stem Nodes Formed	Total No. Observed	% of Plantlet Formation	No. of Stem Nodes Formed	Total No. Observed	% of Plantlet Formation	No. of Stem Nodes Formed
C_{45} Zhengzhou	12	41.6	1.5	24	70.8	1.9	32	90.6	4.8
18-1-5	15	80.0	2.3	15	80.0	2.3	34	94.0	4.6
Chouche	32	81.3	1.8	24	62.5	0.8	33	72.7	2.0
Jin Zao Jin	17	52.9	1.8	3	0	0	31	48.2	1.4
Pearl of Csaba	12	91.6	2.6	13	23.1	1.2	31	84.4	2.7
Monukka	8	50.0	2.0	9	44.4	1.0	32	71.8	3.0
Average	16	66.2	2.0	14.6	46.8	1.2	32	76.9	2.6

rate of plantlet formation for six cultivars was 66.2% after 5 months' storage and that the number of newly developing nodes was multiplied 1.5–2.6 times. When the cultures were stored for 13 months, the average rate of plantlet formation decreased to 46.8%, and the nodal multiplication rate ranged from 0 to 2.3 times (average of 1.2). However, when the cultures were transferred to new medium after they were transferred once, the average rate of plantlet formation was raised to 76.9% and the node multiplication rate ranged from 1.4–4.8 (average of 3.0).

Somatic Embryogenesis

Somatic embryogenesis has the advantages of large-scale production, development of complete embryos, easy plantlet formation, and capacity for fast propagation. It may be the most rapid method for vegetative propagation as well as being the basis for artificial seed production (Mullins and Srinivasan et al., 1976; Luo, 1981). Table 6 lists examples of embryogenesis from different explants in grape. All diploid plants and somatic embryos derived from anther cultures are described in this table. Since Mullins's report in 1976, 30 grape cultivars, hybrids, and wild genotypes have been used for somatic embryogenesis. According to these reports, the induction of somatic embryogenesis in grape depends upon the following factors:

1. Genotype: In vitro somatic embryos were produced from plants of certain genotypes under precise culture conditions. Embryo induction occurs easily in those hybrids whose parent plants have genotypes which favor embryogenesis.
2. Explant source: Successful embryo induction was achieved with female sex organs, such as the unfertilized ovary and nucellus (Srinivasan, 1980; Mullins and Srinivasan, 1976), or male sexual organs, such as anthers (see Volume II, Chapter 15 of the Handbook); embryo induction from somatic cells was only rarely successful.
3. Hormones: 2,4-D was mainly used for callus induction, though NOA and NAA combined with BA were used for a few cultivars. After callus induction, 2,4-D was eliminated from the medium and NAA or NOA in combination with BA was added for embryo induction (Mullins and Srinivasan, 1976; Krul and Worley, 1977; Rajasekaran et al., 1979). Reduced or even eliminated levels of auxin and cytokinin resulted in further development of the embryos and plantlet formation. Somatic embryos have been used to produce planting material in oil palm (*Elaeis quineensis*) and celery (*Apium graveolens*). Investigation of somatic embryogenesis in grape should be continued to permit breakthroughs in propagation.

TABLE 6. Adventitious Embryogenesis via In Vitro Culture in Grape

Plant	Explants	Process of Embryogenesis	Authors
Cabernet sauvignon	Ovule	1. Callus (3 weeks) 2. Callus transfer (2 weeks) 3. Embryo 4. Plant formation	Mullins et al., 1976
Seyval	Stem, leaf, petiole, inflorescence	1. Callus 2. Embryo 3. Normal embryo growth	Krul et al., 1977
Male plant of hybrid Glory from *Vitis vinifera* × *V. ruperstris*	Anther	1. Anther callus (subculture 3 weeks) 2. Embryo (4°C, 4 weeks) 3. Plantlet	Rajasekaran et al., 1979
Savieravi	Anther	1. Callus (1 month) 2. Callus (4–6 months) 3. Embryo cluster (20 days) 4. Plantlet	Cao et al., 1980
Gros Conlman Black Hamburg Muscat Hamburg Muscat of Alexandria	Anther	1. Callus (1 month) 2. Callus (1–2 months) 3. Embryo cluster 4. Plantlet	Zhou et al., 1981
Koshu Sanzaka Neo Muscat Kaiji Askeri Riesling Black Hamburg	Anther	1. Callus (2–4 months) 2. Embryo 3. Plantlet	Hirabayashi et al., 1982
Muscat Hamburg Delaware *V. thunbergii* and 4 hybrids	Anther	1. Callus (2–4 months) 2. Embryo (1–2 weeks) 3. Plantlet	Hirabayashi et al., 1982
12 Cultivars and hybrids	Stem, leaf, inflorescence, and anther	1. Meristematic node 2. Embryo (one from 12 cultivars and hybrids)	Badea et al., 1982

PLANTLET TRANSPLANTATION AND GENETIC STABILITY

Essentials of Transplantation Technique

HARDENING OF PLANTLETS. Two steps for plantlet hardening were carried out. The first step was illumination hardening, which occurred when the cultures were moved from the culture room to the greenhouse. The stopper of the culture tube was removed, and the culture was hardened by exposure to sunshine at an intensity of 40,000 lux for 7–10 days. During this time the temperature of the culture tube had to be closely monitored or the plantlet would overheat by the excessive intensity of noon sunshine. Some of the shoot tips rose above the opening of the culture tube and were hardened in the arid air and intense sunshine, so that their stems gradually turned red and thick. Maintaining the cultures in clean conditions continued to be important for reducing contamination. The second step was sand culture. Plantlets were carefully removed from flasks and medium adhering to the roots was cleaned off. The cleaned plantlets were placed in a tray containing sand and vermiculite. The tray was covered tightly by a plastic cover to maintain humidity at 80%. After that, aeration was increased, temperature was lowered, and intensity of illumination was raised gradually over a period of 8–10 days. The plantlets undergoing hardening by sand culture were kept away from direct sunshine at noon. Through these two hardening steps, the stems and leaves became healthier and the plantlets produced newly regenerated roots with root hairs, laying the foundation for transplantation.

TRANSPLANTING PLANTLETS INTO POTS. The plants which were just hardened could not be transplanted directly into the open field. As an intermediate measure, they were first transplanted into pots containing nine parts of thick sand and one part of nutritive loam mixture. Plants were watered and the pot was covered with plastic which was removed the 10th day after transplanting. The proper soil moisture was maintained and the cultures were protected from the intense sunlight at noon. After growing in the pots for 1 to 2 months in the greenhouse, the plantlets had three to five new leaves and were strong enough to be transferred into the nursery bed or field.

TRANSFER TO NURSERY BED. Plantlets were planted in a ridged nursery bed in the afternoon, in double rows 80 cm apart and with 15-cm spacing between plants in the row. Prior to transplantation, the nursery bed was leveled and decomposed manure was applied as base manure. After transplantation, providing sufficient irrigation and leveling the bottom of the nursery bed were important for promoting plantlet survival. Appropriate field management, such as proper watering, cultivation, insect pest control, and manuring, should be stressed. At the later stages of plantlet growth, the growth was controlled by top pruning to promote maturity of the vine. Four to five months after transplanting, the plantlets grew into strong nursery stocks 30–50 cm in height and with stems 0.5–0.8 cm in diameter.

Characteristics and Genetic Stability of Plantlets

Seedlings from cuttings and from culture were observed for 4 years. The morphology of grapes derived from in vitro cultured plantlets and the shape, color, and taste of the fruit were the same as those of their original cultivar. No variations were found between cultured plantlets, indicating that their hereditary characteristics were stable. The micropropagated plantlets usually had shorter internodes and more branches, so it is necessary to investigate ways to increase the thickness of their vines. Recently, a reliable process for transplantation of in vitro cultured grape plantlets has been established by Wang et al. (Gansu University of Agriculture); 80% or more of plantlets survived after transplantation. Comparisons of the morphology of the stem leaf and fruit between micropropagated plantlets from shoot tip culture and seedlings from seed in the cultivar Rougeon were performed over two growing seasons by Chee et al. (1982). No differences were found, but the plantlets obtained from adventitious buds from callus often revealed distinct variation. The authors achieved rapid grape propagation by continuous multiplication through in vitro culture of apical and axillary buds. The plantlets were genetically stable. Barlass et al. (1982) cut stem apices bearing three to four leaf primodia into pieces and maintained them through successive subcultures for a few years. These cultures retained the capacity for regeneration. Morphological observations of the most recently formed plantlets showed no variation from original plants of the same cultivar.

FUTURE PROSPECTS

The propagation of grape through in vitro culture has clearly advanced and will eventually enter large-scale production, as many researchers have suggested (Chee et al., 1982; Cao et al., 1979; Mashiko et al., 1982). Recently, reliable procedures for production have been established, applied, and popularized. However, certain difficulties and problems remain, such as high cost and identification of viruses. The following investigation should, therefore, be pursued:

1. Industrialization of in vitro rapid propagation for newly bred single plants, rare cultivars, and cultivars, lines, and grafting stock which are difficult to breed by conventional techniques. Simple and rapid processes suitable for each cultivar should be established.
2. Virus elimination by a combination of heat treatment and rapid propagation through in vitro culture of stem apices. Masahiko et al. (1982) obtained intact plantlets through in vitro culture of apices taken from a virus-infected plant of cultivar Zewoji. Of the plantlets 80.4% were

virus-free, a percentage much higher than found for those that underwent heat treatment (27.3%). In that experiment, shoot apices 0.2–0.3 mm long were taken from plants infected by grape vine leaf roll virus and cultured on an aseptic agar medium for 4 months, with four changes in medium components.

3. Induction of somatic embryos and research on artificial seeds.
4. Research on embryonic cell lines for mutagenesis and selection of mutants at the cellular level, protoplast culture, cell hybridization, and other advanced biotechnological techniques.

KEY REFERENCES

Barlass, M. and Skene, K. G. M. 1978. In vitro propagation of grape vine (*Vitis vinifera* L.) from fragmented shoot apices. *Vitis* **17**:335–340.

Cao, Z., Qi, Y., and Guo, C., 1984. In vitro culture of stem apex in grape. *Chin. Pomic*. **4**:26–29 (in Chinese).

Liu, P., and Zhu, L. 1984. In vitro culture of stem apex in grape. *Chin. Pomic*. **2**:43–46 (in Chinese).

Qi, Y., Cao, Z., Wang, J., and Luo, J. 1985. Technique of fast propagation through in vitro culture in grape. *Commun. Agric. Sci. Technol*. **8**:14–15 (in Chinese).

Skene, K. G. M. and Barlass, M. 1982. Micropropagation of grape vine. *Comb. Proc. Int. Plant Propag. Soc*. **30**:564–570.

REFERENCES

Barlass, M., Skene, K. G. M., Woodham, R. C., and Krake, L. K. 1982. Regeneration of virus-free grape vines using in vitro apical culture. *Ann. Appl. Biol*. **101**:291–295.

Bini, G. 1976. Trials on the in vitro culture of apical meristems of *Vitis vinifera. Rivista delia Oroflorofrutticolt. Ital*. **60**(5):289–296.

Chee, R. and Pool, R. M. 1982. The effects of growth substances and photoperiod on development of shoot apices of *Vitis* cultured in vitro. *Sci. Hortic*. **16**:17–27.

Favre, J. M. 1977. Preliminary results on obtaining spontaneous in vitro shoot production in the grape vine. *Ann. Amelior. Plantes* **27**:151–169.

Galzy, K. 1969. Research on the growth of both fan leaf infected and healthy *Vitis rupestris* Scheele cultivated *in vitro* at different temperature. *Ann. Phytopathol*. **1**:149.

Golodriga, R. Y., Zienko, V. A., Butenko, R. G., and Levenko, A. B. 1982. Rapid reproduction of useful grape genotypes. *Subdivision* **3**:24–27.

Guo, Q. 1984. Suggestions on the basis of reference materials of developed countries for developments of science and technology in grape vine of China. *Grape Cultiv. Wine Making* **1**:4–6 (in Chinese).

Jona, K. 1978. Callus and axillary bud culture of *Vitis vinifera* "Sylveuer Riesling." *Sci. Hortic*. **9**:55.

Krul, W. R. and Worley, J. F. 1977. Formation of adventitious embryos in callus cultures of "Seyval": A French hybrid grape. *J. Am. Soc. Hortic. Sci.* **102**:360–363.

Luo, S. 1981. Advancement and application of researches on plant tissue and cell in vitro. *Acta Physiol. Planta* **4**:91–92 (in Chinese).

Mao, M., Tien, H., and Lu, C. 1984. Studies on technology of propagation through tube culturing plantlet in grape. *Shihezi Agric. Coll.* **1**:31–40 (in Chinese).

Masahiko, I., Hideo, S. T., and Katsuyasu, U. 1982. Elimination of grape vine viruses by meristem tip culture. In: *Plant Tissue Culture 1982.* (A. Fujiwara, ed.), pp. 807–808. Japanese Association of Plant Tissue Culture, Tokyo.

Monette, P. L. 1983. Subdivision of proliferating grape vine explants using shoots: Influence of shoot size on proliferation. *Plant Propag.* **28**:13–15.

Morel, G. 1944. Le développement de mildion sur des tissus de vigue cultures in vitro. *Acad. Sci. Paris* **281**:50–52.

———. 1975. Meristem culture techniques for the long-term storage of cultivated plants. In: *Crop Genetics Resources for Today and Tomorrow*. pp. 327–332. Cambridge University Press, Cambridge.

Mullins, M. G. and Srinivasan, C. 1976. Somatic embryos and plantlets from an ancient clone of the grape vine. (cv. Cabernet Sauvignon) by apomixis in vitro. *J. Exp. Bot.* **27**:1022–1030.

Pool, R. M. 1975. The influence of cytokinin on in vitro development of "Concord" grape. *J. Am. Soc. Hortic. Sci.* **100**:200–202.

Rajasekaran, K. and Mullins, M. G. 1979. Embryos and plantlets from cultured anthers of hybrid grape vines. *J. Exp. Bot.* **30**:399–407.

Zhang, X. 1985. Fast propagation in grape cultivar "Kun Yu." *Yunnan J. Bot. Res.* **7**:245–247 (in Chinese).

CHAPTER 22
Hawthorn

Jingshan Wang and *Guang Yu**

INTRODUCTION

Hawthorn, *Crataegus pinnatifida* Bunge, has been cultivated in China for over 3000 years. In recent years, with further understanding of its economic value and especially its medical importance, the demand for hawthorn fruits has greatly increased. Hawthorn cultivation has become one way for Chinese farmers to supplement their income.

Economic Importance

The fruits of *C. pinnatifida* are rich in nutrients. The content of vitamin C, vitamin B, and carotene is 16, 4, and 9 times higher, respectively, than that of apple (Yang et al., 1981). The fruits can be eaten fresh or processed into dozens of specialty foods.

Hawthorn has aroused increasing attention from medical workers for its role in aiding digestion, increasing blood circulation, and curing blood disorders. It has been reported that the fruit is a cardiotonic medicine that inhibits arrhythmia and reduces lipemia, high blood pressure, and glutamic-pyruvic transaminase (Liu Xingzhi and Liu Bin, 1980). It is also

*English translation by Kaiwen Yuan.

a rich source of flavanoids, potential anticancer substances. The total content of flavone in the variety Mianzha is 0.78% by dry weight.

The seeds of *C. pinnatifida* can be ground for medicine, the leaves for animal feed, and the pectin for food processing. The wood is good material for carving. Hawthorn is also an ornamental tree species.

Geographical Distribution

Crataegus pinnatifida Bunge belongs to the family Rosaceae. The genus includes more than 1000 species, widely distributed in the Northern Hemisphere. Many species are grown in North America. In China, about 16 *Crataegus* species are found in the area between 10 and 50 degrees north latitude (Yu, 1979). The cultivated species in China is *C. pinnatifida* var. *major* Br. In recent years, a number of new and useful varieties have been discovered. A cold-resistant variety, Dawang, has been found in northeast China's Jilin Province. It grows vigorously and is resistant to −40°C. This discovery has made it possible to cultivate *C. pinnatifida* in Heilingjiang Province, extending its range over 300 km northward (Hebei Agricultural University, 1980). *C. pinnatifida* varieties with pink or purple-red colored fruit have been discovered in Liaoning, Hebei, Shanxi, and other provinces. The fruits of these varieties are hard; their processed products have beautiful color and do not dehisce. The wild dwarf *C. pinnatifida* species found in the provinces of Henan and Fubei are good parental materials for breeding of dwarf varieties and for close planting. The variety Altai growing in the Altai Mountains is cold-resistant and high-yielding. Its seeds have small kernels and can germinate the first year after formation. It is both a stock wood and a valuable breeding material (Xu, 1984).

Problems in Production

At present, seedling-stock grafting is used to raise seedlings for hawthorn cultivation. Since the seeds contain few kernels and have thick seed coats, they germinate after stratification for two winters. Grafts can only be performed when seedlings mature; raising a seedling usually takes 3 to 4 years, sometimes even 4 or 5 years. To produce a large number of seedlings, the technique requires great quantities of scions, the collection of which inevitably affects the yield of the mother tree. Despite these barriers, a number of valuable newly discovered varieties need to be released in production. Therefore, techniques for rapid production of large quantities of superior seedlings need to be developed for hawthorn cultivation.

LITERATURE REVIEW

Tissue culture work with perennial crops has concentrated on fast propagation of superior clones and seed embryo culture. Great progress has been achieved in recent years (Wang and Zhang, 1975; Horiuchi et al., 1976; Mu et al., 1977; Yang et al., 1982).

Progress has also been made in tissue culture of *C. pinnatifida*. Yu et al. (1981) first regenerated plantlets from shoot tips taken from mature trees and developed many promising clones. They began to study production methods for test tube plants in 1982 and have now established a workable system which will soon be used in practical production.

Young hawthorn embryos have been cultured successfully (Wang et al., 1981, 1982). There are now embryo-derived plants 4 years old growing in the field. Embryo culture has been used to develop green-shoot regeneration systems and also to induce embryos; these uses are significant for both practical applications and theoretical studies.

Wang Jinxuan (1982) and Wang Yuying (1983) cultured mature hawthorn embryos to accelerate propagation of seedling stocks successfully. Hu et al. (1984) developed a technique of rapid plant regeneration, which unfortunately is too expensive at present for commercial use.

SHOOT TIP CULTURE AND RAPID PROPAGATION

Shoot tip culture of *C. pinnatifida* produces self-rooting plants (Table 1). The procedure involves development of superior clones, proliferation of rootless shoots, root initiation, and transplantation.

Development of Superior Clones

CULTURE PROCEDURE. *Explants.* One- or two-year-old healthy shoots were taken from superior mature mother trees. Materials should be collected in winter and spring, for at this period buds are well developed and can be readily regenerated.

Inoculation. The shoots were cut into small segments, each including a bud; soaked in 70% alcohol for 30 sec; immersed in 0.1% mercuric chloride solution for 10 min; and rinsed three or four times with aseptic water. The bracteal leaves were removed under aseptic conditions. The apical point (about 1 mm in size) was picked out and placed onto the medium.

Incubation Conditions. Inoculated explants were incubated in a room with a temperature of 22–28°C. Natural illumination was provided for 8 to 12 hr/day. The light intensity was about 1000 to 3000 lux.

TABLE 1. Comparison of Conventional and In Vitro Propagation of Hawthorn

Conventional Propagation			In Vitro Propagation
	1st year	**1st year**	
1. Collection of wild fruits	Sept.	Oct.	1. Collection of superior scion
2. Removal of fruit flesh	Sept.	Oct.	2. Shoot tip culture
3. Shell cracking treatment*	Sept.	Dec.	3. Regeneration of plantlet
4. Stratification	Oct. to April, 2nd year	Dec.	4. Clonal propagation†
	2nd year	**2nd year**	
5. Sowing	April	Jan.	5. Production of rootless shoots
6. Growth in nursery	May to Oct.	Jan.	6. Root initiation
7. Overwintering	Nov. to March, 3rd year	Feb.	7. Transplantation
	3rd year		
8. Continued growth	April to July	March to April	8. Growth after transplantation
9. Grafting	July to Aug.	May	9. Transfer to nursery
10. Survival after grafting	Aug. to Oct.	May to Oct.	10. Growth in nursery
11. Overwintering	Nov. to March, 4th year	Nov. to March, 3rd year	11. Overwintering
12. Continued growth	April to Oct.	April to Oct.	12. Continued growth
13. Plants ready for sale	Nov., 4th year	Nov., 3rd year	13. Plants ready for sale

*One more year is required for storage if the treatment is not performed.
†After establishment of clones, plantlets can be produced year round in greenhouses.

Proliferation. When the cultured shoot tips had expanded leaves, they were subcultured once. If the original medium was found unfavorable for subculture, it was modified and supplemented with appropriate additives. The regenerated shoots were excised and transferred to fresh medium for proliferation when they were 2 to 3 cm long or had differentiated many buds.

MEDIUM. MS medium was used as the basic medium, supplemented with 2.2 μM BA. Occasionally, 2.3 μM KIN or 0.06–0.57 μM IAA was supplemented as well. Other components included 88 mM sucrose and 0.75–8% agar. The pH of the medium was 5.4 to 5.8. The prepared medium was sterilized under 1.1 kg/cm^2 pressure for 10 min.

MORPHOGENESIS. Shoots were regenerated through direct bud differentiation from cultured shoot tips. Bud differentiation was regulated by cytokinins and was associated with callus formation at the basal parts of inoculated shoot tips.

Callus Formation. Cytokinins such as BA, KIN, and ZEA could stimulate callus formation from inoculated shoot tips, but their effects were different. Concentrations of 2.2 and 22 μM of ZEA, BA, and KIN were tested. The frequencies of callus formation were 61.5% and 100% for ZEA, 35% and 76.2% for BA, and 5% and 7.7% for KIN, respectively. Auxins could also induce callus development. The results varied with the types of auxin and their concentrations. In most cases, 2,4-D was the most effective for callus induction from shoot tip explants, with progressively decreasing effects with NAA, IBA, and IAA.

Bud Differentiation. Bud differentiation of *C. pinnatifida* shoot tips could be regulated and controlled by addition of cytokinins. In the presence of cytokinin, shoot tips could differentiate without any other hormones. Different types of auxin or concentrations of the same auxin had distinct effects on bud differentiation. When the concentration of added auxins was 2.2 μM, BA was best for bud differentiation with a frequency of 70%; ZEA gave 38.5% and KIN gave only 10%. A concentration of 8.8 μM BA or 9.3 μM KIN gave bud differentiation frequencies of 35% and 33.3%, respectively; 76.9% and 23.8% differentiation were achieved with 23 μM KIN and 22 μM BA. Shoot tips did not differentiate at all with either 9.1 μM or 23 μM ZEA. In most cases, many buds could be induced even with low concentrations of BA, but this inhibited the elongation of internodes and affected plantlet development. KIN and ZEA were not very effective for differentiation of multiple buds, but they promoted internode elongation and leaf expansion, making the shoots strong and healthy. The best results were often achieved with a combination of BA and KIN in the medium.

Relationship between Shoot Tip Differentiation and Callus Formation. The inoculated shoot tip died after a period of incubation if little or no callus developed from the base of the shoot tip. If the shoot tip was moderately callused (the callus was similar in size to the vertical section of shoot), it underwent normal differentiation. If the basal part of the shoot tip had overcallused so that the callus was much larger than the vertical section of shoot, the shoot tip developed completely into callus. Buds were difficult to regenerate if the whole shoot tip callused.

To get better results from bud regeneration, suitable hormones and their concentrations, especially cytokinin, should be selected. Too low a concentration of cytokinin may result in no callus formation at the basal part of explants and poor bud differentiation, but if the concentration is too high, too much callus growth results. Therefore, the medium should be modified according to the response of the explants. The appropriate addition of auxin may also be helpful to the culture's growth.

We have used these methods to regenerate plants from shoot tip explants of mature trees in many improved *C. pinnatifida* varieties (e.g.

Dawang, Xifeng-hong, Liaohong, Dajinxing, and Fulihong) and have achieved high regeneration frequencies. Good results also have been obtained in other species using shoot tip explants.

After plantlets are regenerated, a superior clone can be established by further proliferation. To provide a superior clone for production, it is best to have several hundred or even 1000 tubes, each containing 10 plantlets.

Shoot Proliferation

When a superior hawthorn clone was developed, rootless shoots could be continuously produced.

CULTURE PROCEDURE. *Manipulation Techniques.* Shoots taken from established clones were cut into single buds or small segments with buds and placed in differentiation medium. After a period of incubation, the developed buds were separately excised and placed in induction medium to regenerate shoots. The part remaining was transferred into fresh differentiation medium to induce additional shoots. The plantlet regeneration cycle could be repeated. Another method was suitable for plantlet proliferation. When buds cultured in the induction medium grew to 4 to 5 cm long, upper parts of the shoots were cut at a height of 2 to 3 cm for root initiation; the lower parts with buds were placed into the original or fresh medium to induce lateral buds. When the buds grew up, they were cut into two parts and cultured again. Differentiated buds or stem segments with buds could also be repeatedly proliferated in an integrated medium which supported growth and differentiation.

Media. The media adopted for proliferation were similar to those used for shoot tip culture: differentiation medium, growth medium, and integrated medium. In differentiation medium, the concentration of cytokinin (BA) was somewhat higher, usually 4.4 μM, plus an appropriate concentration of auxin. Growth medium contained a lower concentration of cytokinin, about 0.44–2.2 μM, with an appropriate quantity of auxin and GA. The integrated medium was basically intermediate between the other two, but often without GA. Hormones and their concentrations in media were varied, depending on the varieties inoculated. In most cases, good results could be achieved with the integrated medium alone.

Incubation Conditions. The temperature of the incubation room was 18–28°C illuminated 8 to 12 hr/day at a light intensity of 1000 to 3000 lux.

DIFFERENTIATION AND GROWTH OF PLANTLETS. The differentiation of test tube shoots of *C. pinnatifida* is mainly controlled by cytokinin; the growth of buds is determined by both cytokinin and auxin. Differentiation and growth should be balanced: differentiation of too many buds will adversely affect growth, and if buds overgrow, differentiation will be inhibited. The cytokinins used in the differentiation medium discussed previously can stimulate differentiation of lateral buds. Differentiation of secondary lateral buds can occur, so that new lateral buds are again differentiated from the original ones, forming a cluster of buds. The buds should be transferred to growth medium to regenerate rootless shoots. In the integrated medium, good differentiation of lateral buds (without differentiation of secondary lateral buds) can be obtained, and lateral buds can also grow without excision of apices. Therefore, use of integrated medium is a good method for regeneration of rootless shoots. The effects of different plant hormones on differentiation and growth of plantlets are as follows:

BA: A low concentration promoted bud differentiation, but increased concentration (4.4–8.8 μM) produced clusters of buds and inhibited internode elongation.

KIN: A high concentration (9.3–23 μM) stimulated bud differentiation, but bud clusters were not easily produced. However, accelerated internode elongation, leaf expansion, and plantlet formation resulted.

ZEA: A low concentration helped bud differentiation, but not as effectively as BA. Its effect on bud growth was similar to that of KIN.

NAA, IAA, IBA, and GA had no direct effect on bud differentiation, but appropriate concentrations all promoted shoot growth.

Incubation conditions also affected differentiation and growth of shoots. When the temperature was high, growth was faster, but relatively fewer buds differentiated. When the temperature was low, the shoots grew slowly, but more buds could be differentiated. Similarly, low light intensity promoted internode elongation, but the shoots were rather weak. When high light intensity was used, the internodes elongated slowly, but the stems were strong, leaves were expanded, and strong shoots were formed.

Root Initiation

Roots were difficult to initiate from test tube plantlets of *C. pinnatifida*. The achievement of a high rooting frequency is most important in production of intact test tube plants.

ROOTING IN TEST TUBES. *Manipulation Method.* The plantlets used for rooting were 2 cm long with expanded deep green leaves. The plantlets were placed vertically in the rooting medium under aseptic conditions.

Medium. Half-strength MS medium with 4.9 μM IBA, 44 mM sucrose, and 0.75% agar was used for root initiation. The pH was 5.4.

Incubation Conditions. The temperature was 22–29°C. Illumination was provided 10 to 12 hr/day at a light intensity of 3000 to 5000 lux.

Culture Differences. Differences in root initiation under the same conditions have been observed among different materials. For instance, the bud clone Kai 1 from the variety Kaiyuan had a rather high rooting frequency of 62.5% (178/285), although only 21.3% (68/319) rooting frequency was obtained in another clone, 933, from the variety Nailin.

Differences also existed in the speed of root development, size of callus, response to medium, and incubation environment. For example, in a bud clone of the variety Bei 8, only a very low concentration of auxin was required for root initiation.

Auxins. Both IBA and IAA could stimulate root initiation in *C. pinnatifida*. For most varieties, suitable concentrations of IBA were 2.5–9.8 μM. Within this range, a slightly higher concentration favored differentiation of root primordium but inhibited root elongation. In general, it was better to use 5.7 to 8.5 μM. For IAA, effective concentrations were 5.7 to 16.7 μM, with 11 μM the standard concentration. The effects of different concentrations of IBA and IAA on root initiation are summarized in Fig. 1.

Sensitive Period in Root Initiation. The lower part of shoots transferred into rooting medium began to turn red on the second to third day under illumination. No red color appeared when the light intensity was too low. On the third to fourth day, the base of the stem began to swell. On the fourth to fifth day, the swollen part of the stem split open and transparent, loose callus was visible within. On the sixth to seventh day, red spots, which were visible root primordia, appeared on the calli. If rooting conditions were favorable, the red spots soon developed into white roots; if not, calli multiplied and the red spots did not develop further. In that case, new roots slowly developed after another period of culture, or no roots were induced at all. If the developed roots were inhibited for 20 days on the original rooting medium, the root tips turned red, yellow, brown, or even black. The time required for differentiation and growth of roots varied with materials, media, and incubation conditions.

The critical phase in root differentiation is the preparatory period before the root's initial cells undergo their first division (on the second to fourth day after inoculation of the plantlet). During this period, the cultures are very sensitive to temperature, light, and other factors: poor incubation conditions directly reduce rooting frequency.

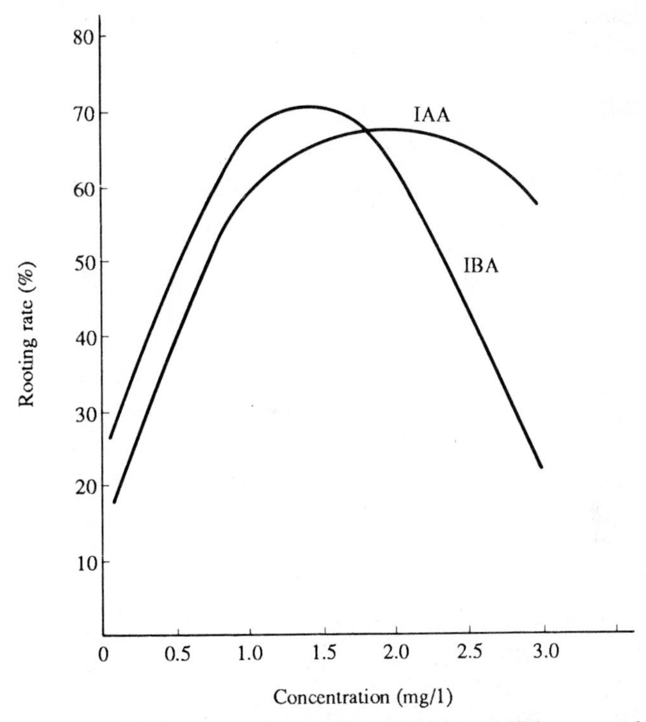

Figure 1. Effects of different concentrations of IAA and IBA on root initiation.

Temperature and light are of prime importance for root initiation. A day-night temperature fluctuation should be used. The day temperature should be controlled at 27 ± 2°C and night temperature at 24 ± 2°C. Night temperatures lower than 18°C seriously affected the rooting. In general, light intensity should be about 4000 lux with an appropriate range of temperature; if the temperature is low, light intensity should be higher and vice versa.

Inhibition of Root Growth. When rooted plantlets continued to be cultured in the original medium, the growth of roots was inhibited. When the inhibited plantlets were transplanted into soil, they showed little growth and the survival rate dropped. The most effective way to ensure good development was to transplant plantlets while their roots were growing vigorously.

A high concentration of auxin was needed for the initiation of root primodium; for root development, a low auxin concentration was required. The effects of auxin on root development in 22-day-old rooted plantlets were determined in three different media: the original medium with 11 μM IAA, medium with low auxin concentration, and one without hormone. Root elongation continued to decrease in the original medium but was slightly promoted in the medium without hormone; the medium with low auxin concentration not only stimulated the root elongation but also raised the total rooting frequency. This occurred because low auxin concentrations encouraged the development of root primordia. To promote root development, the plantlets were transferred to a medium with a low auxin concentration.

ROOTING IN VESSELS. *Manipulation Techniques.* After light hardening, the rootless plants were cut at the base of the stems and placed in a substrate of river sand containing 2.5 μM IBA (or NAA). The containers used for root initiation were glass vessels or 200-ml flasks containing 1 to 2 cm of substrate saturated with auxin solution. One vessel contained 20 to 30 plants covered with plastic which was perforated for aeration. The plants were sprayed with water every day to compensate for evaporation until they were rooted.

Roots can also be initiated in a rooting bed, but the plantlets should be treated with a high concentration of auxin. The rooting temperature in the daytime was 27 ± 2°C, and at night 24 ± 2°C. Natural light was provided.

Effect of Auxin on Root Initiation. When concentrations of 2.7 μM NAA and 2.5 μM IBA were used with the variety Bei 8, the rooting frequency was above 80%. If IAA was used, a higher concentration was required. Similarly, root development varied with types of auxin. With NAA, white calli were first formed at the base of the stem, and then roots grew out slowly through the calli. In the treatment with IBA, calli appeared on some of the plantlets; in others, roots developed directly from the base of stems without the formation of callus. With IAA, very few plantlets formed calli. Instead protuberances first appeared at the stem base, then roots broke through the epidermis and developed. Although different auxins appear to have similar functions in root initiation, their mechanisms are different.

Different concentrations of the same type of auxin had different effects in root initiation. The auxin concentrations needed for rooting in vessels had a wider range than for test tube rooting.

COMPARISON OF THE TWO ROOTING METHODS. Root development was determined 5 days after transplantation in plants which had been rooted in test tubes and in flasks (Table 2). Roots were initiated rapidly in test tubes, usually within 7 to 10 days. The rooted plantlets could be transplanted in about 20 days. However, the incubation environment had to be strictly controlled. Changes of temperature, light, etc., could easily affect rooting or even prevent it. In vessels or flasks, root differentiation occurred 15 to 30 days, resulting in a higher rooting frequency and a greater average root length, because the incubation conditions were less strict.

TABLE 2. Performance of Plantlets Rooted in Test Tubes or Flasks, 5 Days after Transplantation

Rooting	Increase in Average Number of Roots per Plantlet	Average Root Length (cm)	Percentage of Plantlets Showing Root Growth
Plantlets rooted in flasks	2.6	4.14	100
Plantlets rooted in test tube	0	0.05	10

The environmental conditions provided for vessel rooting were similar to that for plantlet transplantation. The plantlets rooted in vessels could be transplanted into soil while their roots were short, and they continued to develop new roots. As their growing conditions will be greatly changed, test tube rooted plantlets should be hardened before transplantation.

In summary, rooting in vessels is superior to rooting in test tubes. It is simple to manipulate and has a low cost and high survival rate. The main problem is that the plantlets are susceptible to rot before they are rooted. The solution includes regeneration of strong rootless plantlets, establishment of suitable rooting conditions, and use of effective disease control.

Transplantation

TRANSPLANTING TECHNIQUES. The test tube rooted plantlets were taken out and the medium adhering to the roots was washed away. They were planted first in nutritive pots and then transplanted to soil in pots after the plantlets were stronger. The nutritive soil was a 1:1 mixture of peat and sand. The plantlets rooted outside the test tube could be planted into pots first and transplanted in the nursery if they survived. If the plantlets were strong, covers were not needed. The transplanted plantlet should be kept in a moist environment, avoiding direct noon sunlight. After 3 to 5 days, ventilation, light, and water spraying should be properly controlled. The temperature should be 15–30°C; low temperature appeared to be better. If plantlets were weak, they could be incubated in water-sprayed sand for a period of time and then transplanted into nutritive pots.

SOME IMPORTANT FACTORS IN TRANSPLANTATION. *Selection of Plants*. Plantlets for transplanting should have strong stems, green color, roots over 0.5 cm long, and height of about 1.5 cm. Plants with too short roots do not survive. In our experiment when plantlets with 0.5-cm and 1-cm roots were transplanted, the survival rate was over 80%; only a 58% survival rate was obtained when test tube rooted plantlets with 0.1-cm roots were transplanted.

The plantlets had roots of three colors: white, red, and yellow. When those with white and red roots were transplanted, the survival rate reached over 90% and they recovered growth rapidly. The plantlets with yellow roots recovered growth slowly and had a low survival rate after transplantation. The yellow roots formed when the culture conditions in the rooting medium were unfavorable.

Transplantation Time. The optimum transplanting time was 15 to 20 days after plantlets were rooted; roots at this time were growing vigorously.

Aerated Soil. Regenerated *C. pinnatifida* plantlets should be transplanted into good aerated soil. We tested three kinds of soil: peat, river sand, and a 1:1 mixture of peat and river sand. The best growth rates were achieved in river sand, because it has good aeration. However, it does not retain water well, and if the plantlets were not watered in time, they would readily die, especially in summer. It is, therefore, better to use a peat and river sand mixture.

Temperature, Light, and Humidity. Low temperature and poor aeration of the soil lead to rotting of roots and stems. High temperature, high humidity, and insufficient light can cause damping off of the transplanted plants. If the air is dry and temperature and light are too high, the leaves may readily wilt, sometimes causing burn damage or even death. Therefore, temperature, light, humidity, and moisture content of soil should be properly controlled so as to achieve a high survival rate of transplanted plants. In general, the transplantation temperature should be about 18–25°C. The light intensity should be 2000 to 3000 lux in the early days of transplantation and increased gradually. After the first watering, the plants are watered again when the soil surface is dry.

This plantlet regeneration system is based on proliferation of axillary buds. Some of the bud clones developed in our laboratory have been cultured for 5 years and for more than 50 generations. One bud has produced several hundred thousand plantlets with no decline in proliferation ability. This technique of rapid propagation can be applied to commercial production.

YOUNG EMBRYO CULTURE IN *Crataegus Pinnatifida*

Culture Procedure

MANIPULATION TECHNIQUES. Thirty-day-old seeds were picked out from young fruits, soaked in 70% ethanol for 30 sec, immersed in

0.1% mercuric chloride solution for 10 min, and rinsed with sterile water three to five times. Under aseptic conditions, one-third of the seed near chalaza was removed with tweezers, then inoculated with the microphyle side of the seed into embryogenic development medium. When embryos grew out from the seed coats, they were transferred to differentiation medium to regenerate intact plants. The developed embryos could also be transferred directly into plantlet differentiation medium to induce rootless plantlets. Intact plantlets were then regenerated through root initiation.

MEDIA. *Embryogenic Development Medium.* The basal medium was N6 (or MS) medium supplemented with 11 μM NAA, 2.3 μM ZEA, 500 mg/l LH, and 146.7 mM sucrose (with or without) 500 mg/l placenta powder.

Embryo Differentiation Medium. MS medium, either hormone-free or with 0.57 μM IAA, 300 mg/l LH, and 88 μM sucrose, was used.

Plantlet Differentiation Medium. MS medium was supplemented with 2.2–8.8 μM BA, 0.06–0.57 μM IAA, and 88 mM sucrose.

Rooting Medium. One-half strength MS medium with 4.9 μM IBA and 44 mM sucrose was used.

The preceding media, with pH at 5.4 to 5.8, were solidified with 0.75% agar and sterilized under 1 kg/cm^2 pressure for 10 min.

INCUBATION CONDITIONS. The culture temperature was 22–28°C. Light was provided 10–12 hr/day at an intensity of 1000–3000 lux.

Morphogenesis

EMBRYO DEVELOPMENT. In the embryogenic development medium, young embryos usually grew from the seed coats in 15 to 20 days. The earliest ones appeared in only 6 days, the last in over 30 days. Most embryos developed from breakage at the chalaza; the two cotyledons were visible at this time. Some embryos grew out from the micropyle end in sizes from 1 mm to several millimeters. In some cases endosperm dedifferentiated into callus, but more calli were formed on embryos growing out from the micropyle end. The key for successful culture of young embryos was to supplement the medium with substances with endosperm characteristics or to simulate an embryo sac environment (Cytological Lab, Shanghai Institute of Plant Physiology, Academia Sinica, 1978). In our experiment with *C. pinnatifida*, we inoculated young embryos together with some of their endosperms into the medium, thus guaranteeing normal development of the cultured embryos. When embryo growth was affected by contraction of the medium and aging of the seed coat, fresh medium was added to the original medium or the embryos were removed and subcultured. This is a simple and effective method of culturing young embryos: 70% of the young embryos usually grow normally.

EMBRYO INDUCTION. When the embryos grew out from seed coats, they were transferred to differentiation medium. A large number of embryos could be differentiated in this way.

Microscopic observation has proved that the embryos originate from epidermal cells of the cotyledons. Embryos have a bipolar structure. Embryos can produce secondary embryos, but plantlets cannot be easily regenerated from secondary embryos. The differentiating ability of the embryos was reduced as incubation went on.

DIFFERENTIATION OF GREEN PLANTLETS. Both developed young embryos, and differentiated preembryos could produce green plantlets. As plantlets were continuously excised for subculture, their axillary buds also continued to give shoots, so thousands of rootless plants could be regenerated from one young embryo. The root initiation method was similar to that used for shoot tip culture.

FLOWER BUD DIFFERENTIATION

Culture Techniques

MANIPULATION METHOD. Healthy branches were taken from a matured mother tree in winter. If the branches were dormant, they could be cultured in water in a warm room to stimulate germination. Terminal buds or upper parts with one to three lateral buds were placed into medium under sterile conditions. After flower buds were differentiated, they were excised and cultured in fresh medium under increased light intensity to stimulate flowering.

MEDIUM. MS medium was supplemented with 0.06 μM IAA, 2.2 μM BA, and 2.3 μM KIN.

CULTURE CONDITIONS. The temperature was 15–25°C. Light was 12–14 hr/day at an intensity of 1000–3000 lux.

Morphogenesis

The flower bud of *C. pinnatifida* is of mixed structure, including shoot, leaf, and flower primordia. There are four types of flower bud differentiation in the test tube: (1) Only the flower bud differentiates. (2) Only the leaf bud differentiates. (3) The flower bud differentiates first, followed by the leaf. The leaf bud develops rapidly and the flower bud gradually disappears. (4) The leaf bud differentiates at first, then develops into a strong plantlet (the original medium should be modified). These plantlets have many thick and large leaves.

In the first three cases, it was found that development of flower buds and leaf buds was mutually exclusive under in vitro conditions. After a period of time in culture, only flower buds or leaf buds remained. The two could not coexist in one tube, because of competition for nutrients. A similar situation occurred among flower buds: one flower bud usually differentiated into many flower buds, but only a few could undergo further development. During culture, flowering could be induced from cultured flower buds by removing the leaf buds.

In the fourth case, it seemed that the flower bud observed was not the flower primordium. The flower primordium had differentiated before in vitro culture. Instead, it was transformed from a leaf bud during the culture process. The physiological conditions under which the leaf bud may be transformed into a flower bud have not been elucidated. The differentiation of flower buds and the actual flowering of *C. pinnatifida* in the test tube indicate that studies of flowering physiology in woody trees can be conducted under in vitro conditions. This system has the great advantage that the experimental conditions can be easily controlled.

FUTURE PROSPECTS

In *C. pinnatifida*, the proportion of abortive seeds is high and frequency of seedling production low. Young embryo culture can greatly facilitate hybridization. As hawthorn seeds have a long dormant period, 2 to 3 years is necessary to produce hybrid plants after hybridization using conventional breeding methods; however, using young embryo culture, hybrid plants can be obtained in the same year. Embryo culture can also be used to preserve hybrid materials. Large quantities of plantlets can be produced to be propagated whenever needed. Especially in distant hybridization, this method can overcome difficulties due to abortions of hybrid embryos and possibly create new species. Tissue culture has broad prospects in hawthorn breeding. In the future, anther culture, unpollinated ovary culture, test tube fertilization, and combinations of these new techniques with conventional breeding method will be carried out. Breeding efficiency will be greatly increased.

The establishment of fast propagation systems for test tube plantlets has laid a foundation for seedling production in hawthorn. If some technical procedures are further improved, this technique can undoubtedly be used in practical breeding. Compared with conventional seedling propagation, test tube plantlet regeneration has the following advantages:

1. Rapid proliferation: One bud can produce thousands of plants in one year by tissue culture; far fewer can be obtained by the conventional method.
2. Short breeding cycle: Plants can be obtained in 1 to 2 years in contrast to 3 to 5 years by traditional breeding.
3. High efficiency: Test tube plantlets can be produced in an industrialized way with high efficiency. Using traditional methods, the seedlings are raised in the field with low efficiency.

Rare and valuable materials can be propagated in large quantities within a short period; this is completely impossible by the conventional method. The technique will have even greater significance if it can be combined with virus elimination to regenerate virus-free plants. For this technique to be useful, methods for reducing its cost are needed in coming years.

In conclusion, *C. pinnatifida* is an economically important fruit tree, but its seedlings are difficult to propagate. The application of in vitro propagation of plantlets to this species should result in even greater social and economic benefits.

KEY REFERENCES

Wang, J. X., Li, S. Z., and Li, B. W. 1982. Induction of stock plantlets of *Crataegus pinnatifida* via in vitro culture of seed embryos. *Acta Hortic. Sin.* **9**(4):7–9 (in Chinese).

Wang, J. S. and Yu, G. 1981. Young embryo culture of *Crataegus pinnatifida. Plant Physiol. Commun.* **4**:42–43 (in Chinese).

——— and Yu, G. 1982. Induced embryoid from young embryo in *Crataegus pinnatifida. Hereditas* **1**:29–30 (in Chinese).

Wang, Y. Y. and Wang, F. X. 1983. Effect of IAA and 6-BAP on the rapid propagation of *Crataegus pinnatifida* in vitro. *Sci. Agric. Sin.* **1**:24–28 (in Chinese).

Yu, G. and Wang, J. S. 1981. Shoot tip culture of matured *Crataegus pinnatifida* tree. *Plant Physiol. Commun.* **4**:43 (in Chinese).

REFERENCES

Cytological Lab, Shanghai Institute of Plant Physiology, Academia Sinica. 1978. *Plant Tissue and Cell Culture*. pp. 200–217. Shanghai Science and Technology Press, Shanghai (in Chinese).

Hebei Agriculture University. 1980. *Cultivation Science of Fruit Trees*. pp. 60–62. Agriculture Publishing House, Beijing.

Hu, L. S. and Huang, B. H. 1984. Studies on techniques of rapid propagation of

Crataegus pinnatifida. Shanxi Guoshu **3**:20–24 (in Chinese).
Liu, X. Z. and Liu, B. 1980. *Hawthorn*. pp. 1–2. Liaoning People's Publishing House, Shenyang (in Chinese).
Mu, X. J., Liu, S. Q., Zhou, Y. K., Qian, N. F., Zhang, F. S., Xie, H. X., Zhang, P., and Yan, Z. L. 1977. Induction of plantlets from the hypocotyl and cotyledon of the young embryo. *Acta Genet. Sin.* **4**(2):140–145 (in Chinese).
Wang, D. Y. and Zhang, J. R. 1975. Artificial culture of embryos in *Citrus. Acta Bot. Sin.* **4**(2):140–145 (in Chinese, English abstracts).
Yang, D. K., Yang, D. B., and Yin, L. C. 1981. *Hawthorn*. pp. 1–6. China Financial Publishing House, Beijing.
Yang, N. B. 1982. List of test tube plants. *Plant Physiol. Commun.* **4**:61–80 (in Chinese).
Yu, D. J. 1979. *Taxonomy of Fruit Trees in China*. pp. 147–150. Agricultural Publishing House, Beijing (in Chinese).
Xu, J. W. 1984. A discovery of some precious *Crataegus pinnatifida* varieties. p. 2, *Guangming Daily*, 30 December 1984 (in Chinese).

CHAPTER 23
Hawthorn: Embryo Culture

Yuying Wang, Xinyi Gao, and *Kai Fu*

INTRODUCTION

The studies on embryo culture of *Crataegus pinnatifida* Bunge, hawthorn, and multiplication of nursery stocks were started by us in 1980 (Wang et. al., 1982). Experiments showed that seed dormancy could be broken in the year of harvest. In that way seeds which had not been treated with low-temperature stratification germinated into young shoots and the terminal, lateral, and axillary buds grew into shoots without roots. The shoots were cut off for induction of rooting. Finally, the rooted shoots were transplanted into the soil. The surviving plants have now grown 2 m in height and 3 to 4 cm in diameter. The experiments on hormone combinations were carried out in order to enhance plantlet induction. As a result, the multiplication rate reached 10.5 times within 40 days (Wang et al., 1983; 1983b).

TISSUE CULTURE TECHNIQUES

Procedures and Media

SEED GERMINATION AND PRIMARY CULTURE OF BUD EXPLANTS. The embryo with or without seed coat could not germinate after 1 month of culture in sand. This suggested that such seeds were still in a state of dormancy which the sand culture was not able to break. In addition, seed with coats did not produce buds, even after culture on medium for 30 days. Only seeds without coats could germinate.

MEDIUM. Mature embryos or buds were inoculated onto MS medium with 4.4–8.8 μM BA, 2.9 IAA, 58.4–146 mM sucrose, and 0.7–0.8% agar. The pH ranged from 5.8 to 6.2. The sucrose concentration was raised to 146 mM if large hawthorn shoots were cultured. After 2 weeks the buds swelled and started to differentiate. On the other hand, the buds on the medium with low sucrose concentration yellowed and finally browned and died.

It was discovered that, for bud differentiation, BA was better than KIN at the same concentration, and IAA at certain concentrations was preferred to NAA. NAA caused many buds to form callus, but when IAA was used, very little callus and many buds were formed. In order to improve the differentiation frequency and to enhance the growth of plantlets, different combinations of IAA with BA were tested (Table 1).

TABLE 1. The Effect of Combinations of IAA With BA on the Differentiation and Growth of Hawthorn Shoots

BA (μM)	Item	IAA (μM)					
		0	0.57	2.9	5.7	11.0	22.0
0	No. of differentiated buds and shoots	25	20	28	20	20	20
	Increase	1.3	1.0	1.4	1.0	1.0	1.0
	Average shoot height (cm)	2.0	3.3	4.5	4.0	2.5	2.0
2.2	No. of differentiated buds and shoots	75	100	113	95	90	90
	Increase	3.8	5.0	5.7	4.8	4.5	4.5
	Average shoot height (cm)	3.9	3.3	4.8	4.6	2.5	3.3
4.4	No. of differentiated buds and shoots	175	165	210	143	95	154
	Increase	8.8	8.3	10.5	7.2	4.8	7.7
	Average shoot height (cm)	3.5	5.9	6.5	4.6	4.1	2.1
8.8	No. of differentiated buds and shoots	200	153	170	155	95	62
	Increase	10.0	7.7	8.5	7.8	4.8	3.1
	Average shoot height (cm)	3.3	3.0	3.0	2.3	2.0	1.9
17.6	No. of differentiated buds and shoots	126	120	65	105	—	74
	Increase	6.3	6.0	3.3	5.3	—	3.7
	Average shoot height (cm)	3.3	3.9	2.9	3.0	—	2.8

As summarized in Table 1, the bud differentiation and proliferation could not take place on medium with IAA but without BA. Without IAA, the average shoot height was 2 cm, double what it was at inoculation. With increasing IAA concentration, shoot height increased. The optimal concentration of IAA was 2.9 μM. When IAA concentration was increased to 11 μM, the growth of shoots was inhibited and substantial callus was formed.

BA had a strong effect on differentiation and proliferation of hawthorn buds and shoots. When explants were inoculated onto media containing BA at different concentrations, after 2 weeks several buds emerged from the basal part of axillary buds; we refer to those over 1 cm tall as "shoots" and those less than 1 cm tall as "buds." Proliferation of buds and shoots depended on the concentration of BA in media.

Table 1 shows that bud differentiation depended on the BA concentration when the IAA concentration ranged from 0 to 5.7 μM. When BA was at 4.4–8.8 μM, bud differentiation was good and the number of buds could increase 7 to 10 times within 40 days. IAA at 2.9 or 5.7 μM and BA at 2.2 or 4.4 μM gave the greatest proliferation. The number of the buds over 1 cm tall increased from 20 to 80; those of less than 1 cm tall increased to 130, and the total number (210) was 10.5 times that of the control. At the same time, the buds grew and developed very well. Thus, this combination of IAA with BA was found to be best for differentiation and growth of hawthorn cultures.

ROOT INDUCTION. Stem sections with three to five leaves cut from the basal part of hawthorn buds were inoculated onto MS medium (½ macro elements) plus 2.9 μM IAA or 2.5 μM IBA and 58.4 mM sucrose for root induction. Roots emerged after 2 to 3 weeks. Our experiments showed that in medium with IAA few calli occurred at the cut end of shoots and the roots which formed were long and thin; with IBA a large amount of callus formed and the roots were thick and short. The formation of callus adversely affected the uptake and transportation of nutrients and water because it often blocked the link of conductive tissue between stem and root, which, in turn, prevented successful transplantation of plantlets. IAA was therefore better than IBA for root induction. Moreover, it was not easy to induce roots if the concentration of the growth hormone was too low, and callus would form if the concentration was too high. Usually, the most favorable IAA concentration ranged from 2.9 to 5.7 μM. Under these conditions the rooting rate of small hawthorn shoots was more than 90%; that of large ones was lower.

Critical Variables

SEED STERILIZATION. First, the seed was taken out of the hawthorn fruit and disinfected for 15 min with a saturated solution of bleaching powder and for 10 min with 0.2% mercuric chloride. It was then rinsed four times with sterilized water. The seed coat was removed in a sterile environment and inoculated onto the medium.

ADULT BUD STERILIZATION. Buds should be collected from branches just germinated in the spring. Buds were washed for 1 hr with running water. After the bud scales were removed, the buds were sterilized 8 to 10 min with 5% Antiformin plus a drop of Tween 80 and then disinfected 4 to 5 min with 0.1% $HgCl_2$. Buds were rinsed four times with sterilized water and inoculated on the medium.

CULTURE TEMPERATURE AND ILLUMINATION. Cultures were placed in a chamber at 25 ± 2°C during the day and 16 ± 2°C at night and illuminated at 2000 lux for 10 hr. Alternatively, they were put into a chamber with natural light at 15–30°C at an intensity of 1000 to 5000 lux.

Transplantation of Plantlets

The best time for transplantation of plantlets was when they had one to four roots that were 2 cm long, roots were white, and root hairs had not yet formed. If they were transplanted too early, when their roots, stems, and leaves had not fully developed, the delicate plantlets were vulnerable to infection. If transplanted too late, the root aged and became discolored. At this time rooting capability was bad and the transplanted plantlets survived with difficulty.

Attention should be directed during plantlet transplantation to the following:

1. Flasks were opened and the plantlets were hardened under scattered light for 1 to 2 days.
2. Young plantlets taken out of flasks were rinsed with lukewarm water then transplanted into sandy soil or vermiculite.
3. After watering, the plantlets were covered with plastic film or placed in a plastic container, in which the relative humidity was over 90% and temperature was 25 ± 5°C. Direct sunlight should be avoided.
4. To keep the young plantlets living, ventilation and watering should be increased a little each week.
5. The plantlets could not be transplanted into the field until the last 10 days of April when the air temperature was stable at about 10°C. The plantlets should be transplanted on ridges where the soil temperature is high, aeration is good, and watering is convenient.
6. It was difficult for plantlets to survive summer in the greenhouse because of overheating. A seedbed might be set up outdoors with automatic watering facilities in sunlight. One month after transplanta-

tion in these conditions, a robust root system would develop. The plants were transferred to the field and the survival rate was over 90%.

REFERENCES

Agricultural University of Hebei. 1982. Hawthorn. In: *The Special Pomology*. p. 319. Higher Education Press, Beijing (in Chinese).

Li, P. 1984. The historical testimony and plentiful plant resources. In: *The History of Cultivated Plants in China*. pp. 3–10 Science Press, Beijing (in Chinese).

Wang, Y. and Gao, X. 1982. The embryonic culture of hawthorn and its propagation. *Plants* **2**:50 (in Chinese).

———, Wang, F., and Gao, X. 1983a. The experiment of the embryonic culture in vitro of hawthorn and its multiple propagation. In: *Collected Papers of Botany*. Vol. 1, p. 103. Science Press, Beijing (in Chinese).

———, Wang, F., and Gao, X., 1983b. Effect of IAA and 6-BAP on the rapid propagation of *Crataegus pinnatifida* in vitro. *Sci. Agric. Sin.* **1**:24–28 (in Chinese, English abstract).

Xie, Y., Dai, L., Guo, M., Xue, M., Yin, W., and Zi, W. 1981. The chemical analysis of fruit of large-fruited Chinese hawthorn and comparison of the fruits of *Crataegus* species in China. *Acta Bot. Sin.* **23**(5):383–388 (in Chinese).

Yu, G. and Wang, J. 1981. The shoot-tip culture of the adult tree of hawthorn. *Plant Physiol. Commun.* **4**:43 (in Chinese).

CHAPTER 24

Black Currant: Shoot Tip Culture

Jiangyun Wu and *Dingqiu Huang**

INTRODUCTION

Importance in Cultivation

The black currant, *Ribes nigrum* L., is a small shrub, belonging to the family Saxifragaceae. Cultivated black currant, introduced by Russians who resided in China, has been grown in Heilongjiang province for more than 70 years. Northeast China contains about 10 wild species of the genus *Ribes*, distributed in mountainous and semimountainous areas. These species are very useful for improving cultivars of *Ribes nigrum* L.

Black currant is a fruit tree with small berries which have high nutrient and economic value. The berries are especially rich in vitamins A, B, and C; the content of vitamin C is 4 times that of orange and 40 times that of apple, about 130 to 224 mg/100 g fresh weight of fruits. The RDA of 40 to 60 mg of vitamin C per day can be met by only 50 g of berries. In addition, there are about 1.7 g protein, 0.1 g fats, 13.1 g carbohydrate, and 1.7 to 3.9 g organic acids per 100 g fresh fruit. The fruits are used to make candy, wine, jam, juice, gelatin, and many kinds of drinks, besides being eaten fresh. The seeds are 16% oil.

*English translation by Yuxiang Jing.

Problems in Production

Heilongjiang Province is the center of Chinese currant cultivation. The conventional methods of cutting and layering used for currant propagation are insufficient to meet the needs of commercial production. In addition, because propagated shoots are expensive, most seedlings are propagated by low-paid laborers. Genotype differences among seedlings are significant, making their cultivation and management difficult, and the yield of each seedling varies greatly, further decreasing productivity. Culturing shoot tips from elite mother trees provides an alternative means of rapid propagation to obtain uniform high-quality nursery stock.

All the varieties of *Ribes nigrum* L. currently in use are old ones, propagated for up to 70 or 80 years, and are highly heterozygous, with low productivity. They urgently need to be purified and restored. Many fine cultivars have been selected inside and outside China. Individual plants with good qualities should be separated and propagated rapidly via shoot tip culture.

Shoot tips have already been successfully cultured from some plants with small berries, e.g., strawberry (Boxus, 1977) and blackberry (Broome and Zimmerman, 1978). Strawberry has been propagated by using tissue culture on a large scale (Boxus, 1977).

PROTOCOLS

Cultural Procedures and Media

DONOR PLANT SELECTION. The tested variety was thin-skinned black currant, which is cold-resistant. Buds were taken for inoculation when dormant or just starting to germinate. The most suitable time for bud inoculation in Heilongjiang is from the last 10 days of January to the first 10 days of April. Terminal and lateral buds could both be used for inoculation, but terminal buds were much better. The shoots developed from them were stronger, and the proliferation rate was significantly higher than that for lateral buds.

MEDIA FOR DIFFERENTIATION AND SUBCULTURE. The best minimal medium was MS, on which the survival rate of tips was more than 90% and the shoots developed very well. The media B5 and ER were also suitable. Although the shoot tip survival rate on White medium reached 81%, the shoots were weakly developed and the differentiation frequency was low. These results show that low levels of nutrients in the medium were insufficient for shoot tip development.

Each tip growing on MS medium supplemented with 8.8–13.2 μM BA and 2.9–4.4 μM GA for 25 days produced on average 4.2 shoots. After the shoots were subcultured by transfer onto MS medium containing 4.4–6.6 μM BA and 5.8–8.8 μM GA, each shoot tip produced an average of 3.18 shoots every 20 days.

We counted the number of shoots twice a month and found that a single cultured shoot tip produced 71,625 shoots, each with an average height of 4 mm and two to four small leaves each, over an 8-month period (Huang et al., 1983).

ROOT INDUCTION. Shoots 1 cm in height bearing several leaves were transferred onto a medium for root induction. The ½ macro MS medium supplemented with 1.7–4.0 μM IAA gave the best results: each of 601 shoots cultured on this medium produced one to three new roots after 20 to 25 days. The next best medium was ½ macro MS medium with 2.5–4.9 μM IBA, on which 30% of the shoots produced roots. If the concentration of IAA was more than 5.7 μM, the rate of root induction was lower.

Sterilization and Culture Conditions

1. Shoots were cut into segments (each with a bud); placed in 70% ethanol for several seconds, then into 0.1% $HgCl_2$ for about 8 min; rinsed with sterile distilled water three times; and finally placed on a piece of filter paper in a petri dish.
2. Shoot tips 0.5–1.5 mm in length with growing points and several leaf primordia were excised from the segments and inoculated onto medium.
3. After clustered shoots grew from shoot tips, they were separated into groups of three to four shoots with attached callus and transferred onto the subculture medium for proliferation.
4. During the whole culture process the shoots were illuminated with fluorescent lamps at a light intensity of 1000 lux for 10–12 hr/day. The temperature was 25 ± 2°C by day and 15 ± 2°C at night.

Shoot Morphogenesis

PROLIFERATION OF BUD PRIMORDIUM. During the shoot tip culture of *Ribes nigrum* L., shoot proliferation mainly depends on growth of the bud primordium. When 1–1.5-mm shoot tips were cultured in the dark on MS medium containing 4.4–13.2 μM BA for 3 days, the basal region of the shoot tips became wider because of division of procambium cells and the primordium of the axillary buds grew larger. On the 5th day, as a result of shoot tip differentiation, two to three primordia of axillary buds emerged. The procambium and ground meristem divided continuously to make the tip tissue longer and wider. By this time, the axillary bud primordium, which had initially differentiated at the tip bases, developed leaf primordia. By the 9th day, four to five axillary buds formed and the basal buds had differentiated four to five leaf primordia. Meanwhile, the early buds differentiated as the tip bases developed rapidly, as a result of contact with the medium and greater distance from shoot apices, which weakened the influence of apical dominance. By the 25th day, each shoot tip had grown four to seven shoots to form a cluster. After 25 days callus formed at the base of the cut shoot. In subsequent subcultures new shoots were differentiated from axillary buds.

DEVELOPMENT OF ADVENTITIOUS BUDS FROM CALLUS. The shoots with callus were transferred to MS medium supplemented with 13.2 μM BA and 2.9 μM GA for 30–35 days of subculture. At this point the callus at the shoot bases had usually aged and browned, but a few were still green. New callus formed from cells which divided rapidly. After several cell divisions a meristematic cell zone formed, swelling into the shape of a horseshoe, which gave rise to one or several adventitious buds. During shoot subculture, the callus generated adventitious buds continuously.

Thorpe (1978) concluded that organs developed from callus parenchyma cells going through successive cell divisions formed a pseudomeristem in which the cells were small and stained darkly. The pseudomeristem gave rise to roots or bud primordia.

Wang et al. (1981) found that in cultures of Chinese larch (*Larix potaninii*) and olive the individual parenchyma cells in callus could be differentiated into embryonic cells. After cell division, they produced tumorlike pseudomeristems and finally formed adventitious buds, roots, and other tissues. Our observations are in basic agreement with this model.

It was clear that there were two ways to proliferate shoots, via callus or via axillary bud proliferation, but the latter is preferred. Plantlets produced from callus might show variation, although very few adventitious buds were produced from callus with our culture methods. Some measures could be taken to prevent variation, including possible mutations, and to keep genetic traits stable. The adventitious buds on the callus formed only from the 30th to 35th day after the shoot tips had been transferred to new medium. The most productive period for axillary bud formation was the first 20 days after transfer. If the shoot tips and bud clusters were transferred promptly (from 20th day to 30th day), the formation of adventitious buds could be prevented. If adventitious buds still formed, they could be removed individually.

CRITICAL VARIABLES

Selection of Shoot Tips

The proliferation rate of inoculated terminal buds was significantly higher than that of axillary buds, larger explants (1 to 1.5 mm) were better than smaller ones (0.5 mm), and both dormant and germinating buds could be used as material for inoculation of culture.

TERMINAL BUDS AS EXPLANTS. The shoot tips from terminal and axillary buds were selected as explants. Terminal buds were stronger and better developed than axillary buds. Experiments comparing inoculation were performed with the medium and culture conditions described in Table 1.

Table 1 shows that the number of shoots produced by terminal buds was significantly ($p < 0.01$) higher than that of axillary buds and that the shoots from the former were markedly taller and stronger than those from the latter (Fig. 1). For currant shoot tip culture, terminal buds were the best material because most axillary buds in the upper part of basal shoots were flower buds and the leaf buds in the lower part were not as strong as terminal buds, which are quite strong. Therefore, when terminal buds were used, propagation was very rapid, but the growth of the mother tree was not affected.

SELECTION OF LARGE EXPLANTS (1 TO 1.5 MM) FOR INOCULATION. In *Ribes nigrum* L. shoot tip cultures the survival rate and shoot number produced from inoculated large shoot tips were higher than those from small ones. These results were related to the content of nutrients and hormones and the proportion of injured explants. If small shoot tips (0.5 mm), which have only a shoot apical meristem, are inoculated, they take longer to differentiate leaf bud primordia. In contrast, the large shoot tips (1 to 1.5 mm) already have two leaf bud primordia, which grow out rapidly after inoculation. Use of large shoot tips for inoculation can thus increase the proliferation rate.

TABLE 1. Comparison of Shoot Number Produced from Shoot Tips of Terminal and Axillary Buds*

Explant	Shoot number ($\bar{x}$ ± SE) in Flask
Terminal bud	5.63 ± 0.14
Axillary bud	4.12 ± 0.40

*Three shoot tips in a flask, 10 replicates, cultured for 25 days.

Figure 1. Rosette shoots. (*Left*) Produced from shoot tip of terminal bud. (*Right*) Produced from axillary bud.

INOCULATION OF BOTH DORMANT AND GERMINATING BUDS. When growing apple shoot tips were inoculated, the frequency of shoot tip differentiation was increased by more than 100% compared with that of dormant buds (Jacobs, 1979). The efficiency of inoculated dormant buds of *Ribes nigrum* (in January) was similar to that of growing buds (in April). This is a practical advantage because the test tube shoots are transplanted in May. If it is possible to start with dormant buds, 3 extra months is needed for subculture to produce more plantlets for transplantation in a single year.

Inoculations were performed at four different times: January 28 and March 23, when the buds were still dormant; April 7, when the buds had started to grow; and April 21, when young leaves on the buds had emerged from the bud scales. In all treatments, shoot tips of the axillary bud were used and cultured in the dark on MS medium containing 13.2 μM BA (Table 2).

Table 2 shows that the numbers of shoots produced from shoot tips inoculated at the three earlier times were significantly higher than for the last time (April 21)($p < 0.01$). There was no significant difference among

TABLE 2. Comparison of Shoot Number from Buds Taken at Different Times of Development*

Stage and Date of Inoculation	Shoot Number ($\bar{x}$ ± SE) in Flask
Dormant	
28 January	4.28 ± 0.39
23 May	4.18 ± 0.38
Germinating	
7 April	3.96 ± 0.40
Spreading	
21 April	2.46 ± 0.25

*Three shoot tips in each flask, 10 replicates, cultured for 25 days.

the earlier three times. Clearly, late inoculation was not beneficial to differentiation of shoot tip lateral buds, and there was no significant difference between inoculations at the dormant stage and the early stage of bud growth.

Complex Effect of Hormones and Light

In *Ribes nigrum* shoot tip cultures, BA was better than other cytokinins for bud differentiation; the best choice was a mixture of 4.4 to 13.2 μM BA with 2.9 to 4.4 μM GA. For subcultures, the concentration of hormones should be altered (4.4 to 6.6 μM BA and 5.8 to 8.8 μM GA). Light culture in media with these concentrations of BA and GA gave significantly higher numbers of shoots than dark culture.

SUPERIORITY OF BA TO OTHER CYTOKININS. The effect of different hormones was carried out using 21 treatments: BA, KIN, and ZEA, each at three concentrations (ranging from 0.44 to 23 μM), and NAA, IAA, IBA, and GA, each at three concentrations (ranging from 0.3 to 8.6 μM). Each treatment had 10 flasks and each flask had three shoot tips. All the flasks were cultured in the dark for 25 days. Shoot tips proliferated in all the treatments with BA and KIN; however, the best response was obtained with BA (Table 3).

From Table 3 it is evident that the treatments with 4.4 or 13.2 μM BA were better than that with 22 μM BA ($p < 0.01$). The former two treatments result in four shoots from each flask after culture for 25 days. KIN also had an effect on shoot tip proliferation but was weaker than BA.

ZEA induced a large quantity of callus without shoot proliferation. All growth hormones used alone did not result in shoot tip proliferation but only induced callus formation. Among them, NAA induced much more callus than did IBA. IAA also did not have any effect on shoot tip proliferation. GA could promote leaf elongation.

TABLE 3. Effect of Concentrations of Growth Regulators on Shoot Number

Growth Regulator	Concentration (μM)	Shoot Number in Flask ($\bar{x} \pm$ SE)
BA	13.2	4.08 ± 0.35
BA	4.4	3.93 ± 0.28
BA	22	2.06 ± 0.34
KIN	23	1.44 ± 0.17
KIN	13.8	1.43 ± 0.15
KIN	4.6	1.09 ± 0.04

SUPERIORITY OF LIGHT CULTURE IN MEDIA WITH BA AND GA TO DARK CULTURE. Dark culture on MS medium with 13.2 μM BA for days induced each tip to produce four shoots. The shoots produced in dark culture were etiolated. They became weaker as the number of subculture generations increased and, consequently, the proliferation rate decreased. It was thus necessary to alternate dark and light culture. Although in all of the treatments described, light inhibited shoot tip proliferation significantly, BA promoted formation of axillary buds and GA and IBA stimulated leaf growth.

The complex effects of hormones and light on shoot proliferation were investigated using MS medium with 13.2 μM BA. GA at 1.45, 2.9, 4.4 μM and IBA at 2.5, 4.9, 7.4 μM were both used. Each combination of auxin and cytokinin was tested in light (1000 lux) and in darkness (24 hr). There were 12 treatments altogether, each with 12 flasks in which three shoot tips were inoculated. Shoots were counted after 25 days of culture. Statistical analysis of data showed that the shoot number in all treatments under light culture was higher than under dark culture, and the number in treatments with GA and BA was higher than with IBA and BA. There was an interaction between illumination and hormones: the GA and BA treatment under light produced the largest number of shoots. The shoot number in GA and BA treatments tended to increase with increasing GA concentration, but in IBA and BA treatments the number increased with a decrease in IBA concentration.

Comparison of Table 4 with Table 3 shows that the maximum average number of shoots produced per tip (four) was the same in both experiments. However, the shoots developed under light were strong, with green expanded leaves on elongated petioles (Fig. 2).

When shoot tips of perennial crops, such as apple, are grown and proliferated in vitro (Jacobs 1979; Wang et al., 1980; Liu et al., 1978),

Figure 2. Shoots on medium.

TABLE 4. Effect of Growth Regulators and Light or Dark Culture on Shoot Number

Growth Regulator (μM)		Shoot number in flask	
		In Light	In Dark
GA	1.5	4.02	2.59
	2.9	4.03	2.62
	4.4	4.16	3.38
IBA	2.5	3.30	2.42
	4.9	2.97	2.25
	7.5	2.78	2.43

results are similar to our observations on currant shoot tip cultures: etiolated shoots produced under dark culture become weaker with each generation of subculture. When GA was used in combination with other hormones and dark culture was changed to light culture, shoot etiolation and the decrease in proliferation rate was prevented. In the combination of GA and BA, BA promoted the formation of axillary buds and GA reversed the inhibition of growth of axillary buds by light (especially red light) and stimulated them to elongate (Jacobs, 1979). Hence, the light culture could also enhance the proliferation rate of shoot tips, and at the same time the illuminated cultures could carry out photosynthesis to synthesize nutrients and hormones in vivo. Shoots grown under such conditions were strong enough for continuous subcultures.

At present, combinations of auxins with cytokinins are used to enhance the efficiency of shoot tip culture in many plants.

REGULATION OF BA AND GA CONCENTRATIONS AND THEIR RATIO TO OBTAIN STRONGER SHOOTS. Differences in growth and development existed among individual shoots. After many rounds of subculture, some shoots had many axillary buds which grew in clusters but did not elongate, and some had only a few axillary buds which were elongated. The concentrations of BA and GA and their ratio in media should be adjusted for specific experiments.

An experiment was performed with nine treatments in which MS was used as basal medium with combinations of BA at 4.4, 8.8, and 13.2 μM and GA at 2.9, 5.8, and 8.7 μM. The difference between treatments was not significant. The best treatment was the combination of 4.4 μM BA + 8.7 μM GA in which each tip developed an average of 3.48 shoots after 20 days of culture. But when the concentration of BA in medium was higher than that of GA, more axillary buds formed but rarely elongated. In contrast, few slender axillary buds formed on media with a low concentration of BA and a high concentration of GA.

Suitable Time Interval for Subculture and Proliferation

A suitable period for subculture is important, especially for rapid proliferation of rare elite plants or species. Culture times of 20, 25, and 30 days were compared, using MS medium containing 4.4 μM BA and 8.7 μM GA and the protocols described.

The average numbers of shoots per flask were 3.48 ± 0.45, 4.14 ± 0.51, and 4.46 ± 0.55 for subcultures of 20, 25, and 30 days, respectively. The proliferation efficiency of the initial 30-day subculture was higher than that of the 20- or 25-day subculture. However, the efficiency of successive subcultures was significantly higher for the shorter intervals than for the 30-day subcultures. The most suitable interval for subculture was 20 days. When the interval was less than 20 days, the growth of shoots after transfer was not restored quickly enough and the shoot proliferation rate decreased. Anderson (1984) suggested that the most suitable interval of time between two transfers should be determined empirically to enhance the proliferation efficiency.

A single shoot tip cultured from April to December 1981 proliferated 71,625 shoots 4 mm long, bearing two to four leaves each. Every month we took random samples of the tip cultures and 500 shoots were examined after proliferation. The number of shoots increased on average 500% per month during the first 8 months (from April to November), and by 370% per month in the later 4 months (from December to May). Therefore, one shoot tip could proliferate up to 73,209,414 shoots in a single year.

The shoots proliferated during subculture were mainly produced from axillary buds. Most of the shoots formed primary and secondary branches within 20 to 25 days. New tips on a few of the shoots with very short internodes were wheel-shaped. In some of the axillary buds there are two to four new shoot tips. Our results show an efficiency of shoot tip proliferation for *Ribes nigrum* that is seldom seen in tissue culture of fruit trees. It has great future potential.

TRANSPLANTATION OF TEST TUBE SHOOTS AND MANAGEMENT OF SURVIVING PLANTS

Rooting outside the Test Tube and Transplantation Techniques

Generally speaking, woody plantlets transplanted from test tubes do not readily survive, but *Ribes nigrum* plantlets have a high survival rate: over 90%. Even shoots with one root could survive transplantation. We performed rooting experiments in 1984. Rootless shoots produced from shoot tip culture of *Ribes nigrum* were grown in various pots with vermiculite, fine sand, or perlite. Vermiculite proved to be the best medium for rooting: it had a rooting rate as high as 90–98%, and plants had a high survival rate after transplantation.

Flasks containing shoots with roots 1.5 cm long were opened for 3 days of hardening. The shoots were then transplanted from the flasks into boxes for 2 weeks to grow well-developed root systems. Finally, they were moved into pots or seedbeds. With this protocol the survival rate reached more than 90% (Fig. 3).

Investigation on Transplanted Cultured Plantlets

After the shoots were taken from flasks into the field (Fig. 4), their growth and development were compared with those of cuttings set at the same time (Table 5). It can be seen from Table 5 that the average number of basilar shoots on test tube plantlets was 9.9, which was significantly higher than for cutting shoots. This phenomenon could be observed even before transplantation, consistent with the multiple center phenomenon of tissue cultured plantlets described by Murashige (1978). This feature has practical implications. In general, two to three plantlets of *Ribes nigrum* must be set together for transplantation, but it is enough to set only one to two plantlets obtained by shoot tip culture. Multiple centers lead to many basilar shoots and to early formation of bushes, resulting in high early yields. Among 48 bushes produced from shoot tip culture, 27

Figure 3. Test tube plant in pot.

Figure 4. Test tube plantlets grown in seedbed.

started to bear fruit in the first year of setting; this does not occur with conventional cuttings.

Genetic Stability of Test Tube Plantlets

Shoots which had been subcultured for more than 1 year were induced to root. We took 100 root tips of 50 randomly chosen plantlets for chromosome observation. All were diploids ($2n = 16$), the same as five cuttings. We also compared the morphological characteristics of the shoots obtained by shoot tip culture for leaf shape and color, petiole, stem, and fruit form and color with shoots derived from cuttings and found that there was no variation in the cultured plantlets.

PROSPECTS

The new technique of rapid propagation by tissue culture is an area of increasing interest. Use of this new method for production rests on two prerequisites: mastery of reliable techniques and demonstration of economic benefits of this new technique compared with conventional propagation of shoots. Our experiments over 3 years show that shoot tip culture is a reliable technique for rapid propagation of *Ribes nigrum* shoots. One shoot tip could be used to produce more than 1000 plantlets within 9

TABLE 5. Growth Comparison of Shoots from Culture with Cuttings Planted in the Field in the Same Year

Shoots	No. of Rosettes Investigated	Average No. of Shoots per Rosette	Average Length of Shoots (cm)	Average Height of Rosette (cm)
Culture	48	9.9	34.4	57.8
Cutting	48	5.3	39.0	50.7

months. The proliferation rate was 6000-fold higher than with conventional breeding. If 1000 shoot tips (from 100 basilar shoots) are cultured for 9 months, 1 million plantlets will be obtained.

The year-round production cost for shoots from *Ribes nigrum* shoot tips was compared with that of cuttings. The average cost per cultured plantlet is 0.038 yuan in China. Since there are twice as many basilar shoots on test tube plantlets as on cuttings, the effective price is halved to 0.019 yuan; the average cost of conventional plantlets is 0.054 yuan. Hence, tissue culture plants are approximately one-third the cost of conventionally propagated *Ribes*. We feel that this tissue culture approach can be used for production and will aid the cultivation of *Ribes nigrum* L.

KEY REFERENCES

Huang, D., Wu, J., and Zhou, S. 1983. Shoot tip culture of *Ribus nigrum* L. and its cytological observation. *Kexue Tongbao* **5**:319 (in Chinese).

Jones, O. P. and Vine, S. J. 1968. The culture of Gooseberry shoot tip for eliminating virus. *J. Hortic. Sci.* **43**:289–292.

Wang, K., Zhang, P., Ni, D., and Bao, C. 1981. Callus formation and organ regeneration in tissue culture of woody plants. *Acta Bot. Sin.* **23**:97–103 (in Chinese, English abstract).

REFERENCES

Boxus, P. B., Quoirin, M. and Laine, J. M. 1977. Large-scale propagation of strawberry plants from tissue culture. In: *Applied and Fundamental Aspects of Plant Cell, Tissue, and Organ Culture*. (J. Reinert and Y. P. S. Bajaj, eds.), pp. 130–143. Springer Verlag, Berlin.

Broome, O. C. and Zimmerman, R. H. 1978. In vitro propagation of blackberry. *Hortic. Sci.* **13**:151–153.

Jacobs, W. P. 1979. Hormones other than auxins and abscisic acid. In: *Plant Hormones and Plant Development*. pp. 263–271. Cambridge University Press, Cambridge.

Liu, S., Chen, W., Wang, H., and Yang, S. 1978. Shoot tip culture of apple seedling and rootstocks in vitro. *Acta Bot. Sin.* **20**:337–340 (in Chinese, English abstract).

Murashige, T. 1978. The impact of plant tissue culture on agriculture. *Front. Plant Tissue Cult.: Proc. Fourth Int. Congr. Plant Tissue Cell Cult.*, (T. A. Thorpe, ed.), Calgary, 15–26.

Papachatzi, M., Hammer, P. A., and Hasegawa, P. M. 1981. In vitro propagation of *Hosta decorata* "Thomas Hogg" using cultured shoot tips. *J. Am. Soc. Hortic. Sci.* **106**:232–236.

Thorpe, T. A. 1978. *Front. Plant Tissue Cult.: Proc. Fourth Int. Congr. Plant Tissue Cell Cult.* (T. A. Thorpe, ed.), Calgary, 49–58.

Wang, B., Cao, Y., and Huang, L. 1980. Stem tip culture of teak in vitro. *Acta Bot. Sin.* **22**:200–201 (in Chinese, English abstract).
Wang, J. and Sui, C. 1981. In vitro culture studies on the proliferation of apple rootstock. *Sci. Agric. Sin.* **6**:33–37 (in Chinese).
Wochok, Z. S. 1980. In vitro propagation and establishment of wax currant (*Ribes inebrians*). *J. Hortic. Sci.* **55**:355–357.

CHAPTER 25
Citrus: Anther Culture

*Zhenguang Chen**

INTRODUCTION

Importance of Cultivation

Citrus belongs to Acrantioideae, Citreae, Citrinae, Rutaceae, and occupies an important position in fruit tree cultivation around the world. Its yield is exceeded by only that of grape. Its fruit is popular because of its numerous varieties, good taste, long availability of supply in the marketplace, and nutritional value. It also has significant economic value. Presently, *Citrus* is cultivated on a large scale in Asia, Europe, America, Africa, and Australia.

Citrus originated in China and has been cultivated for more than 4000 years. The numerous excellent *Citrus* varieties include *Citrus sinensis* var. *sunwuitincheng, Citrus reticulata* var. *tangerina, Citrus reticulata* var. *tankan*, and *Citrus reticulata* var. *unshium*. These varieties have a fine reputation in China and in the world market.

Geographical Distribution

Citrus varieties are widely distributed over 18 provinces in China. The main *Citrus* growing areas in China are Sichuan, Guangdong, Fujian,

*The author thanks Shenkun Zheng for discussion, and Yongfu Wang and Youli Hu for technical assistance. English translation by Xiaxian Zhou.

Jiangxi, Hunan, Zhejiang, Hubei, Guangxi, Yunnan, Guizhou, and Taiwan. China has many *Citrus* germ plasm resources, such as the recently discovered resources of Da Yi in Yunnan Province, *Citrus ichangensis* in Hubei Province, Shan Jin, and others. According to historic records, *Citrus reticulata* (*Citrus nobilis*), *Citrus reticulata* (*mandarin*), and *Citrus sinensis* were introduced from China into the Mediterranean countries and then to other countries in the world. *Citrus* was introduced into America in 1665. The famous *Citrus reticulata* var. *unshiu* was developed in Kusu Kakoshima, Japan, after being introduced from Tian Tai Mountain, Zhejiang Province, by a Japanese monk 500 years ago. Presently, most *Citrus* cultivation is in the United States, Japan, Brazil, Spain, Italy, and Israel. *Citrus* is also cultivated in Morocco, South Africa, Egypt, Algeria, Argentina, Mexico, Greece, and other countries.

Problems of *Citrus* Breeding

Problems of *Citrus* production are as follows:

1. The basic goal of *Citrus* cultivation is the selection and breeding of good varieties of high and stable productivity.
2. The period of maturity of most *Citrus* cultivars in China is winter. Very few varieties mature in summer. Therefore, it is necessary to breed varieties with both early and late maturity and to improve preservation techniques so that there is a continuous supply of *Citrus* to the market.
3. Selection and breeding of productive varieties with good processing, juice making, and storage qualities are important to the canning and juice industry.
4. Breeding of disease- and cold-resistant varieties is necessary.
5. Selection and breeding of *Citrus* stocks with high resistance and good compatibility and of dwarf types, etc., are important.

Literature Review

The role of tissue culture in *Citrus* breeding has received attention in recent years. Embryo culture of *Citrus* has been reported since the 1960s (Rangan et al., 1969). However, there have not been satisfactory results for in vitro culture of immature embryos. Elimination of virus by microbud grafting in the mid-1970s was an important step in breeding virus-free plants. *Citrus* plantlets were recovered by endosperm culture, single cell culture, and protoplast culture. The induction of anther culture of *Citrus* still lags behind the culture of other explants. The author began anther culture in *Citrus* to induce pollen-derived plantlets in 1974. Plantlets were obtained and transplanted successfully to the field in 1979. Hidaka (1979) also induced anther-derived plantlets successfully in *Poncirus trifoliata* (L) Raf. in 1979 and in 1982 reported the production of young plantlets from anther culture of *Citrus aurautium* L. Iwamasa reported in 1983 that young plantlets were successfully induced from anther culture in the variety *Citrus microcarpa*.

ANTHER CULTURE TECHNIQUES IN CITRUS

Culture Procedure and Media

The first step was initiation of microspore division to form embryo. The optimal medium was N_6 medium (Zhu et al., 1977) supplemented with 4.4 μM BA, 0.22–0.45 μM 2,4-D, 147–293 mM sucrose. This step took 40 to 80 days.

After this embryos grew and developed into plantlets. The optimal medium was modified MS medium (Rangan et al., 1969) supplemented with 1.1 μM BA, 0.57 μM IAA, 500 mg/l LH, 147–293 mM sucrose. This took 30 to 40 days.

Finally, the plantlets were grown on a medium to induce roots. Modified MS medium was used with 0.49–0.98 μM IBA and a low concentration of sucrose. This took 20 to 30 days.

The media were solidified by the addition of 1% agar and the pH was adjusted to 5.8.

Major Points of Culture Techniques

1. *Citrus* varieties used for experiments should be able to undergo androgenesis to produce embryos, such as *Citrus microcarpa* and Su Gan.
2. Pollen of the inoculated anther should be at mid- to late-uninucleate stage.
3. Flower buds are placed at 3°C for 5–10 days.
4. Flower buds are sterilized with 0.1% mercuric chloride for 10 min and rinsed with aseptic water five times prior to inoculation.
5. Anthers are removed for inoculation under aseptic condition, and 40 anthers are placed in each flask.
6. The medium is autoclaved to a pressure of 15 lb for 20 min.
7. The temperature in the culture room is controlled at 20–25°C. In general, embryo induction is carried out in the dark.
8. Cultures are kept in light and dark for 12 hr alternately so that organ differentiation can occur. The light intensity is kept to 500 to 800 lux.

KEY POINTS FOR SUCCESSFUL INDUCTION OF POLLEN-DERIVED EMBRYO

Screening of Materials Capable of Induction

Variations in induction frequencies exist among the types of materials inoculated. Somatic cell callus is readily formed in *Citrus* anther culture at the surface of the anther or at the breakage of the filament. High temperature in the culture room and high concentrations of plant hormones favored callus formation from somatic cells. The development of pollen into embryos was seriously inhibited. A comparative experiment with nine different varieties indicated that varietal differences existed in callus formation from somatic cells in anther culture, among different varieties studied. Frequencies of callus formation from somatic cell were recorded as follows: Wu Yi 36%, Si Bu Xiang 29.14%, *Citrus sinensis* var. *sedkkan* 27.76%, *Citrus limon* Burn 27.24%, *Citrus reticulata* var. *tangerina* 18.16%, Lu Gan 13.11%, Su Gan 10.64%, *Citrus microcarpa* 6%, and *Citrus grandis* var. *pinshanyu* Hort. (with a high degree of pollen sterility) 0.5%, the lowest. Pollen-derived embryos were obtained from only two varieties, *Citrus microcarpa* and Su Gan, with low induction frequencies of 2.21% and 1.10%, respectively. *Citrus microcarpa* Bge. was usually propagated by seeds and Su Gan was a natural hybrid. Thus, it seems likely that pollen grains of *Citrus* propagated sexually for a long time readily developed into embryos by anther culture and those of vegetatively propagated varieties had difficulty forming embryos.

Variation in embryo induction by anther culture was also demonstrated by using anthers at different flowering periods. For example, anthers of early flower buds of Su Gan were induced into embryos by anther culture, but anthers of the later flower buds produced no embryos. Variation also resulted from the age of the tree. Flower buds taken from a vigorously growing tree of *Citrus microcarpa* Bge. had a higher frequency of embryo induction than flower buds taken from older trees.

Effect of Low-Temperature Treatment of Flower Bud on Embryo Formation

Flower buds of *Citrus microcarpa* were treated at 3°C for 5, 10, 15, 20, and 25 days prior to culture in vitro, and untreated flower buds were used as control. N_6 basal culture medium was used with 8.8 μM BA, 0.45 μM 2,4-D, and 440 mM sucrose. Culture temperature was kept at 20–25°C and pH was 5.8. The results of these experiments are shown in Table 1.

Table 1 shows that 0.94% of the control anthers produced embryos. Induction frequencies of embryos in the low-temperature treatments for

TABLE 1. Effect of Low-Temperature Treatment (3°C) on the Formation of Anther-Derived Calli and Embryos

No. of Days of Treatment	No. of Anthers Inoculated	Callus of Somatic Cells		Pollen-Derived Embryo	
		No.	%	No.	%
Control	848	5	0.59	8	0.94
5	723	8	1.11	12	1.66
10	902	9	1.00	15	1.66
15	846	18	2.13	5	0.59
20	470	5	1.06	2	0.43
25	434	3	0.69	3	0.69

5 and 10 days were somewhat higher than those of controls. However, low-temperature treatment for an extended period negatively affected pollen development. Frequency of embryo induction decreased sharply when low-temperature treatments were given for more than several days. Optimal low-temperature treatment for *Citrus* flower buds was a temperature of 3°C for 5 to 10 days.

Optimal Medium and Plant Hormone Combinations

Cytological observation of microspore development in *Citrus* demonstrated that equal divisions of pollen grains were often seen in anthers pretreated at low temperature. *Citrus microcarpa* Bge. was used in a 1981 experiment in which N_6 basal medium was supplemented with different concentrations of cytokinin and auxin to study the effect of plant hormone combinations on embryo differentiation. The experimental results are shown in Table 2. Calli of somatic cells were formed when BA concentration was 4.4 or 8.8 μM. Induction frequency of somatic cell-derived callus increased with increasing concentration of 2,4-D. However, frequency of embryo induction decreased to zero when the 2,4-D concentration increased. This indicated that plant hormone combinations suitable for the callus formation inhibited the development of pollen-derived embryos.

The absolute level of 2,4-D in the medium, rather then the ratio of BA and 2,4-D, played a decisive role in *Citrus* embryo induction. Table 2 shows that 0.23–0.45 μM 2,4-D was favorable to embryogenesis. If the concentration was greater than 4.5 μM 2,4-D only callus, and no embryo, was formed. Thus, the optimal plant hormone combination for induction of pollen-derived embryo was 4.4 μM BA and 0.45 μM 2,4-D.

TABLE 2. Effect of Different Plant Hormone Combinations on Embryo and Callus Formation

Plant Hormone Combinations		No. of Anthers Inoculated	Callus from Somatic Cells		Pollen-Derived Embryo	
BA (mg/l)	2,4-D (mg/l)		No.	%	No.	%
2	0.05	481	0	0	4	0.83
	0.1	612	1	0.16	2	0.33
	0.5	678	10	1.47	0	0
	1.0	569	8	1.41	0	0
	2.0	667	18	2.70	0	0
	4.0	622	34	5.47	0	0
1	0.05	524	12	2.29	6	1.15
	0.1	751	9	1.20	10	1.33
	0.5	607	9	1.48	1	0.16
	1.0	477	3	0.63	1	0.21
	2.0	562	16	2.85	0	0
	4.0	570	20	3.51	0	0

*Flower buds inoculated were treated at 3°C for 10 days.

Selection of Optimal Culture Conditions for the Development of Microspores

Four environmental conditions were tested for anther culture of *Citrus* (Table 3). Embryos did not appear at 26–30°C in either light or dark but appeared at 20–25°C. Suitable temperature was an important factor for pollen embryogenesis of *Citrus*. A culture temperature of 20–25°C was the best, but no embryo appeared above 25°C. This is in accordance with the optimal temperature for *Citrus* growth. Therefore, optimal temperature for induction of fruit tree anther cultures is probably related to op-

TABLE 3. Effect of Light and Temperature on Callus and Embryo Formation*

Culture	Conditions	No. of Anthers Inoculated	Callus of Somatic Cell		Pollen-Derived Embryo	
			No.	%	No.	%
20–25°C	Light	583	12	2.0	2	0.34
	Dark	680	44	6.47	15	2.21
26–30°C	Light	490	0	0	0	0
	Dark	661	38	5.75	0	0

*Medium: N_6 + 2 mg/l BA + 0.05 mg/l 2,4-D + 10% sucrose, pH 5.8; flower buds inoculated were pretreated at 3°C for 10 days; light illumination for 12 hr; light intensity: 500 to 800 lux.

timal ecological condition of the tree species. The highest induction frequency of pollen-derived embryos (21.1%) was obtained in the dark at 20–25°C.

KEY TO ORGAN DIFFERENTIATION

Plant Growth Regulator Concentrations in Differentiation Medium

Green embryos were transferred to the modified MS basal medium supplemented with 1.1 μM BA, 0.57 μM IAA, 500 mg/l LH, 58.7 mM sucrose, pH 5.8. They were placed at 20–25°C under 12 hr of light. Some of the embryos formed plantlets. However, organ differentiation was inhibited in the majority of embryos, and they did not develop into plantlets but produced pseudoplantlets or formed abnormal embryos with bugle- or beltlike cotyledons. This especially occurred with higher concentrations of BA. It was, therefore, necessary to add low concentrations of plant hormones to induce organ differentiation and thus increase frequency of plantlet formation.

Root Inducing Medium

An experiment was performed to increase rooting of plantlets because roots of pollen-derived plantlets in *Citrus* are usually weak and few. Plantlets developed whenever 0.49–0.98 μM IBA was added to the basal medium or basal parts of the plantlet were soaked in 245 μM IBA for 2 hr. When plantlets were transferred into medium with 11 μM NAA, a large mass of callus formed at the plantlet base and the plantlet died. A few roots were formed when 5.4 μM NAA was used. Therefore, adding 0.49–0.98 μM IBA to the medium or soaking the basal part of the plantlet in 245 μM IBA was good for root formation (Fig. 1).

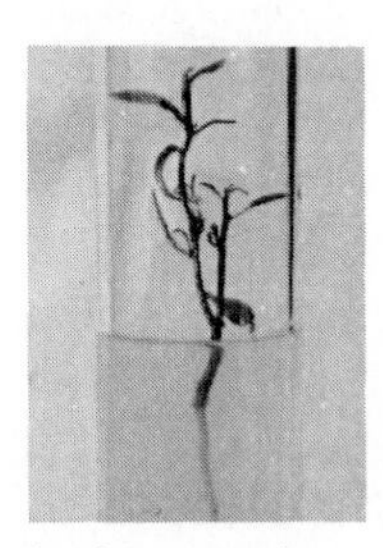

Figure 1. Pollen-derived plantlet in test tube.

TRANSPLANTING TECHNIQUES FOR CULTURED PLANTLETS

Anther-derived plantlets of *Citrus* were extremely weak. They usually died after being transferred from the test tube to soil because of the sudden change of environmental conditions. However, the author was successful in transferring large numbers of plantlets from the test tube into the soil (Fig. 2). Attention should be given to several points.

Growth of Strong Healthy Plantlets

In general, weak plantlets have difficulty after transplantation. Therefore, young plantlets should be soaked in 245 μM IBA for 2 hr and then transferred into wide-mouthed bottles for culture. Plantlets grew well-developed root systems under these conditions, and survival rates were increased after transplanting.

Prevention of Plantlet Infection

Newly transplanted young plantlets have low resistance. They occasionally die from rot caused by infection of the root neck. This is apparently the main cause of early death of plantlets after transplantation. It is, therefore, necessary to wash off the medium attached to the root at the time of transfer, to cut off the old root, and then to smear the root neck with fungal solution to prevent rot.

Preparation of Soil for Transplanting

Transplantation soil should be prepared from peat, sand, and garden compost in ratios of 2:1:1. This mixed soil is suitable for transplantation of test tube plantlets.

Figure 2. Plantlet transplanted from test tube into soil.

Humidity Control

Transplanted plantlets are covered with glass to maintain an appropriate level of humidity. Plantlets were watered once a day or once every 2 days. The glass cover was partially removed after 2 weeks and aeration was increased gradually from then on. The glass cover was removed after 50 to 60 days and the plantlets were then adapted to natural environmental conditions for cultivation.

After young plantlets have survived, dilute nutrient solution may be used to accelerate their growth. The root neck should be examined often after transplantation; if it is seriously infected, the stem above the soil could be cut and it might survive. In addition, control of both temperature and light is necessary for normal growth of transplanted plantlets.

HISTOLOGICAL AND CYTOLOGICAL OBSERVATION OF DEVELOPMENT OF MICROSPORES

Observation of Microspore Development into Embryos

To investigate pollen development of *Citrus*, anthers were fixed with Carnoy's fluid after culture periods of 0, 10, 20, 30, 40, 50, 60, 80, and 100 days and sectioned by using the standard paraffin method. Development of microspores into embryo was observed over time.

ANTHER DEVELOPMENT PRIOR TO IN VITRO CULTURE. Pollen grains were inoculated at the mid- and late-uninucleate stages and the anthers were light yellow. The pollen grains were surrounded by the tapetum, which could be seen distinctly by cross section. However, different anthers from the same flower bud or different pollen grains from the same anther varied in developmental stages. Anthers should be inoculated when most of their pollen grains are at the late-uninucleate stage.

VARIATION OF ANTHER AFTER IN VITRO CULTURE. After the anther was cultured for 20 to 30 days in suitable medium under optimum culture conditions, the first mitosis of a few pollen grains was seen and binuclear or tetranuclear pollen grains formed. Meanwhile, the tapetum disappeared.

In anthers cultured for 40 to 60 days, fewer pollen-derived embryos and binuclear, tetranuclear, or multinuclear pollen grains were observed. It was demonstrated that development of microspores in *Citrus* was extremely varied even when they were derived from the same anther. Cell masses or embryos of different sizes could be observed in the sections of anther cultured for 80 to 100 days. There were also microspores in various developmental stages. Examination of pollen development from single cell to multicell through embryo stages demonstrated that the *Citrus* plantlet derived from anther culture originated from pollen. Observation further demonstrated that the development of *Citrus* microspores in in vitro culture took a long time. It has been reported that in *Triticum aestivum* L. anther culture, two or three cells were formed from individual pollen grains 5 to 7 days after inoculation. After 2 days, mitosis reached its peak. However, binuclear or tetranuclear cells were seen in *Citrus* anthers 20 to 30 days after culture, and cell mass began forming 20 to 30 days later. Even in anthers inoculated for 80 to 100 days, some pollen grains were still observed at the binuclear or tetranuclear stage, although some pollen grains had already formed embryos.

CHROMOSOME NUMBER EXAMINATION. To identify the ploidy level of the chromosome of the differentiated plantlet, root tip tissue or embryo was taken in the morning, fixed with Carnoy's fluid (alcohol/acetic acid: 3:1) for 24 hr, washed with alcohol, hydrolyzed with 1 N HCl at 60°C for 15 min, stained with Schiff's fluid and safranin, and squashed for chromosome examination. The results revealed that the chromosome number of embryo and root tip cells was haploid (n = 9), demonstrating that the plantlets were haploid and developed from pollen.

REFERENCES

Chen, Z., Wang, M., and Liao, H. 1980. The induction of *Citrus* pollen plants in artificial media. *Acta Genet. Sin.* **7**(2):189–191 (in Chinese, English abstract).

Rangan, T. S., Murashige, T., and Bitters, W. P. 1969. In vitro studies of zygotic and nuceller embryogenesis in *Citrus. Proc. First Int. Citrus Symp.*, vol. I. (H. D. Chapman, ed.). pp. 225–229.

Rosati, P., Devreux, M., and Laneri, U. 1975. Anther culture of strawberry. *Hort. Science* **10**:119–120.

Hidaka, T., Yamada, Y., and Shichijo, T. 1979. In vitro differentiation of haploid plants by anther culture in *Poncirus trifoliata* (L). Raf. *Jap. J. Breed.* **29**(3):248–254.

Xue, G. and Fei, K. 1981. A preliminary report on plantlet formation of Delicious apple through anther culture. *Zhongguo Guoshu* **1**:40–44 (in Chinese).

———, Fei, K., and Hu, J. 1981. Haploid plantlet of apple obtained by in vitro anther culture. *Acta Hortic. Sin.* **8**(1):9–11 (in Chinese, English abstract).

Zhu, Z., Sun, J., and Wang, J. 1977. Cytological study on androgenesis in *Triticum aestivum* L. In: *Proc. Symp. Anther Cult.*, pp. 113–120. Science Press, Beijing (in Chinese).

———, Wang, J., Sun, J., Xu, Z., Zhu, Z., Yi, G., and Bi, F. 1975. Establishment of a better medium for anther culture in *Oryza sativa* through comparative experiment of nitrogen sources. *Sci. Sin.* **5**:484–490 (in Chinese).

Zou, C. and Li, P. 1981. Induction of pollen plants of grape (*Vitis vinifera* L.). *Acta Bot. Sin.* **23**(1):79–81 (in Chinese, English abstract).

CHAPTER 26

Litchi: Anther Culture

*Lianfang Fu**

INTRODUCTION

Economic Importance

Litchi, *Litchi chinensis* Sonn., is one of the best fruit trees growing in the subtropics. Its fruits, with beautiful color, fragrant smell, and sweet taste, are in demand in both national and international markets. They are nutritious: every 100 ml of fruit juice contains 62.04 mg vitamin C, and the content of soluble substances in fruits reaches 20%. Its nutritional and medicinal functions have been recorded in "Compendium of Materia Medica," an ancient book on Chinese herbal medicine written in the Ming dynasty more than 400 years ago. Besides being eaten fresh, litchi can be used for making dry fruits, canned fruits, wine, and other processed products. The flower, seed, skin of the fruit, and branches can be used as medicine. The flower of litchi is fragrant and provides a good honey source for developing bee culture. The branches and roots, rich in tannin, can be used for extraction of dyes. The trees, hard and with a well-developed root system, can be planted to protect embankments. In short, litchi has a wide range of uses and a very high practical value.

*English translation by Kaiwen Juan.

Geographical Distribution

Litchi trees originated in the south part of China. The wild species is distributed naturally in the mountains in the southwest of Guangdong Province and Hainan Island. There are many primeval forests of wild litchi in Hainan Island. It grows together with *Dryobalanops aromatica* and is one of the major species constituting tropical rain forests in low mountainous regions.

Litchi is an evergreen fruit tree growing in subtropical areas. According to records, China has a history of cultivating litchi for over 2000 years. It is mainly cultivated between 18 and 25 degrees north latitude. Litchi is a long-lived wood species. Ancient litchi trees, over 1000 years old, can be found today. In China the total acreage under litchi cultivation is about 60,000 ha, with an annual output of 75 million kilograms, ranking first in the world. The cultivated litchi varieties are mainly grown in the provinces of Guangdong, Fujian, Sichuan, Taiwan, and Yunnan, and also in Guangxi Zhang Autonomous Region.

The United States introduced litchi to successful cultivation from Fujian Province at the end of the last century, and litchi products are now sold in many countries of America and Europe. Litchi was also introduced directly or indirectly from China to Thailand, Indonesia, the Phillippines, Cuba, Panama, Brazil, and Australia.

Taxonomy

According to systematic botany, litchi belongs to the genus *Litchi* Sonn, under the family Sapindaceae. There are only two species of the genus *Litchi*.

***LITCHI CHINENSIS* SONN.** Originating in the south of China, *Litchi chinensis* Sonn. is a big tree about 10 m high. The leaf has two to four pairs of leaflets. It has terminal panicles, light-yellow flowers, usually seven pistils, an inverse heart-shaped ovary with two to three locules and a short stalk, elliptical or spherical fruits, and long-elliptical dark brown seeds. The fruits, sour and sweet when they are ripe, are edible.

***LITCHI PHILIPPINENSIS* RADIK.** Originating in the Philippines, *L. philippinensis* Radik is a big tree, with one to two pairs of leaflets, long-elliptical fruits with long stipule and short arils. The fruits are inedible.

Problems in Breeding and Cultivation

Litchi is a perennial fruit tree, beginning to bear fruits when 6 to 8 years old. In production, new strains are usually obtained from spontaneous mutation in stems and buds, which are very scarce. Therefore, new breeding techniques should be developed for the following characteristics.

DEVELOPING STABLE STRAINS WITH SMALLER STONES. Some varieties have big fruit stones and little available fruit. It is very important to develop stable strains with smaller stones to improve the quality of fruits.

CREATING A HIGH RATE OF FRUIT SETTING STRAINS. Litchi has a large number of flowers but few pistillate flowers; therefore, fruit yields are low. It is important to increase the yield of litchi production.

IMPROVING STORABILITY. The harvested fresh fruits are highly perishable. They have characteristics of, as a Chinese folk song says, "turning color in 1 day, changing taste in 2 days, and rotting in 3 days." Therefore, storability of fresh fruits must be improved in order to provide more fresh fruits for the market.

ANTHER CULTURE TECHNIQUES

Anther culture of litchi was conducted in 1981 in our laboratory, and haploid plants were obtained the next year. In order to perfect cultural techniques, the studies were carried out in the following 2 years (Fu and Tang, 1983).

Culture Steps

Pollen-derived plants of litchi were achieved through four steps:

1. Calli were first initiated from anthers. The appropriate medium was MS medium with 4.5–9.0 μM 2,4-D, 4.6–9.3 μM KIN, 2.7–5.4 μM NAA, and 88 mM sucrose.
2. The callused anthers were transferred to fresh medium, with three to four callused anthers per test tube, for further induction of callus.
3. Calli were transferred to differentiation medium for embryo induction. The appropriate differentiation medium was MS medium or modified B5 medium with 2.3 μM KIN, 0.54 μM NAA, 500 mg/l LH, 400 mg/l royal jelly, and 88 mM sucrose (Haploid Breeding Group, Institute of Botany, Academia Sinica, 1972).
4. Visible embryos at 2–3 mm in diameter were transferred to plant regeneration medium. Plantlets with stems, leaves, apical buds, and other organs were regenerated (Fig. 1). The appropriate medium was MS medium or modified B5 medium with 2.3 μM KIN, 2.9 μM GA,

Figure 1. Intact plantlet of litchi.

500 mg/l LH, 400 mg/l royal jelly, 10.95 mM glutamine, and 88 mM sucrose.

Ten-milliliter aliquots of medium were dispensed into 18–20-mm culture tubes, pH was adjusted to 5.8, and tubes were autoclaved.

Experimental Materials and Inoculation Techniques

1. Flower spikes were taken for microscopic examination of developmental stages. Anthers of late-uninucleate developmental stage were used for culture. It was observed that most anthers in flower buds that were 3 to 4 mm long were at this stage.
2. The expanded flower buds and new ones that appeared were removed. The spikes were wrapped in sterile gauze.
3. The wrapped buds were first soaked in 95% ethanol for several seconds, then immersed in 0.1% mercuric chloride solution for 10 to 12 min, and finally washed with sterile water three to four times.
4. Anthers were taken from the sterilized buds and inoculated into medium. Each tube was inoculated with about 30 anthers.
5. The inoculated anthers were incubated at 24–26°C, and illumination was provided for 10 hr/day at an intensity of 2000 lux. For plantlet regeneration, the light intensity should be increased to about 3000 lux.

Cytological Examination

Before inoculation, flower buds were stained with glacial acetocarmine for an examination of the pollen's developmental stage.

When embryos and root tips were taken for cytological observation, they were first pretreated in saturated *p*-dichlorobenzene solution (or in 0.1% colchicine solution) for 3 hr, fixed in the solution of alcohol and glacial acetic acid at a ratio of 3:1 for 24 hr, kept in 70% ethanol solution, hydrolyzed for 20 to 40 min, stained with hematoxylin–iron trichloride, and smeared.

ESSENTIALS FOR SUCCESSFUL ANTHER CULTURE

Selecting Appropriate Medium and Supplements

DIFFERENTIATION MEDIA. Two basal media were employed for callus induction: MS medium and modified B5 medium (with macroelements and ferrous salts of B5 and microelements and organic growth substances of MS). The experiments proved that MS medium was more effective for callus induction.

Hormones. Using two litchi varieties, Chenzi and Gushan Jiaohe, as experimental materials, two different hormone concentrations were tested. One was MS medium with 9 μM 2,4-D, 9.3 μM KIN, 9 μM IAA, and 88 mM sucrose; another was MS medium with 9 μM 2,4-D, 9.3 μM KIN, 2.75–5.4 μM NAA, and 88 μM sucrose. In the first treatment 1500 anthers were inoculated, with a frequency of 80% callus formation; 300 anthers were inoculated in the second treatment, without any callus formation. This indicates that in callus induction medium, besides KIN, two auxins should be added. Among the auxins, 2,4-D was essential, and IAA was better than NAA.

Sucrose. Sucrose concentrations of 88 mM and 176 mM were tested using the variety Chenzi as experimental material. The former concentration produced a frequency of 80% callus induction and the latter 52%. This showed that medium containing 88 mM sucrose concentration was more favorable for callus induction (Table 1).

DIFFERENTIATION MEDIA. *Selection of Media.* MS medium, modified B5 medium, 5 × White medium, 2 × MS medium, and White medium were tested for embryo induction. Results showed that young em-

TABLE 1. Effects of Sucrose Concentration on Callus Induction

Sucrose Concentration (mM)	No. of anthers inoculated	Callusing Anthers		Pollen-Derived Callus	
		No.	Percentage	No.	Percentage
88	200	160	80.0	20	10
176	200	105	52.5	12	6

bryos could not grow, but turned brown and died when they were inoculated into White medium (its microelements and supplements were similar to that in the modified B5 medium). While the effect of 2 × MS medium was not very significant for embryo growth, the color of embryos changed from green into deep green. The macroelements in MS, 5 × White, and modified B5 media favored embryo growth. The embryos were enlarged from 2 mm to about 12 mm. Macroelements of White and 2 × MS media were unfavorable for embryo development.

Hormones. Different hormones and concentrations were tested. The following three combinations were effective:

1. 2.3 μM KIN with 0.54 μM NAA
2. 2.3 μM KIN with 0.57 μM IAA
3. 2.3 μM KIN with 2.9 μM GA

In all three combinations, 400 mg/l royal jelly and 500 mg/l LH were added. MS medium containing 88 mM sucrose was the basic medium with pH at 5.8.

After 20 days' inoculation, the calli from anther walls and filaments were transferred into differentiation medium. They continued to proliferate, but no organ was differentiated, and they finally died. However, when pollen-derived calli were transferred into differentiation medium, they first formed transparent small embryos then further developed into embryos in various sizes and shapes.

Different kinds of auxin affected callus differentiation. When granular calli were inoculated into MS, modified B5, and other media with 2.3 μM KIN, 0.57 μM IAA, 500 mg/l LH, 400 mg/l royal jelly, and 88 mM sucrose, embryos were differentiated in 1 week. One cone-shaped callus 8 mm × 6 mm could usually produce 20 to 30 embryos. If the callus was put in a similar medium in which 0.57 μM IAA was replaced by 0.54 μM NAA, it produced 100 to 200 embryos in 1 month. It was revealed from repeated experiments in 2 years that low concentration of NAA greatly stimulated embryo formation, but it was not as effective as IAA in promoting embryo growth (Table 2).

Regenerating Intact Plants

EFFECTS OF SUCROSE CONCENTRATION ON ROOT INITIATION. In rooting medium with 29.3 mM sucrose, roots induced from embryos were thin and delicate; in medium with 88 mM sucrose, the initiated root system was thick and strong. The roots were loose and brittle in medium with 146.5 mM sucrose. From these findings it was concluded that 88 mM sucrose was optimal for embryo culture.

TABLE 2. Effects of NAA and IAA on Embryo Induction*

Year	Treatments (μM)	No. of Embryos per 10 Calli	Size of Embryos (mm)	Time Required for Embryo Induction (Days)
1981	NAA 0.54	100–200	1–3	30
	IAA 0.57	20–30	2–15	7
1982	NAA 0.54	100	1–4	30
	IAA 0.57	30–40	1–10	10

*MS medium with 2.3 μM KIN was used for all treatments.

REGENERATION OF NORMAL PLANTS. Most shoots that differentiated from embryos could not develop stems or they grew abnormally with curved stems in MS medium with 500 mg/l LH, 400 mg/l royal jelly, and 0.47 μM KIN. When the quantity of KIN was increased from 0.47 to 2.3 μM and 2.9 μM GA was added, the plantlets showed normal growth. It was concluded that the KIN concentration should not be less than 2.3 μM, and 2.9 μM GA was essential for embryo differentiation.

EFFECTS OF GLUTAMINE ON THE GROWTH OF PLANTLETS. When embryos with developed shoots and roots were transferred to the differentiation medium with 10.95 mM glutamine described previously, they grew rapidly. Glutamine stimulated plantlet growth even in the medium with a low hormone level.

Experiments over several years showed that an appropriate increase of KIN shortened the culture period. For instance, anthers inoculated in April 1981 produced plantlets in May 1982. Altogether 13 months elapsed from anther inoculation to plant formation. After the increase of KIN from 0.47 to 2.3 μM, the culture period was greatly shortened. Anthers inoculated in April 1984 produced plants in October 1984, after only 180 days.

It should be mentioned that the following problems occurred during plant regeneration from embryos; they were solved as described.

Embryos Turning Brown and Callused. In order to promote the normal embryo growth, besides supplementation of hormones and organic substances in MS medium, three different concentrations (100, 200, and 400 mg/l) of royal jelly were tested. A concentration of 400 mg/l was the most effective: it not only inhibited the callus formation from embryos but also prevented the embryos from turning brown and aging.

Effects of Concentrations of Vitamins B_1 and B_6 on Cotyledon Initiation. When concentrations of vitamins B_1 and B_6 were reduced below 0.3 μM and 3.0 μM in MS medium, respectively, embryos could not grow. When concentrations were increased up to 59 μM, cotyledons were differentiated from young embryos about 3 mm long. A high concentration, therefore, of vitamins B_1 and B_6 was indispensable for embryo differentiation (Lou, 1978).

CHARACTERISTICS OF MORPHOGENESIS

Initiation and Differentiation of Callus

Calli were initiated from anther walls 3 weeks after inoculation. They were light brown and irregular in shape and proliferated rapidly. Another type of calli developed from anthers after 2½ months of culture. They were pollen-derived calli, loose in texture, pale-yellow in color, and granular in shape, and they proliferated slowly. Calli derived from anther wall did not differentiate even after passages in differentiation culture, and finally died. The granular pollen-derived calli developed gradually into transparent or milky white embryos in differentiation media; usually the milky ones grew faster than the transparent ones.

Embryo Development and Regeneration of Plants

Cotyledons were regenerated from small spherical embryos in different ways. Some embryos developed one cotyledon first and later another cotyledon arose from the other side. Some embryos developed two cotyledons simultaneously. In some irregular embryos, plumule and radical differentiated gradually when they grew to a certain size. At a low concentration of hormones, the embryos could not develop into intact plantlets.

Embryonic Culture

Embryos were separated when they were at the spherical, heart-shaped, or torpedo-shaped stage. Separated embryos were inoculated in MS or modified B5 differentiation medium with low or no hormones. New embryos or embryonic cells could then be regenerated. Once embryonic cell cultures capable of subculture were induced, they could undergo normal proliferation and differentiation even in medium without auxin and cytokinin. After embryos form through repeated subculture, the concentration of kinetin should be increased gradually in the medium in order to regenerate plantlets.

CYTOLOGICAL OBSERVATION

Root tips from 24 regenerated plantlets were cytologically examined. In the root tip cells that were observed, the chromosome number was

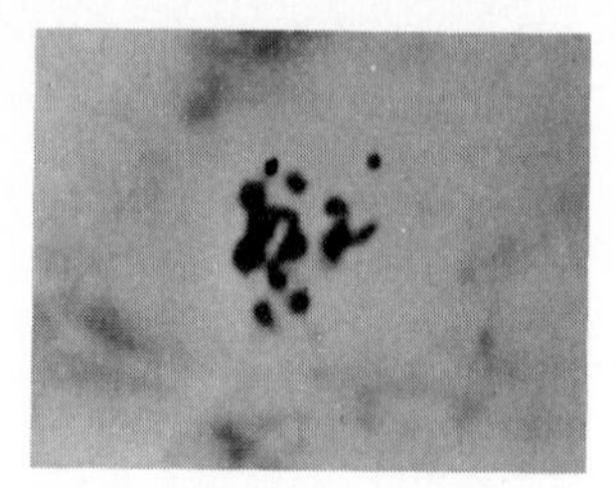

Figure 2. A root tip cell (n = 15).

mostly haploid, $n = 15$ (Fig. 2). Aneuploids and triploids were also observed. The chromosome numbers of aneuploids were usually about 15. In embryos subcultured for long periods, more aneuploids and triploids were found and fewer haploid cells were observed. Most of the root tip cells were found to be aneuploids in the embryos regenerated from embryonic cell cultures; their chromosome number was about 45 or more. If plantlets can be regenerated from triploids in the future, they may have practical value in production. The results of chromosome counting on pollen plants and embryos have shown that they are derived from pollen.

FUTURE PROSPECTS

Pollen-derived plants have been obtained by anther culture in two litchi varieties, Chenzhi and Gushan Jiaohe, which are improved cultivars. The regeneration frequency of pollen plants is being increased. It is possible to develop economically important pure strains from the two varieties by anther culture. Future studies will be directed to observing androgenesis, improving culture techniques, and further enhancing the regeneration frequency so as to perfect the whole culture process and accelerate the use of pollen-derived plants in breeding.

KEY REFERENCES

Fu, L. F. and Tang, D. 1983. Induction of pollen plants of litchi tree (*Litchi chinensis* Sonn). *Acta Genet. Sin.* **10**(5):369–374 (in Chinese, English abstract).

Haploid Breeding Group, Institute of Botany, Academia Sinica. 1972. In vitro culture of haploid plants and its significance in breeding. In: *Collections of Haploid Breeding*. Vol. 1, pp. 1–8. Science Press, Beijing.

Luo, S. W. 1978. Plant tissue and cell culture: Recent advances and application. *Acta Phytophysiol. Sin.* **4**(1):91–112 (in Chinese).

CHAPTER 27
Longan: Anther and Embryo Culture

Wenxiong Wei

INTRODUCTION

Economic Importance and Geographical Distribution

Longan (*Dimocarpus longana* Lour.) is a subtropic evergreen fruit tree. The longan fruit is pleasant tasting and nutritious. In addition to its high sugar content, it has vitamin C (143.1–163.9 mg/100 g) and vitamin K (196.5 mg/100 g). The fresh fruit when dried is called "Gui Yuan" in Chinese. Gui Yuan has long been considered an effective medicine. The canned longan fruits and longan paste are important commercially. The seeds are used for brewing or making high-quality active charcoal. Its roots, flowers, and leaves have also been used in traditional medicines. The root system contains bacteriorhiza, making it well suited to the hilly red loam soil. The longan tree is long-lived, branched, and leafy and has a good shape. It is well known as an ornamental tree and a source of nectar for bees. The wood is hard, bright purple, and closely grained. It is good carving material that is used for production of high-quality furniture.

Longan originated in southern China. There are large areas of wild longan trees growing in the southeast to southwest district in Yunnan

Province and in the Wu Zhi Mountain (Five Fingers Mountain) in Hainan Island, Guangdong Province, China.

According to history recorded in ancient publications, the cultivation of longan in China began about 3000 years ago (Sun, 1983) in the early years of the Han dynasty. At present, longan is distributed in the provinces Fujian, Taiwan, Guangdong, Guangxi, Sichuan, Yunnan, and Guizhou. Longan yield in Fujian Province was 60% of that of the whole of China (Longan Investigation Division, Fujian Provincial Horticulture Society, 1983). Longan is also cultivated in Thailand, India, Vietnam, and the Philippines. Since the 19th century, it has been introduced into most tropical and subtropical zones of Europe, America, and Africa. Most of these countries are still in the introductory stages of cultivation. At the present time, China is the leader in longan production because of the vast cultivation areas, high yield, good quality, and numerous varieties.

Taxonomy

Longan belongs to Sapindaceae, *Democarpus* Lour., Euphoria. There are 10 species in Democarpus Lour. (*Euphoria* Comm.). All of them originated in tropical zones in Asia. Of the 10 species, 3 originated in China; only 1 of the 3 in China was cultivated as a fruit tree, longan (*D. longana* Lour., *E. longana* Lour., *Nephelium longana* Lam; chromosome number $2n = 30$). It is closely related to *Litchi chinensis* Sonn. botanically, as both belong to the same family. Longan is still propagated by seed. For breeding, plants of high quality are gathered from farmers for identification and release. Hybrid breeding is currently in the early stages. Seedlings of excellent varieties are difficult to obtain. Attention is currently focused on the breeding of disease-resistant, cold-tolerant varieties with shrunken fruit stones or kernels.

Present Status and Value of Tissue Culture

Callus and embryos were induced from young cotyledon explants, and intact plantlets were obtained in 1981 (Wei et al., 1981). Since then, haploid plantlets have been produced by in vitro anther culture, and plantlets were grown directly from culture of young embryos. Many plantlets were also formed through the formation of calli or embryos (Yang et al., 1984, 1985). Many of these results have not yet been reported in the Chinese literature. It is not possible to obtain homozygous diploid plants by self-fertilization since longan is a perennial, self-incompatible perennial crop. However, haploid plantlets may be obtained by anther culture, and homozygous diploid lines may be produced by chromosome doubling. Homozygous diploid plants obtained by this method are useful as parents for hybrids and are useful material for studies of longan genetics. It has been shown by examination of chromosomes of anther-derived plants that some cells were aneuploid or diploid, in addition to those that were haploid ($n = 15$), which comprise the majority. This has provided the opportunity for the production of plantlets with variation in ploidy level from progeny of anther-derived plantlets.

The in vitro culture technique of young cotyledons provided a way to propagate hybrid plants vegetatively. It also provided a method for selecting elite plants and breeding resistant strains and mutations.

It should be noted that some rare longan varieties having shrunken stones were discovered among natural varieties. The shrunken-stoned variety was named because the embryo is aborted during development or develops slowly, causing the stone in the mature fruit to become shrunken, with thickened flesh; increase in the edible portion; increase in the soluble, solid flesh component; and general improvement in overall fruit quality. These qualities make the shrunken variety one of the most desirable. It has been assumed that fruit formation with shrunken stone results from incomplete fertilization and underdeveloped endosperm, so that the embryo was not adequately nourished during its development and was aborted or developed slowly. It has also been suggested that this may be due to variation in chromosome structure, inducing abortion (Zhejiang Agricultural University, 1980). Nevertheless, the trait of shrunken stones is an inherited characteristic. To date no homozygous plants have been found among the natural varieties having shrunken stones. Hence, fruit of trees with shrunken stone fruits do not have 100% shrunken stone fruits. Moreover, most plants with shrunken kernels have unstable traits and some undesirable qualities. Therefore, research is aimed at improved screening for better longan varieties with shrunken stones. Since the seeds of shrunken stones have lost their viability, they do not germinate after sowing. The anatomy of embryo development has been well investigated by Yang et al. (1983). It was confirmed that embryos of shrunken stones develop during early developmental stages but are aborted during later development. Therefore, embryo culture has been investigated. To date, more than 30 strains with shrunken stone have been bred successfully by embryo culture.

TISSUE CULTURE TECHNIQUES

Culture Procedure

The induction of longan anther culture and the production of plantlets consisted of the following steps:

1. Anther callus or embryogenic tissue was induced on MS medium supplemented with 4.6 μM KIN, 9 μM/ 2,4-D, 5000 mg/l active charcoal, and 5% sucrose. Thirty to 40 days was required for culture.
2. Embryos were induced by differentiation. It was necessary to reduce the concentration of the inorganic salts. Rhododendron medium (Anderson, 1975, 1978) (Table 1) could be used instead of MS medium. Table 1 shows that most inorganic salt concentrations in rhododendron medium are lower than in the MS medium. Optimal plant hormone concentrations were 2.2–4.4 μM BA, 0.54 μM NAA, with 88 mM sucrose. Thirty to 80 days was required for embryo differentiation.
3. Embryos were subcultured on new medium for stem elongation and leaf expansion to form plantlets. Rhododendron medium or MS medium could be used in this stage. However, rhododendron medium was generally better than MS medium. BA (1.32 μM), IAA (0.57 μM), and 88 mM sucrose were added to the medium. A period of 100 to 130 days was required for plantlet regeneration. Medium was solidified with 0.65% agar and pH was adjusted to 5.8.

Development of longan young embryos into plantlets was similar to that of anther culture. After culture of sexual embryos large quantities of asexual embryos were obtained by vegetative propagation. A sexual embryo may also germinate directly into a seedling.

There are two ways to subculture. One is to maintain integrity of the proembryo for multiplication of similar new proembryos. The other is to dedifferentiate the embryos into callus, which differentiates into new embryos. In longan tissue culture (including anther culture), strong viability of embryo-derived plants has been observed. It has been shown that some asexual embryos subcultured for 12 generations over a 2-year period retained strong propagation ability. The ability to develop into plantlets was maintained for 10 embryo generations.

TABLE 1. Comparison of Concentrations of Macroelements and Ionic Salts in MS Media and Rhododendron Media

Chemicals*	MS Medium	Rhododendron Medium
KNO_3	20.6 mM	5.2 mM
NH_4NO_3	18.8 mM	4.6 mM
KH_2PO_4	1.25 mM	—
$NaH_2PO_4 \cdot H_2O$	—	2.8 mM
$FeSO_4 \cdot 7H_2O$		
Na_2-EDTA	100 μM	200 μM

*Components of MS medium not included in Table 1 are the same in both media.

There were two ways in which embryos proliferated into new embryos: The first was proliferation at a fixed site, in which a young embryo was formed perpendicularly at the concave of the heart-shaped embryo. The second was proliferation at an unfixed site, in which numerous new embryos were formed at any part of the embryo, such as the edge or surface of a cotyledon. The latter was often seen. However, no difference was found between the embryos formed by either way.

Cell Culture Techniques (Yang and Wei, 1984)

ANTHER CULTURE

1. Male flowers or hermaphrodites which were about to flower at different parts of the inflorescence were collected. Pollen development was observed by acetocarmine staining. Closed flower buds at mid- and late-uninucleate stage 3 mm in diameter were selected. The flower head was cold-pretreated for 24 hr or left untreated.
2. The flower surface was sterilized with 0.1% mercuric chloride for 10 min and rinsed with aseptic water four or five times.
3. Anthers were removed aseptically and inoculated in 25 ml of medium in a 50-ml flask. Seventy anthers were placed in each flask.
4. After inoculation, anthers were cultured at 25–27°C. Callus induction was carried out in complete darkness. Embryo and plantlet formation was carried out with light for 10 to 12 hr/day at 2500 lux.
5. Root tips were fixed with a solution of glacial acetic acid and alcohol (3:1), stained with haematoxylin and squashed for observation of chromosomes.

KEY POINTS IN TISSUE CULTURE

Selection of Media

MS medium was successfully used throughout the three steps in the young cotyledon culture of longan. However, in anther culture, MS medium was favored for callus propagation when embryos were differentiated from callus. If rhododendron medium was used, the results were superior to those with MS medium. This is demonstrated not only in the variation in differentiation frequency of the embryos but also in the quality of the embryos (Table 2).

It is shown in Table 1 that inorganic nitrogen and potassium in rhododendron medium is sharply decreased and that iron and sodium are increased compared to MS medium. These differences were correlated with differences in embryo development. In rhododendron medium, the

TABLE 2. Effect of Different Media on Embryo Formation from Pollen-Derived Callus of Longan

Medium*	Total No. of Callus Inoculated	No. of Embryos		Average Frequency of Embryo Induction Based on 9 Repeats (%)	Total
		Yellow Semi-Transparent	White		
Rhododendron	78	1	43	44	52.96
MS	60	3	8	11	17.78

*Includes 2.2 μM BA, 0.54 μM NAA, and 88 mM sucrose.

development of embryo was promoted, embryo growth was vigorous, and embryos were white from the outset. In MS medium, a large amount of calli were produced as well as a large quantity of loose semi-transparent pearl-shaped or heart-shaped embryos of light yellow or yellowish green. The white, vigorous embryos were produced from the embryos only after long-term culture.

In longan tissue culture, oxidation by phenolic compounds produced by the explant were often seen, and toxic substances readily accumulated. It was necessary to add antioxidants to the medium to prevent oxidation and compounds to absorb toxic substances. MS medium was used with 4.6 μM KIN and 9.0 μM 2,4-D in this experiment. The results demonstrated that active charcoal was the best supplement (Table 3).

Walkey (1972) reported that in culture of axillary buds in *Malus pumila*, polyvinyl pyrolidone (PVP) played an important role in preventing cultures from oxidation by phenolic compounds. It is shown that this supplement was far less effective than activated charcoal. Vitamin C was also less effective than active charcoal. Table 4 shows that the order of the effectiveness of these three supplements in anther tissue was similar to those in young cotyledon explant culture. Activated charcoal was 3.5 times more effective than PVP ($p < 0.01$), and vitamin C was far less effective than PVP.

TABLE 3. Effect of Different Medium Supplements on Callus Induction of Young Cotyledons of Longan

Supplements (mg/l)	No. of Explants Inoculated	Callus		Color
		No.	%	
Active charcoal 5000	14	13	92.9	Yellowish green
PVP 5000	12	4	33.3	Light brown
Vc 20.5	16	5	31.3	Yellowish brown

TABLE 4. Effect of Different Supplements on Callus Induction from Pollen in Longan

Supplements (mg/l)	No. of Anthers Inoculated	Formation of Pollen-Derived Callus	
		No.	%
Active charcoal 5000	1083	157	14.5*,†
PVP 5000	992	43	4.3
Vc 5	546	3	0.5

*The difference between active charcoal and PVP treatment was significant, $t = 4.868 > t$ at 0.01 level = 3.355.

†The difference between active charcoal and Vc treatment was significant, $t = 11.322 > t$ at 0.001 level = 8.610.

MS medium was used with 4.6 μM KIN and 9.0 μM 2,4-D, and activated charcoal, PVP, and vitamin C were added separately. The amount and growth rate of the callus from each anther as well as the color and overall quality of the callus were affected by these three supplements (Table 5). In media with active charcoal, about one-third of the callusing anthers produced at least two calli per anther. In media supplemented with PVP, nearly 90% of the callus producing anthers produced only one callus per anther. Thus it is clear that callus induction frequency increased by more than one-third when activated charcoal was added to the medium. By the 40th day after inoculation, calli were 1 mm in length when PVP was added. However, when activated charcoal was added, more than half of the calli were more than 2 mm in length. This demonstrated that androgenesis and callus growth can be rapid. Remarkable differences were found in callus quality. More than 95% of anther-derived longan calli were fresh milky white, milky yellow, or yellowish green in medium containing activated charcoal. In medium supplemented with PVP, nearly 50% of calli were yellowish brown or gray, indicating a low probability that they were capable of further differentiation.

Addition of activated charcoal to medium was first proposed by Nakamura, Itagaki, and Anagnostakis (Maheshwari et al., 1979). Its effectiveness has been shown repeatedly by other researchers. It can be concluded that the beneficial effects resulted from the absorption of substances from the medium which suppress androgenesis and were secreted by the anther. This result demonstrated that toxic substances absorbed by activated charcoal adversely effect androgenesis by decreasing induction frequency and markedly inhibit callus growth, destroying the capacity for further differentiation. In young cotyledon culture, these toxic substances also affected the dedifferentiation of somatic cells and the viability of longan callus.

In the culture of anther and young cotyledons of longan, the pH

TABLE 5. Effect of Different Supplements on the Yield and Quality of Anther-Derived Callus in Longan

Medium (mg/l)	No. of Anthers Inoculated	Number of calli on anthers						Callus diameter* (mm)					
		1		2		3		1		2		3	
		No.	%	No.	%	No.	%	No.	%	No.	%	No.	%
MS medium + active charcoal 5000	41	26	63.4	14	34.1	1	2.4	27	47.4	15	26.3	15	26.3
MS medium + PVP 5000	19	18	94.7	1	5.3	0	0	20	100	0	0	0	0

*Observed 40 days after inoculation.

should be 5.8 and 0.65–0.7% agar should be added to the medium. During the period of callus induction, 4.6–13.9 μM KIN and 4.5–9.0 μM 2,4-D should be added. Cultures should be maintained in complete darkness at 25–27°C. Supplementation to the medium of a combination of 1.32–4.4 μM BA and 0.54–5.4 μM NAA was needed to induce differentiation of embryos from callus. The concentrations of inorganic nitrogen and potassium in the medium in this period should be decreased as previously mentioned. In addition, light illumination at 2500 to 3000 lux was needed for 12 hr/day. In embryo subculture, too high a concentration of plant hormones made the embryo dedifferentiate into callus and turn brown. New embryos were best propagated with lower levels of plant hormones. A combination of 1.1 μM BA and 0.57 μM IAA promoted production of many new embryos. Embryos dedifferentiated into calli if higher BA concentrations or 4.6–9.3 μM KIN was used. Callus proliferated by subculture could be differentiated into new embryos. Light yellow calli gave the best yields.

Five to nine months elapsed from inoculation to callus formation, to embryo induction, and finally to development of embryo into plantlets.

So far, young cotyledon cultures of longan gave better results than anther culture for callus formation and embryo differentiation. The callus formation frequency and embryo differentiation ability reached more than 90%. Success or partial success has been obtained in anther culture in varieties Dong Bi, Oolong Mountain, Hong He Zi, and You Tan Ben. The callus induction frequency ranged from 10 to 15%, and the differentiation frequency of callus to embryo reached 50%. However, embryo germination into a plant takes a long time, but it is believed it can be shortened.

MORPHOGENESIS AND CHROMOSOMAL OBSERVATION OF LONGAN TEST TUBE PLANTLETS

In longan tissue culture, it has been reported that numerous miniball-like pearl-shaped embryos were formed from callus. They were white or semitransparent, soft, and loose in texture. Pearl-shaped embryos were produced, as well as heart-shaped embryo after culture (Fig. 1, upper row center). Most of the asexual embryos of longan were heart-shaped. They could be subcultured and numerous new heart-shaped embryos proliferated or germinated directly into plantlets. Most heart-shaped embryos grew gradually and stopped proliferating when plant hormone levels were reduced and inorganic nitrogen was decreased; then embryos with various shapes were formed (Fig. 1). Adventitious roots or buds could form at the terminals of some irregularly shaped embryos. These

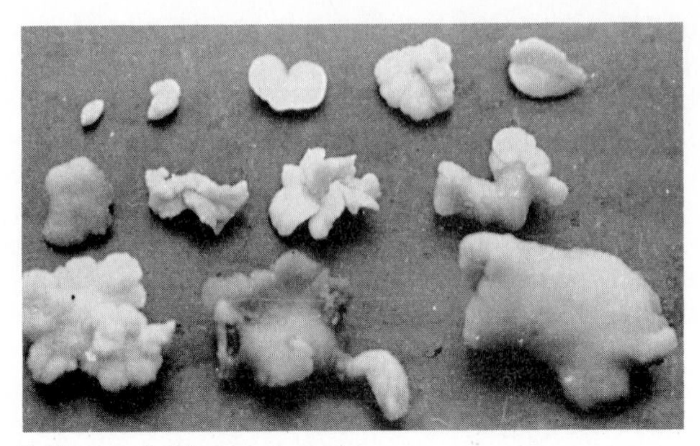

Figure 1. Embryos of different shapes.

were mostly aberrant plantlets which could not grow as healthy test tube seedlings. This phenomenon often occurs in pollen culture. When other heart-shaped embryos grow, two huge cotyledons are formed at the two sides of the original concave. They extend like waves, green in color and in the shape of water lettuce leaves. When the embryos with cotyledons matured, healthy roots germinated downward and buds germinated upward from the central growth point between the two cotyledons (Fig. 2). Later on, the buds grew into normal test tube plantlets. It has been found that in young longan cotyledon culture the water lettuce leaf–like cotyledon could grow very large; the central embryo bud often differentiated into clustered buds, resulting in the germination of shoots instead of a single shoot. This mechanism should be investigated. In anther culture, the heart-shaped embryo might grow homogeneously, retaining its heart shape. After maturity, the healthy root germinated at the concave end

Figure 2. Plantlets grown from cotyledon of embryo.

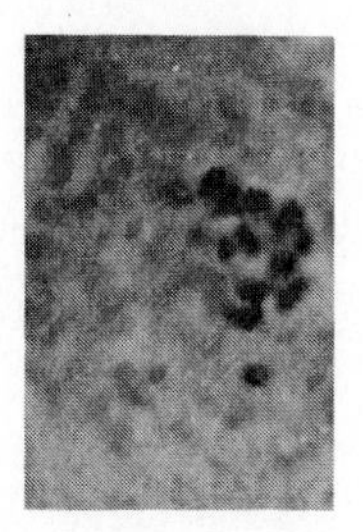

Figure 3. Pollen plantlet of longan, $2n = 15$.

first, followed by the upward germination of the bud. Direct germination of heart-shaped embryos into plantlet has not been observed in longan tissue culture.

The green plantlet differentiated from pollen-derived callus was very weak, with small leaflets (0.5 × 1 mm) on multiple leaves. The leaflets were few, were ellipse shaped with weak stem, and grew very slowly. However, plantlets differentiated from young cotyledon-derived callus grew quickly and healthily, with leaves like the leaves of normal seedlings, leaflets of ellipse shape and large size (1 × 3 mm), and many leaflets in each multiple leaf. The chromosome number of most of the cells of pollen plantlets at the mitotic phase (around 82.7%) was $n = 15$ (Fig. 3). Some cells were irregular aneuploids with chromosome number 19–22. Some root tip cells of some individual plantlets were diploid ($2n = 30$). The root tip chromosomes of the test tube plantlet derived from cotyledon explant were stable (Wei and Yang, 1981).

KEY REFERENCES

Wei, W. X. and Yang, Y. Q. 1981. Induction of cotyledon-derived embryoid and breeding of test tube plantlet of longan. *Fujian Shida Xuebao* (*Nat. Sci.*) **2**:102–106 (in Chinese).

Yang, Y. Q. and Wei, W. X. 1984. Induction of anther-derived plantlet in longan. *Acta Genet. Sin.* **11**(4):288–293. (in Chinese, English abstract).

———, Chen, X. and Chen, Z. F. 1985. Achievements in embryo culture of longan with shrunken stone. *Zhongguo Guoshu* **2**:54 (in Chinese).

REFERENCES

Anderson, W. C. 1975. Propagation of rhododendrons by tissue culture: 1. Development of culture medium for multiplication of shoots. *Proc. Int. Plant Propag. Soc.* **25**:129–135.

———. 1978. Rooting of tissue cultured rhododendrons. *Proc. Int. Plant Propag. Soc.* **28**:135–139.

Liang, E. (ed.). 1979. *Economic Fruit Trees*. Vol. II. Book Series of Youth, H.V. 793. pp. 35–49.

Longan Investigation Group, Horticulture Society of Fujian Province. 1983. Report on investigation of longan production in fujian province. *Fujian Gioshu* **1**:1–15 (in Chinese).

Maheshwari, S. C., Rashid, A., and Tyagi, K. 1979. *Proc. Int. Symp. Plant Cell Cult. Sin*. **11**:288–293.

Sun, Y. W. (ed.). 1983. *History of Fruit Trees and Fruit Trees Resources of China*. pp. 2, 7–9, 20–21, 142. Shanghai Science and Technology Press, Shanghai (in Chinese).

Walkey, D. G. A. 1972. Production of apple plantlets from axillary-bud meristems. *Can. J. Plant Sci*. **52**(6):1085–1087.

Yu, D. J. 1979. In: *Taxonomy of Fruit Trees in China*. p. 321. Agriculture Press, Beijing (in Chinese).

Zhejiang Agricultural University (ed.). 1980. *Fruit Trees Breeding*. p. 397. Shanghai Science and Technology Press, Shanghai (in Chinese).

CHAPTER 28
Loquat: Shoot Tip Culture

*Yongging Yang**

INTRODUCTION

Economic Importance

Loquat (*Eriobotrya japonica* Lindl.) is a well-known fruit of southern China. It has a very good market because it matures in the late spring, when other fruits are still unripe. Its pulp is soft and juicy with a pleasant taste. Besides its use as fresh fruit, the loquat can be processed into jam, candied fruit, fruit wine, and syrup. In addition, it is a popular Chinese medicine for respiratory ailments and good for relieving digestive problems (Huanan Agricultural College, 1981). Its leaves contain volatile oils such as nerolidol and acacia alcohol, which are used to control asthma. Leaf extract is the major component of the Chinese medicine loquat leaf syrup (Tong, 1983). Recently, researchers in the United States and Japan demonstrated that the leaves of loquat contain amydalin (vitamin B_{17}), which reportedly has an effect on cancer control (Li, 1979). The seeds of loquat contain 20% starch, which is used to make wine. The flowers of loquat in the autumn and winter give off a rich perfume. Loquat is an evergreen tree with thick branches and wide leaves. Its crown is compact, so it is also used as an ornamental plant (Yu, 1979; Huanan Agricultural College, 1981).

*English translation by Kezhi Bai.

Geographical Distribution

Loquat grew in central and western China originally. According to a survey of the fruit trees of Sichuan Province in 1958, there were many loquat primeval forests distributed over Luding County and the Daduhe Valley. In the Qingjian Valley and the Yangzi River's tributary in Hubei Province, some wild and semiwild primeval loquats can be found (Zhang et al. 1982). There are also wild loquat forests in Quizhou and Yunnan provinces The cultivars in some places of Japan are called Early Tang, Middle Tang, and Late Tang, implying that they were introduced from China in the Tang dynasty (Liu, 1981; Liu, 1982). About the 18th century, loquat was introduced into Europe from China and Japan. In the late 19th century, it was introduced into the United States (Liu, 1982). Loquat requires moderate climate and soil conditions, so it grows mainly in the warm and moist subtropical zone between 25 to 30 degrees south and north latitude. The areas most suitable for loquat growth include the Mediterranean, northern India, southern Japan, California and Florida in the United States, Mexico, the southern USSR, Argentina, Brazil, and Australia (Ye et al., 1982).

China is the major area of loquat production in the world. Within China, the most important growing areas are in Zhejiang, Fujian, Jiangsu, and Anhui provinces. There are some plantings in Hunan, Hubei, Guangdong, Sichuan, Jiangxi, Taiwan, Guangxi, Yunnan, and Guizhou provinces and Hanzhong District of Shanxi Province. The loquat has been cultivated for more than 2000 years, as shown by the discovery of its seed and buried fruit in an ancient tomb in Jiangling County, Hubei Province (Zhang et al., 1982).

Taxonomy

Loquat (*Eriobotrya japonica* Lindl.) belongs to the family Rosaceae. The genus *Eriobotrya* contains about 20 species all distributed in the warm temperate zone and subtropical zone. There are 11 species in China. Of them, only one species, *E. japonica*, is cultivated. Others, such as *E. prinoides, E. fragrans, E. cavaleriei*, and *E. deflexa*, are still considered wild and can be used as stocks (Yu, 1979). In China, there are about 100 good varieties distributed in three countries, namely, Tangai of Zhejiang Province, Wu of Jiangsu Province, and Putian of Fujian Province. The new selected varieties such as Taichen-4 have big fruit with small seed. Its pulp is 1 cm thick and is yellowish pink, and the edible part of the fruit is approximately 70% of the total fruit. More than 70% of the fruits have only one seed, so this variety is very good for fresh eating and canning (Huang, 1980).

Problems in Breeding

Selection of elite germ plasm, bud mutation, and hybridization is used in loquat breeding programs to solve the following problems:

1. Development of varieties which can be cultivated more widely: Since loquat can only be cultivated in limited areas, through breeding, workers want to select varieties resistant to low temperature and varieties adapted to different soil conditions so that the cultivation area can be expanded.
2. Development of varieties which mature at different periods: The fruit of loquat are perishable and cannot withstand storage and long transportation lines. Breeding programs are geared toward obtaining varieties which ripen early and late to prolong fruit availability.
3. Development of smaller seed or seedless varieties: Compared with other fruits, loquat fruits have seeds which are particularly large and numerous. Development of varieties with smaller seeds or no seeds is needed.
4. Development of techniques for rapid propagation of elite individuals: Routine transplant methods do not satisfy the need for intensive cultivation. It is difficult to introduce a newly developed variety to cultivation rapidly.

Literature Review

Table 1 lists articles about loquat tissue culture published in recent years in China. Embryo culture of loquat may be used in distant hybridization to prevent the abortion of hybrid progeny. Endosperm culture was being used to obtain triploid seedless loquat, but it was unclear whether the plantlets obtained were triploids. However, the method still offers potential for obtaining plants of different ploidy level.

Shoot tip culture can be used to grow buds directly; this is useful for maintaining varietal characteristics in rapid propagation. Plantlet regeneration through callus formation and somatic embryogenesis can increase proliferation rates greatly and at the same time increase the incidence of somaclonal variation which is favorable to obtaining valuable mutants.

TISSUE CULTURE TECHNIQUES

Cultural Protocols and Medium Composition

MATERIAL. The terminal bud is chosen for inoculation when the growth of the branch stops or the young tip is chosen during the growth

TABLE 1. **Studies on the Loquat Tissue Culture**

Explants	Methods	Results	References
Unmatured embryo	Germination	Plantlets	Zhuang and Pan et al., Nov. 1980 (Fujian)
Shoot tip of seedling	Direct propagation	Plantlets and proliferation	Yang and Chen et al. Apr. 1982. (Fujian)
Endosperm	Callus induction	Abnormal organs	Zhuang and Pan et al. Sept. 1982 (Fujian)
Shoot tip of mature tree	Direct propagation	Plantlets and rapid propagation	Yang and Chen et al. May, 1983 (Fujian)
Endosperm	Callus induction and asexual embryo	Plantlets	Chen and Lin et al. Dec. 1983 (Fujian)
Young embryo and shoot tip	Callus induction and asexual embryo	Unknown	He and Zhang et al. Dec. 1983 (Taiwan)

phase of the branch. Our experience indicates that shoot tip culture can be started at any time during the annual life cycle of loquat, but the rate of shoot formation was different in buds taken at different growing stages. For culture of shoot tips from seedlings, the highest rate of shoot formation could be obtained by inoculation of buds in the period from complete cessation of summer shoot growth to germination of autumn shoots. In our experiments, the shoot formation rate of 26 shoot tips was 88.5%, about 50% in shoots with extended leaves. For the shoot tips taken from vigorously growing autumn branches or buds just starting summer dormancy, most shoots would survive, but only a few could extend leaves. This implies that the buds at the end of the dormancy period grew well, perhaps because a large amount of nutrients had been accumulated in the tissue. The buds picked at the end of dormancy and before germination are the best material for inoculation. Shoot tips inoculated in the early spring, when young fruits grow slowly and spring shoots are about to sprout, give the highest rate of shoot formation. For example, when the buds of Taichen-4 were used as inoculum just after fruit setting, the shoot formation rate was 76.9%. But the buds taken in the summer or fruit ripening stages were not good, and their shoot formation rates were only 6.9% and 16.1%, respectively.

INDUCTION OF BUD GERMINATION AND LEAF EXPANSION. When the shoot tips were inoculated on MS medium containing 2 μM BA, 0.5

μM NAA, 0.6 μM GA, and 58.7 mM sucrose, after 1 or 2 months the buds began to grow and young leaves expanded. GA stimulated leaf differentiation and expansion. The concentration of GA in loquat shoot tip culture must be low, about 0.3 to 0.9 μM. The effect of GA was to promote the initial growth of the shoot tip. Prior to use, GA should be filter-sterilized through a 0.45 μm millipore filter and then added to the sterilized medium.

SHOOT FORMATION. The hormone concentration needed for development of shoot tips from young seedlings was different from that of shoot tips from mature trees. After initial growth, the shoot tips taken from young trees or seedlings must be transferred to a medium containing lower concentrations of cytokinin and auxin for further development. If they continued to be cultured in the medium with the same hormone levels, leaf development ceased and the bud die. Medium for shoot tips taken from young trees eventually consisted of MS medium with 1 μM BA, 0.3 μM NAA, 58.7 mM sucrose, pH 5.8. However, when shoot tips were taken from mature trees with fully developed leaves, it was necessary to transfer the cultured shoot tips to MS medium with high concentrations of hormones, i.e., 4.5 to 9.0 μM BA, 1.3 μM NAA, 58.7 mM sucrose, pH 5.8. In this medium, about 75% of the shoot tips developed into shoots.

EMBRYO FORMATION IN CALLUS. It was reported by researchers in Taiwan that shoot tips 1 to 2 mm long were cultured in MS medium with 4.5 to 9 μM 2,4-D, 0.2 to 0.9 μM BA, and vitamins. They formed embryogenic callus first, were subcultured in the preceding medium, and differentiated into embryos. High levels of 2,4-D in the medium were favorable to callus growth, and low levels promoted embryo differentiation. BA promoted embryo development to plantlets. This report showed that the buds picked in spring and winter produced more embryogenic callus (He et al., 1983).

ROOTING. Shoots obtained by culture of shoot tips from young seedlings could be cut when they were 1 to 2 cm long and then transferred to MS medium without hormones. After 20 days they rooted and could be transplanted after hardening. If the shoots came from shoot tips taken from mature trees they were first soaked in 0.5 μM IBA solutions and then transferred to the medium mentioned, after which they rooted rapidly. Some shoots even developed 21 roots in 20 days of culture. It seems possible that the growing points of seedling shoots possess greater ability to form auxin than those from shoot tips of mature trees, which depend on the external hormone supply to a greater degree.

Procedures and Culture Conditions

PROCEDURES

1. Strong terminal buds or apices of young branches about 1 to 1.5 cm long were taken. Young leaves were removed and cut into 0.8-cm segments.
2. Segments were washed 3 to 5 min with supernatant of saturated bleaching powder solution.
3. Segments were transferred to 0.1% $HgCl_2$ solution for 10 min under aseptic conditions.
4. Then they were washed three to five times with sterilized water.
5. Segments were excised carefully into 2–5-cm-long shoot tips with growing points with some leaf primordia and used as explants.

The explant should not be excised too small since loquat has only a few viral diseases and shoot tips shorter than 1 mm oxidize easily and seldom survive. Attention should be directed to the state of the outer young leaves of shoot tips, which supply nutrition to growing points at the initial stage of culture and prevent the shoot tip from browning. After the inner leaves of the shoot tip start to grow, the outer leaves, having exhausted their nutrition, dry and turn brown. The envelope formed by dried outer leaves inhibits the growth of young leaves and should be removed.

CULTURE CONDITIONS. Explants were cultured at 25°C in the dark for the first 20–30 days. After 30 days, dark grown cultures had twice as many calli 8 mm or larger as cultures grown in the light. These calli must be removed at this stage and shoot tips should be transferred to the light (1500 lux, 12 hr) to promote shoot formation. If these calli were not removed, shoot formation and leaf expansion were often inhibited. The survival rate of shoot tip cultures from young seedlings was lower in the dark than in the light if enlarged calli were not removed. In contrast, the survival rates of shoot tip culture from mature trees were the same in light and dark. It is clear that light inhibits callus formation and shoot tip growth in the beginning of shoot tip culture. From these experiments, it is clear that shoot tip culture could not succeed without callus formation. Therefore, dark culture, which encouraged callus formation, was necessary at the initial stage of shoot tip culture.

KEYS TO SUCCESSFUL SHOOT TIP CULTURE. The loquat shoots obtained by shoot tip culture did not readily form adventitious buds. They showed strong apical dominance. It was difficult to circumvent this problem by cutting the terminal bud. Usually, when this was done the first axillary bud under the cut surface sprouted to replace the terminal bud and other buds did not germinate. It is necessary to supply a suitable cytokinin to obtain propagation of shoots.

The optimal BA concentration (6.5 to 9.0 μM) significantly increases the the survival rate. The concentration of BA plays an important role in inducing development of axillary shoots. Fig. 1 shows that as the concentration of BA increased, the apical dominance of loquat shoot tips was broken and a number of axillary shoots grew out (Fig. 2). The relationship between shoot numbers and concentration of BA was quadratic. The maximum shoot formation occurred when the concentration of BA was about 9 μM. Under these conditions, even a 1-cm-long shoot transferred to the medium would be multiplied rapidly to nine shoots of 1 to 2 cm in length in 30 days. At this high hormone level, some shoot tip explants not only formed axillary buds but also developed strong rosette shoots from their basal parts (three to five shoots on each explant). In 9 μM BA, culture growth increased 4.25-fold and 5.3-fold within 30 and 60 days, respectively. Two types of shoots could be seen during shoot tip culture: One was normal, bright green or dark green; leaves were expanded with serrated margins and hairy epidermis; and young stems were not very thick but strong and covered with silver hairs. All of these are typical characteristics of normal loquat shoots and leaves. Another type

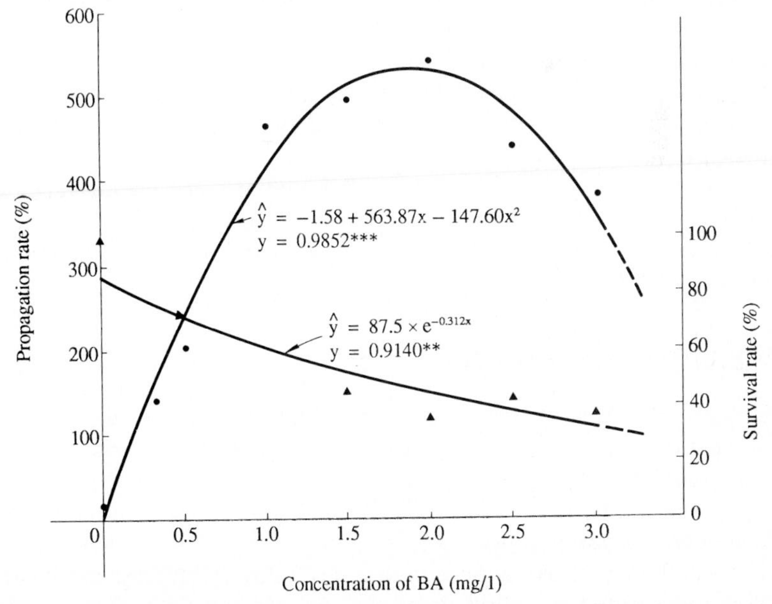

Figure 1. The effects of BA on propagation rate and growth of loquat shoots from shoot tip culture. Circles: normal shoot propagation rate (%). Triangles: No. of normal shoots/total no. of shoots formed (%).

Figure 2. The axillary buds of shoot tip developed into shoots.

was deformed; the shoot and leaf were yellow-green. The newly developed leaf was crimped and had difficulty expanding. Such leaves contained few epidermal hairs and looked translucent. If they were allowed to develop, their viability decreased rapidly and they eventually died. As the concentration of BA increased, the percentage of normal shoots decreased gradually (Fig. 1). The relationship between normal shoots and concentration of BA changed exponentially. The deformed shoots increased as the BA concentration increased over 9 μM. When abnormal shoots were used as explants, the propagation rate decreased sharply to one-third of that of the normal shoots. In summary, the use of very high concentrations of cytokinin and of abnormal shoots as explant material results in decrease in the propagation rate. The optimum concentration of BA was 6.7–9 μM.

PLANTLET TRANSPLANTATION

If the plantlets were strong, it was not difficult to attain high survival rates after transplantation. The present average survival rate is 85.2%, with the maximum up to 95.8%. These survival rates are not lower than those of grafting, which are typically 80–90% during the optimum season (Huang, 1978). Loquat shoot tip culture is not restricted by seasons.

The method of transplantation was as follows: Plantlet roots were washed and planted in sandy loam (or soil). In the first 10 days, they were maintained in high humidity, avoiding intensive light. Humidity was gradually decreased and the seedlings developed (Fig. 3). During trans-

Figure 3. Plants in pots.

planting, the loquat shoot tip seedlings were about 2 to 3 cm in height, and their expanded leaves were about 2 cm × 1 cm in size. After 1 year, the average height of plants was more than 20.5 cm, and the average size of the largest leaf area was 27.5 cm × 8.6 cm. They grew healthy, strong, and uniform and could be transplanted into the fields. On the basis of observations on the plantlets of variety Taichen-4 planted in an orchard, the avarage height of new shoots was 35 cm, some as much as 68 cm, which is double that of plants derived from grafted plants started at the same time. No variation was found in the morphology of plants obtained by culturing shoot tips.

At present, test tube plants of the following varieties have been cultured successfully in Fujian province: Taichen-4, Banhong (Huang et al., 1981), Jiefangzhong, Baili, Jiajiao, and Senwezaosheng introduced from Japan.

REFERENCES

Chen, Z. G., Lin, S. Q., and Lin, Q. L. 1983. A primary report on obtaining plantlets by endosperm culture of loquat. *Acta Fujian Agric. Coll.* **12**(4):343–345 (in Chinese).

He, W. Q., Zhang, Q. S., and Huang, S. N. 1983. Somatic embryogenesis in young embryo and shoot-tip culture of loquat. *Chin. Hortic.* **29**(4):322 (in Chinese).

Huanan Agricultural College. 1981. *Cultivations of the fruit trees*. pp. 199–206. Agricultural Press, Beijing (in Chinese).

Huang, J. S. 1979. The grafting of loquat seedlings. *Chin. Fruit Trees* **2**:24–29 (in

Chinese).
——— and Xu, S. D. 1980. Taichen-4, a new variety of loquat. *Chin. Fruit Trees* **81**:35–37 (in Chinese).
Huang, J. S., Xu, S. D., Cai, G. H., and Cai, J. C. 1981. Banhong, a new variety of loquat. *Chin. Fruit Trees* **2**:25–27 (in Chinese).
Li, H. R. 1979. The loquat has good effect on cancer control. *Dev. Abroad Sci. Technol.* **9**:44–45 (in Chinese).
Liu, Q. 1982. Advances of loquat studies in China. In: Loquat, Collection of the Scientific and Technical Data. pp. 7–11 (in Chinese).
Liu, X. H. 1981. The loquat in Japan. In: Collection of Translations of Fujian Agricultural College. pp. 56–63 (in Chinese).
Tong, P. Y. 1983. *The History of the Fruit Trees*. pp. 210–215. Agricultural Press, Beijing (in Chinese).
Yang, Y. Q., Chen, G. L., and Tang, D. Y. 1982. The shoot-tip culture of loquat seedlings. *Plant Physiol. Commun.* **2**:39–40 (in Chinese).
———, Chen, G. L., and Tang, D. Y. 1983. Studies on loquat shoot-tip culture and propagations. *Acta Hortic.* **10**(2):79–86 (in Chinese, English abstracts).
Ye, L. X. and Xia, Q. Z. 1982. A survey of loquat cultivation researches at home and abroad. In: Loquat, Collections of the Scientific and Technical Data, pp. 1–6 (in Chinese).
Yu, D. J. 1979. *Taxonomy of Chinese Fruit Trees*. pp. 309–312. Agricultural Press, Beijing (in Chinese).
Zhuang, F. C., Pan, W. X., and Wu, J. Z. 1982. Induction of endosperm callus and differentiation of abnormal organs of loquat. *Subtropical Plant Commun.* **2**:11–17 (in Chinese).
———, Pan, W. X., Wu, J. Z. and Cheng, S. H. 1980. Culture of young embryo of loquat. *Subtropical Plant Commun.* **2**:3–7 (in Chinese).
Zhuang, H. Z. and Zhang, Y. D. 1982. The studies of wild loquat in Hubei. Loquat, Collections of the Scientific and Technical Data, pp. 18–23. (in Chinese).

CHAPTER 29
Kiwi Fruit

Zhenguang Huang and *Suying Tan*

INTRODUCTION

Tissue culture has been applied to almost all the major species of fruit trees. This technique has been employed extensively to reproduce nursery stocks of fruit trees on a large scale in countries such as Italy, the United States, and Britain. Tissue culture of *Actinidia* (kiwi fruit) has proved to be a valuable technique for rapid propagation.

Taxonomy and Geographic Distribution

Kiwi fruit is the common name of *Actinidia* of the family Actinidiaceae. According to Liang (1980), *Actinidia* includes 54 species, of which 52 are native to China. Except for a few provinces and autonomous regions in Northwest China, which have not been investigated, it is known that *Actinidia* is distributed in China between Northeast China and Guangdong and Guangxi in the south, and from Zhejiang in the east to the Himalayas in the west.

Of the *Actinidia* species, *A. chinensis* Planch has most economic value because of its large fruit. This species has four varieties: var. *chinensis*, var. *hispida*, var. *setosa* Li., and var. *jingangshanensis* C. F. Liang. The first two varieties are distributed mainly in the Yangtze River Valley, the third exists in Taiwan, and the fourth flourishes in Jiangxi.

Economic Importance

The use of *A. chinensis* Planch by the Chinese people can be traced back 2000 years, as recorded in an ancient Chinese book called "Er Ya." Since then, it has been regarded as a herbal medicine in the Chinese Materia Medica in every dynasty, having tonic and antipyretic effects and promoting secretion of body fluid. Its fruit contains vitamin C at a concentration 20 to 60 times higher than in apple. In 1906, a New Zealander took the seed from China to his homeland, where it was bred through selection into five varieties. In the 1940s it was artificially cultivated and produced commercially. Because of the rapid increase in its cultivated area and the amount of fruit produced, it has become one of the major exports of New Zealand. It has also been introduced from New Zealand directly or indirectly to Britain, the United States, Italy, France, Belgium, India, Japan, and the Soviet Union as one of the most unique fruit trees in the world.

Since the 1970s, a great deal of attention has been paid to the economic value of *Actinidia* in China. Extensive investigations have been carried out on *Actinidia* germ plasm resources, resulting in the selection of a number of good lines and the establishment of bases for cultivation.

Problems of Production

A. chinensis Planch is dioecious. Since it has a long period of natural hybridization, the wild plants exhibit remarkable differences from one another. For example, individual fruits weigh from 30 to 200 g; the color of the pulp can be green, yellow, or red, and the fruit has a variety of tastes ranging from fragrant and sweet to sour and bitter. The vitamin C content of every 100 g of pulp ranges from 30 to greater than 300 mg, and periods of maturity vary considerably. Therefore, good varieties must be selected before *A. chinensis* Planch can be cultivated. Selected superior plants should not be propagated through seeds, because seedlings are highly heterozygous, and about 70% of offspring are male plants. Moreover, traditional cuttings and graftings cannot reproduce a large quantity of nursery stock in a short period because of a limited number of branches. Therefore, tissue culture is an ideal and practical way to propagate *Actinidia* rapidly.

Research History

Harada (1975) was the first to report organ culture of *A. chinensis* Planch. Explants 1.5 cm long and 5 mm in diameter were cut from the stem and root of the plant of var. *hispida*. Both males and females were used as culture material, resulting in the formation of buds, roots, and globular embryos. A few years later, scientists in China and in other countries made an extensive study of *A. chinensis* Planch tissue culture. Plantlets were obtained from cultured shoot tips and stem segments with or without axillary buds of var. *chinensis* (Gui et al., 1979a, 1979b; Huang et al., 1980, 1984; Liu et al., 1981; Standardi, 1981). Wang et al. (1982) and Hong et al. (1981) obtained plantlets from stem segments and terminal buds with leaves with *A. arguta* Mig as culture material. Regenerated plants were obtained from endosperm of var. *chinensis* and var. *hispida* (Huang et al., 1982). Regenerated plantlets were obtained in culture of other explants such as filament cotyledon leaf and embryos (Gui et al., 1979a; Tripathi et al., 1980; Wang et al., 1982; Yu, 1983). The Italians have already reproduced nursery stocks of *Actinidia* on a commercial scale through tissue culture (Gilberto et al., 1981). Presented here are several methods for regenerating plants from different explants of *Actinidia*.

TISSUE CULTURE TECHNIQUES

Regeneration of Plants from Stem Segment Callus

SELECTION OF MATERIAL. Select 1-year-old hard branches or young shoots from healthy individual plants of var. *hispida* and var. *chinensis*. We have also used another related species, *A. eriantha* Benth. This species has vitamin C content of 620 to 1050 mg, or even as high as 2146 mg/100 g of fresh fruit. This is 8 to 20 times higher than that of *A. chinensis* Planch, although its fruit is somewhat small (about 20 g). Recently a number of good individual plants with larger fruit have been discovered.

Culture Procedures and Media

EXPLANT DEDIFFERENTIATION AND BUD INDUCTION. To culture explants of var. *chinensis*, MS medium with 4.6–13.7 μM ZEA and 87 mM sucrose was used. Induction of callus in explants may reach 92.0–100%. Buds may be differentiated directly from the explants without transferring to a different medium (Huang et al., 1980). Culture of stem segments of superior plants of *A. eriantha* Benth required a higher content of ZEA (13.7 μM) in the culture medium. The capacity for bud differentiation from callus was much weaker than that of *A. chinensis*

Planch, although axillary buds elongated faster and developed roots readily. They could even root directly on differentiation medium without any special treatment for rooting.

Liu et al. (1981) believed the following to be the ideal differentiation medium: MS + 0.05 μM NAA + 2.2 μM BA + 4.6 μM ZEA, or MS + 2.7 μM NAA + 8.8 μM BA + 4.6 μM ZEA, with 87 mM sucrose. Others have used N_6, White, MT, or Nitsch medium with 2,4-D as growth regulator. Experiments indicated that these media are not as effective as MS medium. Even though they can induce callus formation in explants when 2,4-D is added, the callus has either no differentiation ability or a lower frequency of bud differentiation (Hong, 1981; Lin et al., 1982; Zhou et al., 1984).

Harada (1975) suggested that the addition of 4.6 μM ZEA promoted bud differentiation. He used MS macroelements with the following microelements and organic components: 100 μM $MnSO_4 \cdot 4H_2O$, 159 μM H_3BO_4, 35 μM $ZnSO_4 \cdot 4H_2O$, 1.0 μM $Na_2MoO_4 \cdot 2H_2O$, 0.1 μM $CuSO_4 \cdot 5H_2O$, 500 mg/l CH, 4.1 μM nicotinic acid, 26.6 μM glycine, 2.4 μM pyridoxine HCl, 1.5 μM thiamine HCl, 1.1 μM folic acid, 2.05 μM biotin, 0.1 mM Fe-EDTA, 58.4 mM sucrose, and 0.7% agar, pH 5.5. He determined that explants of root segments produced more shoots and roots than stem segment explants and that the male plant had stronger regeneration ability than the female plant. Gui et al. also showed that 4.6 μM ZEA was best for promoting bud differentiation (1979a, 1979b).

Harada (1975) discovered that when explants were cultured on media containing 0.45–4.5 μM 2,4-D or 0.54–5.4 μM NAA, many globular embryos appeared. However, if embryos were cultured on this medium continuously, they dedifferentiated rapidly and complete plants were not obtained from these somatic embryos.

SUBCULTURE. Differentiated cultures can be subcultured at 30–40-day intervals. The subculture medium the authors used was the same as the differentiation medium, except when excessive shoot differentiation caused by the accumulation of endogenous hormone in the culture occurred. In this case, ZEA concentration must be reduced to 2.3 μM.

ROOT INDUCTION. It has been reported that young shoots 1 cm in length cut from the stem base are suitable for root induction. The results of our experiments suggested that 3–4-cm-long bud shoots should be used to induce roots so that the rooted plants were more easily transplanted. The basal part of the plant must be soaked in 250 μM IBA solution for 3–3.5 hr and then put into hormone-free MS medium in which the concentration of macroelements has been decreased to 50%, with 29.2 μM sucrose and 0.5% activated charcoal (Harada, 1975; Huang et al., 1980,

1984). After 1 month, good root systems develop and the plants may be transplanted into soil immediately. A rooting rate of up to 93.3% was reported and the number of roots for each plant was 9.4 ± 7.53.

Operating Techniques and Culture Conditions

STERILIZATION OF MATERIAL. Shoots can be cut off for inoculation anytime during the year. First the shoots were cut into segments 10 cm long, surface-sterilized with 0.1% mercuric chloride solution for 15 min, and washed with sterile water three times. Finally, the bud eyes were cut off and the cortex was peeled off to expose the cambium. This reduced the infection rate by eliminating microorganisms parasitic in the eyes and cortex.

INOCULATION. Sterilized shoots were cut longitudinally with a pair of scissors into four to eight pieces and then cut horizontally into pieces 1–1.5 cm long and inoculated into culture medium.

CULTURE CONDITIONS. Cultures were placed in the culture room at 25°C with artificial illumination for 12 hr/day at 850 to 1200 lux.

TRANSPLANTATION OF PLANTLETS. To increase survival rate of transplants derived from tissue culture of *Actinidia*, attention must be directed to several conditions, such as water content in the soil of the nursery beds, air humidity, and illumination. The following steps are keys to cultivating healthy plantlets.

1. Select cultures that have not produced too many shoots for subculture and if necessary reduce the concentration of ZEA to 2.3 μM.
2. Fine sand is recommended as an ideal and economic transplanting matrix, because it is permeable, can maintain a certain amount of water, and need not be sterilized. Either well-rooted or poorly rooted plantlets can be transplanted to a sand nursery bed. Plantlets are then covered with a plastic film and provided with proper shade.
3. During the first 15 days following transplantation, the relative humidity of the air beneath the plastic film should be 100% and the absolute water content in the sand over 10%.
4. Illumination in the nursery bed was kept at about 2500 lux during the summer and at about 5000 lux during the winter.
5. Fifteen days after transplant, the plastic films were opened a little at a time, to adapt plants to normal air and also to reduce the water content in the sand. After another 15 days, the surviving plantlets exhibited three to four newly sprouted leaves. At this time the plastic films could be removed but proper shade should still be provided because *Actinidia* is a shade-tolerant plant.

In 1983, the authors transplanted a total of 1354 rooted plantlets, of which 1104 survived, for a survival rate of 81.8%. Even poorly rooted plantlets may have a survival rate of up to 70% if they are transplanted under suitable conditions. From the results of several years of experiments, it was estimated that nine generations could be subcultured per year, giving a propagation efficiency of 2.5 per generation. After 1-year propagation, a 0.3-m-long branch with three to six buds could theoretically produce up to 30,000 plantlets, showing the extent to which this technique may accelerate the speed of nursery stock propagation.

Plantlets were obtained from callus differentiation. Through subculture of multiple generations over 2 years, only 2% of the recovered plants exhibited obvious changes in morphology. Genetic stability and variability of materials require further study.

Culture of Shoot Tips and Stem Segments with Axillary Buds

The difference between shoot tip and axillary tip culture from stem segments without buds is that plants resulting from culture of meristems are derived from repeated multiplication of meristematic tissue without going through the callus stage.

PREPARATION OF MATERIAL. Culture of shoot tips and stem segments with axillary buds of *A. chinensis* Planch requires a shoot tip of 1–1.5 cm with three to four nodes (Harada, 1975; Gui, 1979b).

EFFECT OF CULTURE MEDIUM. The plant material was inoculated onto MS basic medium supplemented with 2.2 μM BA, 2.9 μM GA, and 4.6 μM ZEA. Approximately 2 weeks later, the apical bud enlarged to 3.5–4 cm, with leaves sprouting in the axillae. One week after inoculation axillary buds of stem segments with lateral buds grew out, sprouting leafs. Two weeks later, the sprouted shoots branched. However, if axillary buds without stem segments were inoculated, the buds enlarged but had difficulty developing and the leaves did not expand (Gui, 1979b). Another type of medium used was Nitsch medium supplemented with 0.29 μM GA, 0.009 μM 2,4-D, 4.4 μM BA, and 58.4 mM sucrose. Using var. *hispida* (Hayward) as material and young shoots with terminal buds as explants, material of the 28th generation and that of the 3d generation were compared. The results demonstrated that there was no difference in rooting ability, length of the roots, or maximum number of roots. In a second experiment, shoot tips and young nodal stem segments with and without terminal buds were used as explants and materials of the 29th generation and the 4th generation were compared. The propagation efficiency of the 29th generation was relatively high, but the length of new shoots and the weight of the callus decreased and no variation occurred in the size and shape of the leaves or in the number of chromosomes during subculture (Standardi, 1981, 1982). In Italy, there are two nursery gardens that reproduce var. *hispida* Hayward in large quantities using shoot tip culture. Tissue culture has become a common method to propagate nursery stocks of *Actinidia* in Italy (Zucherelli and Zucherelli, 1981).

Endosperm Culture of *Actinidia*

The fruit of *Actinidia* is rather small and contains a great number of tiny seeds which taste unpleasant when broken. Therefore, a major goal is to breed high-quality large, seedless fruit. The authors attempted to reach this goal by creating a triploid plant using the endosperm culture method. However, this approach is not certain as it is possible that the fruit could become smaller because the fruit of the triploid plant does not contain seed.

SELECTION AND STERILIZATION OF MATERIALS. At the beginning of August, the seeds in the fruit of wild var. *chinensis* are brown and the inside is white. At this stage, the fruits are sterilized with 1:15 bromogeramine solution for 20 min and then washed with sterile water three times. The fruit is then cut open to obtain the seed and the seed coats are removed to retrieve the yellowish white kernels. Finally, the needle-shaped young embryos are squeezed out and the remaining endosperm is retained for culture.

CULTURE PROCEDURES AND MEDIA

1. The endosperm explants were cultured on MS medium, with 13.7 μM ZEA and 9.0 μM 2,4-D. Several days later white callus appeared (Huang et al., 1982). Callus could also be induced on MS medium supplemented with 13.7 μM ZEA, 2.3–4.5 μM 2,4-D, and 400 mg/l CH (Gui et al., 1984).
2. In one experiment, callus was transferred to MS medium containing only 4.6 μM ZEA. From three pieces of callus, a large quantity of globular embryos developed and grew into cotyledonary embryos. If these cotyledonary embryos had been grown on the previously mentioned medium, their cotyledon and hypocotyl would undergo callus formation, inducing plenty of new embryos. These cultures maintained the ability to differentiate into embryos even after being subcultured for more than 2 years (Huang et al., 1982). If callus was kept on MS medium supplemented with 4.6 μM ZEA and 400 mg/l CH, embryos which grew into complete plants emerged (Gui et al., 1984). If var. *hispida* and var. *chinensis* were used as materials and the

endosperms were put onto MS dedifferentiation medium supplemented with 13.7 μM ZEA and 0.54 μM NAA, papillae developed directly from the endosperm surface and later grew into expanded thalloids (Gui et al., 1982).

3. For regeneration of complete plants, cotyledonary embryos were transferred to medium containing 1.8 μM ZEA and developed into complete plants. On MS medium supplemented with 13.7 μM ZEA and 0.54–2.7 μM NAA, greenish callus occurred. When the callus was transferred onto the preceding medium, the buds differentiated from the callus and roots developed after induction with IBA. Complete plants grew and survived transplantation into the soil.

CULTURE CONDITIONS. The inoculum was placed into a culture room at 25–30°C. During the first 25 days, weak indoor diffused light was used. After 25 days artificial illumination was used for 12 hr/day, with illumination at 850 to 1200 lux.

CYTOLOGICAL IDENTIFICATION. *Observation Technique.* The root tips were cut from endosperm embryos and complete plants in early morning and pretreated with saturated bromonaphthalene solution for 3 hr. The root tips were fixed in Carnoy's solution and stained with haematoxylin to determine chromosome ploidy. The chromosomes may also be observed by F-BSG banding technique. After 2 hr, the plantlet root tips were removed and treated with the following procedure: pretreatment with 8-hydroxyquinoline pretreatment in lower osmolarity, enzymolysis, posttreatment in lower osmolarity, fixing, slicing, staining, and observation. If observations of chromosome banding are desired, it is necessary to perform denaturation and renaturation after slicing and then repeat the staining. This method yields good dispersion of chromosomes and easy counting.

Observations. Root tips of over 300 plantlets and embryos were observed. Among 46 distinct plant samples, there were 19 plants with $2n = 3x = 87$ (41.3%); 12 with $2n = 58$ (26.1%), 10 with $2n = 4x = 116$ (21.7%); and 1 with $2n = 6x = 174$ (2.2%). Four were aneuploids, equaling 8.7% of the total.

Triploid endosperm plantlets differed from diploid parent plants because the hairs on the leaves and stems were long, thick, and hard. The root tip of *Actinidia* has a short metaphase period and a large number of chromosomes, making chromosome observation difficult. Endosperm chromosomes are a rich area for future investigation. Among the endosperm plants obtained from embryos, most are triploids, some are tetraploids and diploids, and there are a few hexaploids and aneuploids. All of them are valuable material for breeding fruit trees.

Culture of Cotyledon, Leaf, and Embryo

Some researchers have already experimented with cultures of the many tissues and organs of *Actinidia*, including anthers, cotyledon, leaf blade, and embryo. Tripathi et al. (1980) reported that when the anther filament of *A. chinensis* Planch was cultured in liquid medium containing 2 μM IAA, roots developed and later plants were produced. Yu (1983) cut the cotyledons from 10-day-old seedlings into 2–3-mm pieces and then induced callus in MS medium supplemented with 4.6 μM ZEA. Buds differentiated, with callus induction rates of 80–90% and differentiation rate of 60–70%. The basal part of 2-cm-tall plantlets was soaked in 250 μM IBA solution for 2 hr and transferred to ½ MS medium supplemented with 29.2 mM sucrose; a week later roots differentiated. The plantlets were transplanted into pots and grew well.

Gui et al. (1979a) cut leaf blades of *A. chinensis* Planch into pieces 2 × 2 cm, 6 × 6 cm, and 12 × 15 cm and inoculated them onto MS medium supplemented with 0.45 μM 2,4-D and 4.6 μM ZEA. These were best for callus induction, but there was little differentiation. When they were cultured either on MS medium supplemented with 0.5 μM IBA and 4.6 μM ZEA or on MS medium supplemented with 4.6 μM ZEA, plants differentiated. Most plantlets were differentiated from the larger explants. Hong (1981) cultured leaf blades of *A. chinensis* Planch and *A. arguta* (Sieb. and Zucc.) Planch × Mig. Medium for callus induction was MS supplemented with 4.5 μM 2,4-D, 4.6 μM ZEA, 2.3 μM KIN, 2.7 μM NAA, and 58.4 mM sucrose. The callus was transferred onto N6 or MS medium supplemented with 9.3–13.9 μM KIN, 1.1 μM IAA, and 58.4 mM sucrose. Induction rates of callus were 16.7% and 93.3% for *A. chinensis* Planch and *A. arguta* (Sieb. and Zucc.) Planch. × Mig., respectively; the differentiation rate was 87.5% and 11.5%, respectively.

Wang et al. (1982) cultivated immature embryos of *A. arguta* (Sieb. and Zucc.) Planch × Mig. to develop a new method of rapid propagation. The embryo was inoculated onto ½ MS medium supplemented with 2.2 μM BA, 2.9 μM IAA, and 150 mg/l LH. Numerous plantlets grew from the embryo buds and the hypocotyl with a differentiation rate of 83.3%. After 5–6 weeks, the numerous plantlets were separated into small clumps, each with five to seven plants. Finally, they were subcultured onto MS medium supplemented with 2.2 μM BA and 300 mg/l LH and separated into small clumps once a month.

It was found from culture of filament, cotyledon, and leaf that the best results were obtained with MS basic medium supplemented with a certain concentration of ZEA. Good results were also obtained when MS basic medium was supplemented with BA.

It is only in recent years that attention has been focused on the production of *Actinidia* in China. However, in Italy, tissue culture has been used for commercial production. This technique has been drawing the interest of researchers, showing that it has entered a stage for practical production and application.

In summary, *Actinidia* is an ideal material for tissue culture, with high induction and differentiation rates of callus. Moreover, because it is dioecious, the number of both female and male plants can be controlled artificially by means of elite individual plants. Further studies should be focused on cell culture techniques and protoplast culture of *Actinidia* to advance the breeding and propagation of superior varieties.

KEY REFERENCES

Gui, Y., An, H., Cai, D., and Wang, J. 1979a. Tissue culture of *Actinidia. Kexue Tongbao* **24**:188–190 (in Chinese).

Hong, S. 1981. Plant regeneration and callus induction of stem segment and leaf of *Actinidia. Hubei Nongye Kexue* **9**:28–30 (in Chinese).

Huang, Z. and Huangfu, Y. 1980. In: *Research Reports on Actinidia* (*1978–1980*), *Zhengzhou Institute of Fruit Trees*. pp. 225–232 Agriculture Academy, Beijing (in Chinese).

———, Huangfu, Y., and Xu, L. 1982. Triploid plant from endosperm culture of *Actinidia. Kexue Tongbao* **27**:247–250 (in Chinese).

———, Wang, Q., Huangfu, Y., and Xu, L. 1984. Future prospect of tissue culture propagation of *Actinidia. Sci. Fruit Trees* **1**:29–33 (in Chinese).

REFERENCES

Gui, Y. 1979b. Studies of callus induction and plant regeneration of stem segment of *Actinidia* in vitro. *Acta Bot. Sin.* **21**:340–344 (in Chinese, English abstract).

———, Mu, X., and Xu, T. 1982. Studies on morphological differentiation of endosperm plantlets of Chinese gooseberry in vitro. *Acta Bot. Sin.* **24**:216–221 (in Chinese, English abstract).

———, Gu, S., and Xu, T. 1984. *Abstracts of International Symposium on Genetic Manipulation in Crops*, p. 82 Science Press, Beijing, China.

Harada, H. 1975. Research note on in vitro organ culture of *Actinidia chinensis* Planch as a technique for vegetative multiplication. *J. Hortic. Sci.* **50**:81–83.

Liang, C. 1980. Taxonomical flora of China *Actinitidia* species. *Guihaia* **1**:30–40 (in Chinese).

Liu, M. and Shu, J. 1981. The report on tissue culture of stem segments of *Actinidia* in vitro. *Collect. Bot. Res.* **4**:34–35 (in Chinese).

Standardi, A. 1981. Micropropagazione dell' *Actinidia chinensis* Planch mediante coltura in vitro di apical meristematic. *Frutticolt* **43**:23–27.

———. 1982. Effects of repeated subcultures in shoots of *Actinidia chinensis* (P1)

In: *Plant Tissue Culture, 1982*. (A. Fujiuwara, ed.), pp. 737–738. Japanese Association for Plant Tissue Culture, Tokyo.

Tripathi, B. K. and Saussay, R. 1980. Sur la multiplication vegetative de l'*Actinidia chinensis* Planchen, "Chinese gooseberry" par culture de racines issues de filets staminaux. *Comptes Rendus Acad. Sci.* Series D **291**:1067–1069.

Wang, J. and Li, S. 1982. Propagation tissue culture of *Actinidia. Liaoning Nongye Kexue* **1**:32–34 (in Chinese).

Yu, S. 1983. The plantlets differentiated from cotyledon of *Actinidia* by tissue culture. *Plant Physiol. Commun*. **2**:37–38 (in Chinese).

Zhu, X., Shao, X., and Chen, X. 1982. The experiment on the segment culture of *Actinidia. Fujian Guoshu* **2**:1–4 (in Chinese).

Zhou, Z., Zhou, F., and Liu, G. 1984. Callus induction and plant regeneration of shoot segment of *Actinidia. Acta Agricult. Univ. Rekinesis* **1**:83–84 (in Chinese).

Zuccherelli, G. and Zuccherelli, G. 1981. *Llactinidia Pianta da Frutto e da Giardino*. p. 50. Officine Grafiche Calderini.

PART D
Extractable Products

CHAPTER 30
Oleaster

Jingli Pan, Puxuan Wang, Rulan Gao, Hui Fan, and *Lizhu Tu*

INTRODUCTION

Economic Significance

The narrow-leaved oleaster, *Elaegnus angustifolia* L., is a wood species with significant economic value. Possessing nitrogen fixing bacteria, its roots are able to supply nitrogen to the plant. The plant contains tannin, and black and brown dyes can be extracted from the bark and leaves. The wood is useful for manufacture of musical instruments, handicraft articles, and furniture and for construction. The gum of the bark has many uses as a hue-enhancing agent for printing and dying textiles. Similar to the nutritional constitution of alfalfa leaves, the leaves have 15.7% protein and 6.5% lipids and can be used for refined fodder. From the flower, perfume essences and spices can be recovered, and the tree is a nectar source and ornamental plant. The fruit, rich in sugar, vitamin C, protein, and other nutrients, is edible and may be used in brewing and vinegar making and as a food or beverage. In addition, medicines made with the flower, fruit, shoot, and leaf are used as cures for chronic bronchitis, dyspepsia, neurasthenia, and other ailments. *E. angustifolia* is resistant to cold, drought, wind, and salt so that it is often the chief tree of protective forests in arid areas, along boulevards, and in gardens in cities and countrysides.

Geographical Distribution

E. angustifolia is naturally distributed in arid and semiarid regions in north and northwest China and cultivated elsewhere in China. It is also cultivated in the south and west parts of the USSR, and in Spain, Italy, and some Central European countries (Northwest Institute of Botany, 1977; Compiling Commission of Tree Annals of China, 1978). As far as we know, there are no other reports on tissue culture of *E. angustifolia* except this study on the morphological characteristics of its stem and leaf in vitro. Oleaster somatic cells were cultured and the cultures developed into structures similar to the zygotic embryo. This shows that the somatic cell of *E. angustifolia* possesses totipotency.

METHOD FOR TISSUE CULTURE

Culture Procedures and Media

SELECTION OF MATERIALS. A flowering shoot collected from *E. angustifolia* was placed in water at room temperature for 40 hr. The ovary, stem, and leaf were removed and placed at 3–5°C for 48 hr. Explants were prepared from the top part of the ovary, 3 mm long, stem segments, and pieces 5 × 5 mm of young leaves.

INDUCTION OF CALLUS. Explants were inoculated onto the following medium: MS with 2.7 μM NAA, 4.6 μM ZEA, 300 mg/l CH, 88 mM sucrose, and 0.6% agar. After 15 to 30 days, callus appeared on the wounded surface of the cuttings in three different colors: milky, yellow-green, and green. The effects of different combinations of plant growth regulators on callus induction are summarized in Table 1.

CALLUS DIFFERENTIATION. After the induced callus has been transferred several times to subculture medium of MS plus 0.91 μM ZEA and 8.8 μM BA, most calli, except those of milky color, differentiated into buds. However, differentiation continued to occur after subculture onto MS medium supplemented with 9.3–18.6 μM KIN with 1.0 μM IBA. Callus which had been subcultured for 2 years still differentiated. Furthermore, embryonic cell lines and embryos could be induced on this medium.

INDUCTION OF ROOTLESS SHOOTS. To obtain intact plantlets, three types of auxins at concentrations of 250–290 μM were used to treat basal parts of rootless shoots (Table 2). The treatment periods were 1, 2, or 4 hr, respectively. The best of the three hormones tried was IBA. Induction frequencies of rooting were 100% with treatments of 1 or 2 hr. In contrast, rooting was greatly inhibited by a 4-hour treatment with IBA

TABLE 1. Induction of Callus with Different Combinations of Plant Hormones

Growth Regulator (μM)	Explant	No. of Explants	No. of Callus	Induction Frequency (%)	Color of Callus
2.7 NAA + 4.6 ZEA	Leaf	11	11	100	Opaline
	Stem	11	11	100	Yellow-green
	Top part of ovary	36	36	100	Yellow-green, green, opaline
2.7 NAA + 4.6 ZEA + 9.0 2,4-D	Leaf	13	13	100	Yellow-green
	Stem	5	5	100	Opaline
2.3 2,4-D + 8.8 BA	Leaf	9	7	77.8	Yellow-green
	Stem	12	12	100	Opaline
Control	Leaf	18	0	0	—
	Stem	8	0	0	—

TABLE 2. Root Induction with Different Auxins

Auxin	Treatment Time (hr)	No. of Shoots Cultured	No. of Shoots Rooted	Rooting Frequency (%)	Total No. of Roots	Average No. of Roots
IBA (250 μM)	1	5	5	100	20	4.0
	2	3	3	100	19	6.3
	4	4	1	25	1	1.0
IAA (290 μM)	1	5	3	60	14	4.7
	2	5	0	0	0	0
	4	3	0	0	0	0
NAA (290 μM)	1	5	1	20	3	3.0
	2	5	0	0	0	0
	4	6	0	0	0	0

and even more with a 1-hour treatment using IAA and NAA. In general, IAA and NAA were not effective for induction of roots from rootless shoots. For xylophyta tissue cultures, root induction is a key problem in cloning plantlets (Fossard, 1978). Root formation in *E. angustifolia* demonstrated that placing the basal part of the shoot in 250 μM IBA for 2 hr was the most effective method for root induction. The rooting rate reached 100% and the root number per shoot was highest.

Optimal Growth Regulator Combination: A Key for Induction of Organ Differentiation

Cui (1952) believed that bud and root formation were conditional on regulation of the cytokinin/auxin ratio. Bud differentiation occurred and root formation was inhibited when the ratio was high. In contrast, roots differentiated and buds were inhibited when the ratio was low. Experimental results with *E. angustifolia* are in accord with this. If the concentration of IBA is very high, rooting instead of shooting occurs. Chen et al. (1980) found that with a poplar hybrid, KIN was ineffective for shoot differentiation. In our study on the tissue culture of *E. angustifolia*, however, KIN was indispensable, and at a concentration of 18.6 μM it played a major role in shoot and embryo differentiation. High concentrations of BA (26.4 μM) improved the differentiation frequency, although abnormal shoots predominated over normal ones. Thus the use of high levels of BA is not recommended.

THE EFFECT OF IBA ON DIFFERENTIATION. Adding IBA alone to medium was not good for shoot differentiation as it increased the number of abnormal shoots. At least 50% of the shoots were abnormal even though the concentration of IBA was 0.98 μM. Over 80% of the total normal shoots were abnormal if IBA was higher than 4.9 μM (Table 3). The combination of KIN at 9.3 μM and IBA at 0.98 μM was most effective for induction of normal shoots, improved growth rate, and development of the root system. However, with increasing IBA concentration, the percentages of abnormal shoots rose from 70% to 90% or even higher (Table 4). Since the callus in Group I had been treated with 250 μM IBA for 2 hr before it was transferred onto the differentiation medium, the percentage of abnormal shoots was very high even with an optimal ratio of KIN to IBA (9.3:1). High concentrations of IBA were unfavorable for differentiation of normal shoots.

THE EFFECT OF KIN ON DIFFERENTIATION. KIN did not induce callus to differentiate into shoots until the concentration was more than 4.6 μM. Higher concentrations of KIN (9.3 to 37.2 μM) seemed to induce normal shoots. Multiple shoots sometimes occurred (Fig. 1), and 18.6 μM appeared to be optimal since the percentage of abnormal shoots is decreased to 13.3% (Table 3). Normal shoots and the highest percentage of calli with green shoot bud spots were formed with 19.6 μM IBA; 18.6 μM KIN is the optimum concentration whether used alone or in combination with 1.0 μM IBA, and 1.0 μM IBA caused shoots to develop into intact plantlets.

THE EFFECT OF BA ON DIFFERENTIATION. BA is effective for callus differentiation of *E. angustifolia*. A concentration of 2.2 μM BA caused differentiation, and when it was increased to 26.4 μM, the differentiation rate was optimal. The difference in the effects of BA and KIN on differentiation is that BA causes at least 60% of the shoots to be ab-

TABLE 3. Effect of Various Growth Regulators on Differentiation

Item	KIN (μM)							BA (μM)						IBA (μM)				
	0.93	2.3	4.6	9.3	18.6	27.9	37.2	0.88	2.2	4.4	8.8	17.6	26.4	0.98	2.5	4.9	9.8	19.6
No. of calli transferred	16	16	17	17	23	30	17	20	20	20	20	20	20	14	14	14	15	14
No. of shoots formed	0	0	2	25	60	44	25	0	5	10	19	43	111	4	7	9	9	5
No. of shoots per 100 calli	0	0	11.8	147.1	260.9	146.7	147.1	0	25.0	50.0	95.0	215	555.0	28.6	50.0	64.3	60.0	35.7
Normal shoot no.	0	0	1	21	52	37	21	0	2	3	4	13	33	2	2	1	1	1
%	0	0	50.0	84.0	86.7	84.1	84.0	0	40.0	30.0	21.1	30.2	29.7	50.0	28.6	11.1	11.1	20.0
Abnormal shoot no.	0	0	1	4	8	7	4	0	3	7	15	30	78	2	5	8	8	4
%	0	0	50.0	16.0	13.3	15.9	16.0	0	60.0	70.0	78.9	69.8	70.3	50.0	71.4	88.9	88.9	80.0

TABLE 4. Effect of Different Combinations of KIN and IBA on Differentiation

Item	I					II					III						
	KIN:IBA (μM)					KIN:IBA (μM)					KIN:IBA (μM)						
	9.3: 1.0	**9.3: 2.5**	**9.3: 4.9**	**9.3: 9.8**	**9.3: 19.6**	**0.9: 9.8**	**2.3: 9.8**	**4.6: 9.8**	**9.3: 9.8**	**18.6: 9.8**	**0.93: 1**	**2.3: 1**	**4.6: 1**	**9.3: 1**	**18.6: 1**	**27.6: 1**	**36.8: 1**
No. of calli transferred	40	40	40	40	40	40	40	20	40	30	25	24	26	21	20	25	13
No. of shoots formed	64	46	14	33	36	13	9	9	33	8	0	0	6	20	36	21	2
No. of shoots per 100 calli	160.0	115.0	35.0	82.5	90.0	32.5	22.5	45.0	82.5	26.7	0	0	23.1	95.2	180.0	84.0	15.4
Normal shoot no.	19	7	1	4	3	1	0	1	4	0	0	0	3	14	30	16	1
%	29.7	15.2	7.1	12.1	8.3	77	0	11.1	12.1	0	0	0	50.0	70.0	83.3	76.2	50.0
Abnormal shoot no.	45	39	13	29	33	12	9	8	29	8	0	0	3	6	6	5	1
%	70.3	84.8	92.9	87.9	91.7	92.3	100.0	88.9	87.9	100.0	0	0	50.0	30.0	16.7	23.8	50.0

Figure 1. A great number of normal shoots occur on the medium supplemented with 18.4 μM KIN.

normal. KIN, therefore, and not BA should be employed for callus differentiation of *E. angustifolia*.

DISCUSSION OF EMBRYONIC CELL MASSES AND EMBRYOS

Callus differentiation of *E. angustifolia* can occur in two ways. One is organogenesis, in which a plantlet is formed via an adventitious bud with single polarity. The second route is embryogenesis, which is similar to formation of the zygotic embryo, in which an embryo develops into the intact plantlet. In organogenesis, the original cells come from the division of epidermal cells and parenchymatous cells. In embryogenesis there are three sources of the original somatic embryo cells. First, a single parenchymatous cell under the epidermis of the callus changes into an embry-

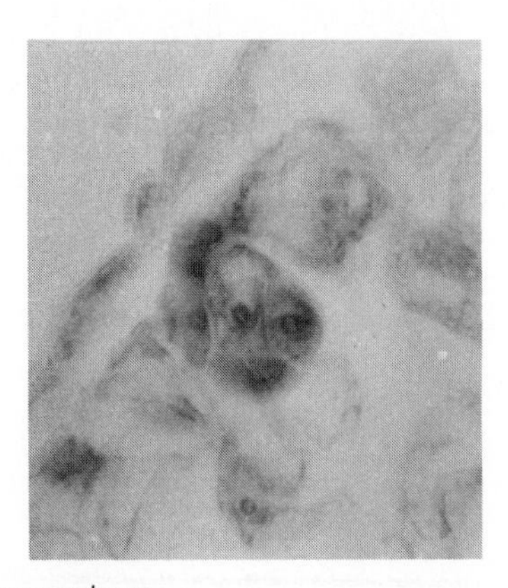

Figure 2. A four-cell proembryo.

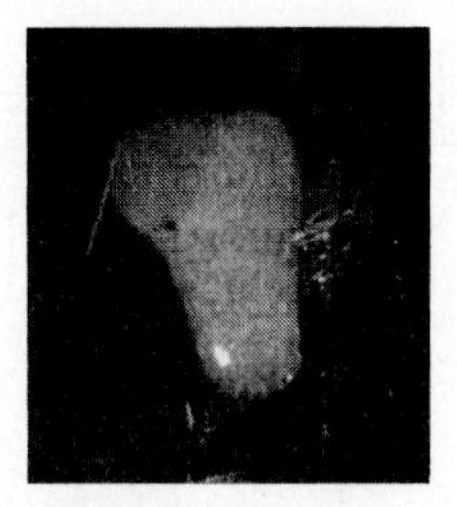

Figure 3. A torpedo-shaped embryo.

onic cell. Second, they are formed through the transverse division of epidermal cells. Third, the epidermal and adjacent parenchymatous cells divide simultaneously and form an embryo. Unlike the early division in the zygotic embryo, these embryonic cells first divide into two cells, then divide into three or four cells (Fig. 2), and finally into a globular proembryo. In the early cell division, the typical basal cell and terminal cell are not formed, and, therefore, there is a lack of the typical structure for the suspensor. Under some conditions, there is a suspensor-shaped structure. The globular proembryo develops from a heart-shaped embryo, into a torpedo-shaped embryo (Fig. 3), and then into an embryo with two cotyledons. During development of the embryo, the epidermal cells adjacent to it are ruptured (Fig. 4), and the parenchymatous cells around the embryo disintegrate (Fig. 5). Finally, an individual embryo is formed and emerges from the callus surface. Occasionally, the aged parenchymatous cells within the callus are able to recover the ability to divide and to form meristematic cell masses, which develop into calli and embryos with time. Further study in this area is needed.

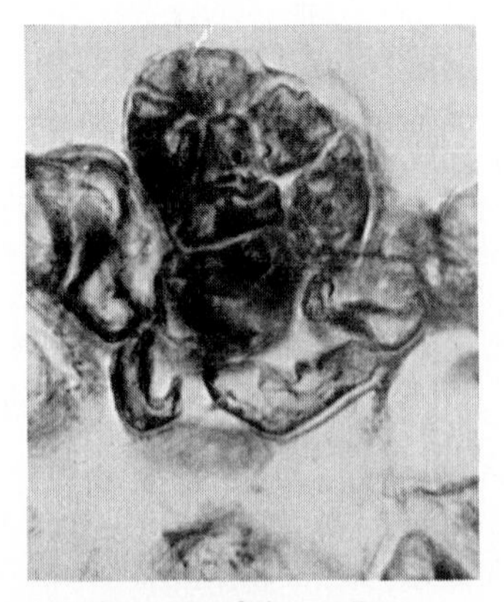

Figure 4. A proembryo growing out of the callus.

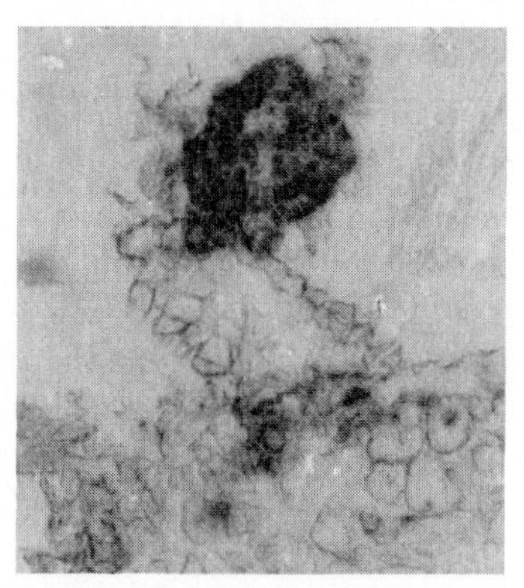

Figure 5. The parenchyma endocytes around the embryo are exhausted and disintegrated.

Steward et al. (1964) noted that a cell within a structural tissue has to be separated from the cells around it in order show its inherent trend to undergo embryogenesis and develop like a zygote into an embryo. In spite of this, there are many experiments in tissue culture demonstrating that a typical embryo comes from an embryogenic cell linked to its neighboring cells. These cells are located mainly at or below the surface of calli. Konar (1972) observed in *Ranunculus sceleratus* L. that the embryo emerged from the transverse division of an epidermal cell, which formed two cells, one an embryonic cell and the other a suspensor cell. Yan (1980) also observed that the epidermal cell of *Camellia oleifera* Abel formed embryos through cell division. In stem cultures of *Tylophora indica*, Rao (1972) discovered that embryos were produced from tissues deep within the callus. Zhang et al. (1981) found a similar pattern with poplar tissue cultures forming embryos. During callus differentiation of *E. angustifolia*, however, most of the embryos originated from the parenchymatous cell close to the epidermis, although sometimes epidermal cells also were involved with embryogenesis or the epidermal cells and aged cells deep within the calli formed embryos.

Pareek et al. (1978), Radojevic (1979), and Reinert et al. (1978) all believed that an embryo arose from a single cell possessing double polarities and that the embryo's vascular system was not derived from the maternal plant. The significant difference between adventitious buds with solitary polarity and embryos is that embryos are embryonic intact plants from the outset of development. This was believed to be true of the embryo of *E. angustifolia*, but the probability that an embryo came from some adjacent cells simultaneously dividing was not excluded. *E. angustifolia* somatic cell embryos are similar in origin, growth, and development to zygotic embryos. The cells may absorb nutrition from the tissues around them and form an independent plantlet separate from the maternal plant. This may be the method of diversification for the adventitious bud.

KEY REFERENCES

Cui, C. 1952. Chemical effect on organ formation in plant. *Acta Bot. Sin.* **1**:151–162 (in Chinese, English abstract).

Pan, J., Wang, P., Gao, R., and Tu, L. 1985. Morphological and anatomical observation of *Elaeagnus angustifolia* L. in vitro. *Acta Bot. Bor. Occ. Sin.* **5**:132–136 (in Chinese, English abstract).

Yan, M. 1980. Induction in vitro of the embryo of *Camellia oleifera* Abel. *Linye Keji Tongxun* **12**:18 (in Chinese).

REFERENCES

Chen, W., Guo, D., Yang, S., and Tsui, C. 1980. Organogenesis of leaf explant of *Populus davidiana* × *P. bolleana* L. hybrid and effect of growth regulators on organogenesis. *Acta Bot. Sin.* **22**:311–315 (in Chinese, English abstract).

Compiling Commission of Tree Annals of China. 1978. *The Afforested Techniques of the Major Tree Species in China*. pp. 1267. Agricultural Press, Beijing (in Chinese).

Fossard, R. A. 1978. Tissue culture propagation of *Eucalyptus ficifolia* F. Muell. *Proc. Symp. Plant Tissue Cult.* Science Press, Beijing, 425–438.

Konar, R. M., Thomas, E. and Street H. E. 1972. Origin and structure of embryos arising from epidermal cells of the stem of *Ranunculus sceleratus* L. *J. Cell. Sci.* **11**:77–93.

Northwest Institute of Botany and Lanzhou University. 1977. *The Oil Plants in Northwest China*. pp. 168–170. Shaanxi People's Publishing House, Xian (in Chinese).

Pareek, L. K. and Chandra, N. 1978. Somatic embryogenesis in leaf callus from cauliflower (*Brassica oleracea* var. *botrytis*). *Plant Sci. Lett.* **11**:311–316.

Radojevic, L. 1979. Somatic embryos and plantlets from callus culture of *Paulownia tomentosa*. *Z. Pflanzenphysiol.* **91**:57–62.

Rao, P. S. and Narayanaswami, S. 1972. Morphogenetic investigations in callus cultures of *Tylophora indica*. *Physiol. Plantarum* **27**:271–276.

Reinert, J., Bajaj, Y. P. S. and Abell, E. 1978. Aspects of organization—organogenesis and embryogenesis. In: *Plant Tissue and Cell Culture*. 2nd Ed. (H. E. Street, ed.), pp. 389–427. Blackwell, Oxford.

Steward, F. C., Mapes, M. O., Kent, A. E. and Holsten, R. D. 1964. Growth and development of cultured plant cells. *Science* **143**:20–27.

Zhang, L., Zhang, Y., Wang, F., and Liu, S. 1981. The induction and development of the somatic embryo of poplar. *Linye Kexue* **19**(4):426–427 (in Chinese).

CHAPTER 31
Walnut

Biwen Han and *Shulan Liu*

INTRODUCTION

History and Economic Importance

Walnut is widespread in the world and is cultivated in the area from the equator to 35 degrees north latitude. It has been more than 2000 years since China began to cultivate walnut. Historically, the growing area has extended from the west to the Yellow River Valley. Walnut is now distributed very widely in China, especially in Yungui Plateau and provinces in the north, except for severely cold regions, as well as the middle and lower reaches of the Yangtse River.

Walnut is an important tree economically. The nut tastes delicious, is nutritious, and is used in Chinese traditional medicine. According to the record in *Compendium of Materia Medica*, walnut can reinforce vital energy and nourish the blood, moisten dryness, eliminate phlegm, cure the lungs, and moisten the intestines. There is a very high oil content in walnuts, generally about 60% or even as high as 75% or 80%. The oil contains 94.5% unsaturated fatty acid. Walnut contains 63% lipid, 15% protein, and vitamins, including 1.7 mg carotene, 3.2 mg thiamine, 1.5 mg riboflavin, and 10 mg nicotinic acid per 100 g nut. The walnut also provides wood of excellent quality for high-grade furniture.

Taxonomy

Walnut (*Juglans regia* L.) has a catkinlike inflorescence; unisexual flower; diverticillate flower, with or without perianth; an inferior ovary; and is pollinated by wind. Walnut belongs to family Juglandaceae.

Walnut (*Juglans regia* L.) is a deciduous tree usually 10 to 20 m tall, at most 30 m, with a long life span, 200 to 300 years. The genus *Juglans* L. consists of more than 20 species growing in the temperate and subtropical zones in Asia, Europe, and North America. There are 6 species of walnut in China. Many good varieties of walnut have been produced by superior selection. China has more than 350 varieties of walnut, among which the most famous ones are thin-shelled walnut, exposed walnut, and soft walnut.

Problems in Breeding and Cultivation

In the past there were late varieties of walnut which began to bear fruits after 8 to 10 years and early varieties which bore fruit after 2 to 4 years. Selection of early varieties with good quality and high yield is very important for commercial production. When plants are propagated from seed, excessive variation of many traits, including the size and weight of nuts, is observed. In order to prevent variation among seedlings, grafting may be adopted, but the limited number of scions excised from a single plant adversely affects the propagation efficiency. Tissue culture techniques allow more rapid propagation and preserve the good characteristics of a variety.

Walnut contains tannin compounds which inhibit growth, so vegetative propagation methods such as leaf cutting or shoot cutting are difficult in walnut cultivation. However, by in vitro culture methods the tannin synthesis could be controlled and bud and root formation as well as plantlet regeneration would be easier.

LITERATURE REVIEW

A few papers about walnut tissue culture have been published. Rodriguez (1982a) reported that ½ K medium (Cheng, 1975) (Table 1) plus 4.05–40.5 μM 2,4-D and 1.86–9.3 μM KIN induced callus formation from walnut cotyledons. After some time the callus browned and growth stopped. Root differentiation was achieved in culture when 3.9 μM IBA or 8.8 μM NAA and 1.86 μM KIN or 1.76 μM BA were added. NAA was more ef-

TABLE 1. Composition of Media for Walnut Tissue Culture

Compound	DKW (pH = 5.2)	WPM (pH = 5.5)	K Medium (pH = 5.4)
NH_4NO_3	17.7 mM	5 mM	20.6 mM
$Ca(NO_3)_2 \cdot 4H_2O$	8.2 mM	2.3 mM	—
KNO_3	—	—	18.8 mM
KI	—	—	5 μM
K_2SO_4	8.95 mM	5.68 mM	—
$CaCl_2 \cdot 2H_2O$	1 mM	0.66 mM	3 mM
KH_2PO_4	1.9 mM	1.25 mM	1.25 mM
H_3BO_3	71.4 μM	100 μM	—
$Na_2MoO_4 \cdot 2H_2O$	1.6 μM	1 μM	1 μM
$MgSO_4 \cdot 7H_2O$	3 mM	1.5 mM	1.5 mM
$MnSO_4 \cdot H_2O$	148 μM	100 μM	100 μM
$ZnSO_4 \cdot 7H_2O$	—	30 μM	37 μM
$Zn(NO_3)_2 \cdot 6H_2O$	46 μM	—	—
$FeSO_4 \cdot 7H_2O$	120 μM	100 μM	108 μM
$CuSO_4 \cdot 5H_2O$	1 μM	1 μM	0.1 μM
$CoCl_2 \cdot 6H_2O$	—	—	0.1 μM
Na_2-EDTA	120 μM	100 μM	97 μM
Vitamin B_1	6 μM	3 μM	15 μM
Vitamin B_6	—	3 μM	6 μM
Nicotinic acid	8.12 μM	4.06 μM	8.12 μM
Glycine	26.6 μM	26.6 μM	26.6 μM
Myo-inositol	5.55 mM	555 μM	2.78 mM
Calcium D-pantothenate	—	—	2.1 μM
CH	—	—	100 mg/l
Sucrose	88 mM	58.6 mM	58.6 mM
Agar	—	0.6%	0.5%
Gelrite	200 mg/l	—	—

fective than IBA in root induction. Root formation was inhibited by 6.12 μM KIN or 5.72 μM BA. Rodriguez (1982b) also reported that when seeds were sown on ½ K medium containing 39.6 μM BA in the presence of different levels of IBA, 39.6 μM BA and 0.49 μM IBA was optimal for supporting a good rate of bud production. Driver and Kuniyuki (1984) reported that shoots were excised from seedlings of walnut hybrids to induce apical and lateral bud elongation. At least five buds containing 4.4 μM BA and 0.005 μM IBA were induced from each node on DKW medium (Table 1). The effect of IBA and NAA on root induction after several subcultures was tested, and 29.4 μM IBA was found to be the most effective. The shoots also rooted with 3 μM NAA, but the vascular bundle between the newly formed roots and shoots was poorly developed. On DKW medium the buds elongated into shoots in a short time. Before the shoots were transplanted into pots, their lower part was dipped in 0.005 μM IBA solution. The pots were placed in a greenhouse with natural light and 85% relative humidity. Rooting occured in 10 to 14 days. Ninety percent of the plantlets survived transplanting to the field. This was the only report describing walnut plant regeneration in vitro.

The main difficulty in walnut tissue culture is browning. The callus always turned brown, could not differentiate, and died. The key to successful walnut tissue culture is the prevention of oxidation and browning.

TECHNIQUES FOR TISSUE CULTURE

Callus Induction

MATERIAL SELECTION AND PRETREATMENT. The thin-shelled walnut from Xinjang Province was used as material. Leaves, shoot tips, petioles and veins, etc., excised from the seedlings or cotyledons pretreated at low temperature (5 ± 2°C) for 12 to 24 h were used as explants.

MEDIUM. The WPM (McCown and Lloyd, 1981) (Table 1) medium supplemented with 22 μM NAA, 44 μM BA, and 5.8 μM GA was used to induce callus from shoot tip, leaf, petiole, and vein. The modified MS and Cheng medium (K medium, Table 1), with 27 μM NAA, 2.2 μM BA, and 5.8 μM GA, were used for callus induction from cotyledon. The WPM medium and the modified MS and Cheng medium were supplemented with 5 ml/l of 20% sodium thiosulfate solution.

PROCESS. In our past reports (Liu Shulan and Han, 1984), the callus was derived from various parts of the walnut plant with different morphological characteristics.

1. *Callus Induced from Shoot Tip.* Callus was of two types: a loose type which was milk-white in color and a compact one which was greenish. During culture, the green callus produced anthocyanin and differentiated vascular tissue.
2. *Callus Induced from Leaf.*

   Young leaf: When the abaxial surface was in contact with the medium, dark green compact callus formed and granularlike structures with anthocyanin developed. The callus could be maintained for 40 days. When the adaxial surface of the leaf was in contact with the medium, the callus was loose and colorless and survived only 25 days.

   Mature leaf: Callus was only produced at the cutting edge.

   Parts of leaf: When the leaf was cut into three sections that were cultured separately the callus was induced most rapidly from the middle section of leaf (7 days later), less readily from the leaf tip (10 days later), and least readily from the proximal part of the leaf (14 days later).

Petiole and vein: Callus produced from epidermal vein under natural light was compact and light green. The petiole callus was derived from cells inside the swelling petiole and was loose and colorless.

Cotyledon: Callus induced from cotyledons in semisolidified medium appeared nodulated and opaque and in the shape of a cauliflower. The callus became greenish after transfer to solidified medium under light.

In our experiments, plantlet regeneration and root differentiation from cotyledon callus were unsuccessful. However, Rodriguez (1982a) achieved root induction from cotyledon callus. It seems possible that shoots, roots, and intact plantlets could differentiate from cotyledon callus under proper conditions.

Shoot Tip Culture

Though it was difficult to get roots and buds from walnut callus, bud differentiation was obtained from shoot tip explants.

MATERIAL SELECTION. The shoot tips of seedlings 2 to 3 years old or young branches of mature tree (var. Tangshan Soft Walnut) were used as explants. The scale, young leaves, large leaf primordia, and other tissues around the shoot tip were excised.

MEDIUM. MS was used as basal medium, plus 2.7 μM NAA, 4.4–8.8 μM BA, 1.45 μM GA, 2.3 μM folic acid, 2.1 μM calcium D-pantothenate, 100 mg/l LH, 58.4 mM sucrose, 7 g/l agar, 5 mg/l 20% sodium thiosulfate solution, pH 5.4 to 5.6

Multiple Bud Formation

In the medium described, shoot tips and newly formed leaves began to elongate after 1 week. Three weeks later, they were 2 to 3 cm in length and axillary buds began to form at the base of leaf primordia, 5 to 10 axillary buds forming on each shoot tip. When the axillary shoots were 2 to 3 cm long, they were cut and used for root induction. The rooting medium was ½ MS supplemented with 2.7 μM NAA, 243 μM boric acid, or 29 μM IAA or 29.4 μM IBA; other components were the same as in the medium for bud induction. Root primordia were formed after 2 weeks.

EMBRYO CULTURE

Material Selection and Pretreatment

The Fuping spicate Walnut was used as material. Walnuts with shells were washed with tap water for 12 to 24 hr. The embryos were excised and washed with tap water for 8 to 12 hr and then refrigerated (5°C) for 12 to 24 hr.

Medium

MS medium was used as basal medium and supplemented with 2.7 μM NAA, 8.8 μM BA, 1.45 μM GA, 2.3 μM folic acid, 2.1 mM calcium D-pantothenate, 100 mg/l CH, 5 g/l agar, 5 ml/l 20% sodium thiosulfate solution, pH 5.6 to 5.8. The embryos were first inoculated in semisolid medium containing 5 g/l agar and cultured in the dark for 1 week, then transferred to solid medium (containing 7 g/l agar).

Multiple Bud Induction

The sterilized embryos were inoculated (immersed) in semisolidified medium for 1 to 2 weeks. The emerging hypocotyl elongated and developed into an opaque flesh cone on which two lines of lateral buds were formed. The cultures were then transferred to solidified medium under 16-hr photoperiod. The buds turned green and began to elongate. Buds were cut when they were about 1 cm in length and transferred to the shoot tip medium mentioned to grow shoots. Roots could be induced from shoots 2 to 3 cm in length.

Sterilization Procedure and Culture Conditions for Various Explants

STERILIZATION

1. Explants: Explants were soaked in 70% ethanol for 0.5–1 min, subsequently sterilized in saturated bleaching powder solution (with several drops of Tween-20) for 20 mins, then in 0.1% $HgCl_2$ solution for 10 min, washed four to five times with sterile water, and inoculated on culture medium.
2. Shoot tip: Shoot tips without leaves and scales were washed with running tap water for 12 hr, rinsed in 70% ethanol for 0.5–1 min, then sterilized with 5% antiformin solution or saturated bleaching powder

solution for 20 min, and washed again four to five times with sterile water. Shoot tips 0.3 to 0.5 cm long were cut and inoculated on culture medium for bud induction.

3. Embryo: The pretreated embryos were rinsed in 70% ethanol for 1 min and in saturated bleaching powder solution for 20 min, then immersed into 0.1% $HgCl_2$ (with 2 drops of Tween-20 added) for 20 min, washed five to six times with sterile water, and cultured in proper medium.

CULTURE CONDITION. The inoculated explants were cultured at a 16-hr photoperiod at 25°C in day and 18°C in night and 2000-lux light intensity.

METHODS TO PREVENT WALNUT CULTURE FROM BROWNING

There are large amounts of tannins contained in walnut cells, among them phenolic compounds which are easily oxidized to quinones, causing the tissue to turn brown. These compounds permeate the medium and inhibit the growth of cultures, eventually causing death. Preventing phenolic oxidation and tissue browning is essential for successful walnut tissue culture. This problem was resolved by addition of antioxidants to the medium for cotyledon and embryo cultures.

Degree of Browning of Explants of Different Age and Stage

Various explants oxidize and brown to different degrees. Usually, the explants from mature trees are affected more seriously than those from seedlings. Explants taken in summer are more susceptible than those taken in winter, early spring, and autumn. Because the buds in winter remain in deep dormancy and are difficult to grow, it is better to take explants in early spring and autumn.

Antioxidant's Role in the Medium

The effects of addition of 5 ml 20% sodium thiosulfate solution, 5 mg active carbon, thiothreitol, phloroglycinol, and vitamin C to 1 l of medium were tested separately. Results showed that addition of 5 ml 20% sodium thiosulfate or 0.4 mg dithiothreitol was most effective. Beneficial effects of other antioxidants were not evident. We suggest the use of sodium thiosulfate rather than dithiothreitol, as the latter is more expensive.

Pretreatment of Explant

Cotyledons, embryos, and shoot tips (especially shoot tips from aging trees) were washed with running water and refrigerated at low temperature (5 ± 2°C) for 12 to 24 hr. After sterilization, the explants were inoculated on agar medium containing sucrose for 5 to 7 days to remove phenolic compounds. Explants were then rinsed with 0.1% bleaching powder solution for 10 min and inoculated on appropriate culture medium. Such pretreatment reduced the degree of browning and helped explants develop and differentiate normally.

Role of Medium Consistency in Preventing Browning

When different explants were inoculated on solidified (8 g/l agar), liquid (with filter paper bridge as support), and semisolid (5 g/l agar) medium, the callus could be induced from embryo and cotyledon only in semisolid medium. Also, bud formation could occur only in semisolidified medium in a low-oxygen environment. On solidified and liquid media bud induction failed because of browning.

FUTURE PROSPECT

Techniques for walnut tissue culture should be developed in the following directions:

1. Cell suspension culture, which is very important for studying the relationship of embryo differentiation, media, and environmental factors as well as secondary metabolites and their utilization
2. Mutant induction and somaclonal variation, which will provide material for walnut breeding
3. Multiple bud induction from shoot tip and embryo, which may lead to the development of a new method for rapid propagation
4. Low-temperature preservation of cultures, which is very helpful for preserving walnut germ plasm

Tissue culture techniques for walnut developed later than those for other fruit trees and are much more difficult. Our work is just beginning, but we have full confidence that greater developments in walnut tissue culture will be achieved in the near future.

REFERENCES

Cheng, T. Y. 1975. Adventitious bud formation in culture of Douglas fir (*Pseudotsuga menziensii* (Mirb.) Franco). *Plant Sci. Lett.* **5**:97–102.

Driver, J. A. and Kuniyuki, A. H. 1984. In vitro propagation of Paradox walnut rootstock. *Hortic. Sci.* **19**:507–509.

Liu, S. and Han, B. 1984. Callus induction in walnut. *Plant Physiol. Commun.* **4**:38 (in Chinese).

McCown, B. H. and Lloyd, G. 1981. Woody plant medium (WPM)—a mineral nutrient formulation for microculture of wood plant species. *Hortic. Sci.* **16**:453 (Abstract).

Rodriguez, R. 1982a. Callus initiation and root formation from in vitro culture of walnut cotyledons. *Hortic. Sci.* **17**:195–196.

———. 1982b. Stimulation of multiple shoot bud formation in walnut seeds. *Hortic. Sci.* **17**:592.

Shaanxi Institute of Pomology (ed.). 1980. *Walnut*. pp. 1–29. Chinese Forestry Press, Beijing (in Chinese).

CHAPTER 32
Oil Palm

Yuanfang Cui and *Zheng Gong**

INTRODUCTION

History and Economic Importance of Oil Palm Cultivation

Oil palm, *Elaeis guineensis* Jacq, of the Palmae family, is a perennial woody crop which grows in tropical zones and originated in moist forests in Africa. The major areas of palm production include West Africa, Central Africa, Indonesia, Malaysia, and Zaire. Oil palm was introduced from Southeast Asia into China and planted experimentally on Hainan Island between 1926 and 1949 and has been cultivated commercially since 1960. In China, oil palm currently grows in the south and northwest of Hainan Island, and on a small scale in Xishuangbana of Yunnan Province.

Oil palm blossoms 2 years after planting, and harvest begins in the third year. The full production period begins in the sixth or seventh year. The economic life span is 20 to 25 years; the natural life span of the plant is 80 years or more. The oil output may reach 5250 to 10,500 kg/ha in the major producing regions of Southeast Asia. In some experimental plots, yield may be 15,000 kg/ha. Acreage of oil palm has exceeded 1.33 million ha in the world. Malaysia ranks first in both production and export of

*English translation by Kaiwen Yuan.

palm oil. With development of oil palm plantations, the total output of palm oil in the world has increased by a large margin; in 1981, palm oil accounted for 15% of total oil products. Hybrid varieties under cultivation are being increased such as the hybrid tenera, which was developed as a cross between a homozygous thick-shelled variety and a homozygous shellless one. The thickness of pericarp is a characteristic determined by a single gene. The hybrid has a very thin endocarp.

Oil palm yields two major products. One is palm oil extracted from the fruits, of which 50% is oil. The other is the kernel oil extracted from seeds, which is also 50% oil. Palm oil, which is red-brown, is an undried oil containing carotenoid and vitamin E. The kernel oil, similar to coconut oil, is rich in lauric acid. As well as being an edible oil, consumed by people in tropical areas, it is an important oil for industry. The oil residues can also be used as animal feed.

Biological Characteristics of Oil Palm

Oil palm is a monocot plant with only one apical dome without tillers. If the apical cone is cut off, the plant stops growing. Every leaf has a flower bud in its axil. Flower bud and leaf bud formation take place simultaneously. However, differentiation and growth of the leaf bud are much faster than that of the flower bud. The inflorescence usually appears at the leaf axil 9 to 12 months after leaf blade formation. Differentiation of the inflorescence occurs 2½ to 3 years before the blossom. Oil palm is monoecious with male and female inflorescences. The male and female both have their own inflorescence. Mixed inflorescences of both male and female flowers are also occassionally observed. Pistillate inflorescences are composed of many spikelets. Each spikelet contains between 4 and 40 pistillate flowers, with one spike bearing 1000 to 1500 fruits and weighing 10 to 15 kg. Staminate inflorescences are composed of many finger-shaped spikelets. Each spikelet contains 700 to 1200 staminate flowers.

Oil palm varieties are classified according to the characteristics of their fruit into dura, tenera, and pisifera varieties. Tenera are economically the most important for oil production. The tenera was derived by crossing the dura and the pisifera varieties. It has medium-sized fruits with a thin shell and thick mesocarp and thus is high-yielding. The oil content from panicles is 22–25%.

Problems of Production

The key to oil palm cultivation is to obtain planting material of good quality with uniform phenotype. Oil palm is a cross-pollinated plant and only propagates sexually. Therefore, progeny show great variation. Most asexual propagation techniques do not work in the propagation of oil palm. Hence, characteristics of high yield and stress tolerance of the mother tree cannot usually be transmitted to the next generation. Good planting material is obtained by conventional sexual propagation, which usually takes several generations; 8 to 10 years constitutes one generation. If tissue culture techniques are used to propagate a mother tree with desired characteristics, it is possible to eliminate the problems of conventional propagation. Good clones can be easily established through tissue culture. Successful application of tissue culture will also yield new methods for breeding oil palm. Using tissue culture techniques, it is possible to select and quickly propagate good planting material and to replace low-yielding varieties with high-yielding varieties having stress resistance and high adaptability.

Review of Oil Palm Tissue Culture

Tissue culture of oil palm has a history of more than 20 years. Progress has been achieved in England, France, Malaysia, and other countries, and research continues worldwide on oil palm tissue culture. Callus has been induced from apical meristem (Staritsky, 1970), and plantlets have been regenerated from embryonic callus (Rabechault and Martin, 1976) and from shoot tips of seedlings (Jones, 1974). Jones thought that plantlets could be regenerated directly from callus through explant culture or from culture of divided meristems without callus formation. Jones was able to obtain plantlets by the former method, but the plantlets did not survive after transplantation. The first batch of asexually propagated plants obtained in England were transported successfully to Malaysia in 1977; all these plants were regenerated from explants of unselected seedling plants (Corley et al., 1977).

Malaysia Agricultural Research and Development Institute (MARDI) began to study oil palm tissue culture in 1971. Callus was induced from in vitro cultured root tips. However, they grew very slowly and could not be grown in liquid culture (Ong, 1975). Karthigasu (1982) has reported that proliferation of oil palm by in vitro culture has been achieved in Malaysia.

Similar progress on oil palm tissue culture has been achieved in a plant physiological laboratory of Office de la Recherche Scientifique et Technique d'OutreMer (ORSTOM) of France (Ahee et al., 1981). The study concentrated on improving Rabechault's method (1976) of regenerating plantlets from young leaf tissues, attempting to increase induction frequency and shorten the duration of each culture stage. Fast-growing cultures were obtained by culturing the primary callus and then inducing embryos and plantlets. Ahee and coworkers shortened the time needed for culturing fast-growing cultures from 2 to 3 years to 6 to 12 months. Many of the plantlets obtained from leaf tissue by this technique have been planted and survive in an oil palm experimental station in the Ivory Coast. In 1984 the authors reported that embryos could proliferate when induced from embryonic calli and the matured embryos could develop further into plantlets.

TISSUE CULTURE TECHNIQUES USED IN CHINA

Culture Procedure and Medium

CULTURE PROCEDURE. Regenerated oil palm plants were obtained through induction and culture of embryos produced from callus. This method can be divided into three steps. The first step is callus induction from the explant. The callus grows in solid medium or as cell suspensions cultured in liquid medium. The second step is to transfer the callus into a new medium for proliferation. The third step is plantlet differentiation. Embryos regenerate from callus, then proliferate, and develop into intact plants. Buds can also grow directly from callus and will form intact plants. Callus formation can be prevented by using apical meristems, which are split as explants. One bud can be grown from each split meristem and will root. The advantage of the former technique is that culture can be carried out on a large scale; in the latter method, the number of cultures is limited and plant growth is likely to be stunted. In fact, most of the plants obtained were produced through callus formation (i.e., explant, callus, embryo, plantlet). Because the plantlets developed complete connections between their stem and root vascular tissues, they survived.

MEDIA. The basic media for tissue culture of oil palm were MS medium (Murashige and Skoog, 1962), modified MS medium, and VW medium (Viacin and Went, 1949). Table 1 presents compositions of the modified MS medium. Agar (0.7–0.8%) should be added to solidify medium. The pH of the medium was maintained between 5.6 and 5.8. The prepared medium was autoclaved for 20 min at a pressure of 0.8 to 0.9 kg/cm^2.

Culture Conditions and Manipulation

STERILIZATION AND INOCULATION. *Seed Embryo.* The seed of oil palm has a shell (endocarp). When seed embryos were used as explants, the seed shells had to be cracked. The kernel was picked out,

TABLE 1. Composition of Modified MS Medium

Constituent		Constituent	
KNO_3	5.25 mM	$Co(NO_3)_2$	0.14 μM
KH_2PO_4	1.84 mM	Glycine	26.6 μM
$(NH_4)_2SO_4$	3.73 mM	Inositol	555 μM
$MgSO_4 \cdot 7H_2O$	1 mM	Nicotinic acid	4.1 μM
$Ca_3(PO_4)_2$	0.65 mM	Pyridoxine chloride	2.43 μM
$MnSO_4 \cdot H_2O$	179 μM	Thiamine chloride	0.3 μM
H_3BO_3	100 μM	$Na_2EDTA(FeSO_4 \cdot 7H_2O)$	100 μM
$ZnSO_4 \cdot 4H_2O$	30 μM	Sucrose	5.87 mM
KI	5 μM	Glucose	5.55 mM
$(NH_4)_2MO_2O_7 \cdot 4H_2O$	1 μM		
$CuSO_4 \cdot 5H_2O$	1 μM		

rubbed with bleaching powder, and immersed in bleach solution (0.2 g/l effective chlorine) for 12 hr. The kernels were then put into a beaker under aseptic condition, soaked in 75% ethanol with agitation for 10 min, placed in a distilled-water solution containing bleach (0.4 g/l effective chlorine), stirred for 20 min, and finally rinsed three times with distilled water. After these treatments, the seed embryos had to be aseptically excised and inoculated onto the medium.

Shoot Tip. Young shoot tips were taken from seedlings at least 5 cm long. Leaves were removed until the apical meristem appeared, and then the central part of the apical tissue was taken for sterilization.

Inflorescence. The inflorescence was removed from the leaf axil near the top of the plant. Leaves were removed and the growing young inflorescence excised. Both stem tip and inflorescence were sterilized by immersing in a solution containing 10% sodium hypochlorite for 15 min, followed by three rinses with sterile water.

Root Tip. Root tips were removed for culture when the roots of sprouting seeds grew to 1.5 cm. Root tips were rinsed with 75% ethanol for 2 min, washed with a solution containing 0.1% mercuric chloride for 2 min, then surface-sterilized with 10% sodium hypochlorite and rinsed three times with sterile water.

Leaf Blade. Shoot tips were cut about 40 cm below the apical dome of matured trees or 2- to 3-year-old seedlings. The leaf tips were cut off and the lower 30 cm was left. Under aseptic conditions, the outside leaf sheaths were removed one layer at a time. Every whorl of the leaf sheath was cleaned with 75% alcohol. Since the outer layers of leaf sheath are very compact, the inner leaves do not need to be sterilized. Leaf sheaths were cut into pieces of 2 cm^2 and put into culture. A 2-year-old (calculated from sprouting) seedling can give 400 explants; a mature tree can provide 1000 explants.

CULTURE CONDITION. Light may have some inhibitory effects on callus growth. Therefore, dark culture was used for inducing callus formation from leaf and in vitro embryos. Light is not a factor in periods of formation, proliferation, and maturity of embryos. Temperature of the culture room was 28 ± 0.5°C, and relative humidity was 60–70%. Light was needed when buds were induced from embryos. Light intensity was 2000 to 3000 lux for 12 to 14 hr/day.

Key Points in Plantlet Regeneration from Embryos

HEAT TREATMENT OF SEED EMBRYO. The frequency of callus induction could be markedly improved with heat pretreatment of seeds. Heat-treated seed embryos have enhanced viability, and the number of swelled and split embryos increases greatly during culture. This is favorable to embryo differentiation. Heat treatment was given by placing the cleaned seeds at 38–40°C. Seeds should be kept moist and removed for air drying after 40 to 45 days. This treatment may affect the time required for callus formation and callus induction frequency. The final frequency of callus induction of heat treated seeds was increased by 10%.

CALLUS FORMATION AND HORMONE REGULATION. Embryos began to swell after 1 week of in vitro culture in appropriate medium. The swollen embryos gradually split and grew into callus after 30 days' culture (Fig. 1). The callus was yellow or pale yellow. The size of healthy growing callus is 5 to 10 times larger than that of embryos after 30 to 60 days in culture. Calli are then subcultured and 2 months later cultured for differentiation. Time needed for embryo development from callus is rather long: from 30 to 200 days or longer.

2,4-D had a positive effect on dedifferentiation of embryos; if 2,4-D was used in combination with NAA, the effect on induction was even

Figure 1. Calli derived from plumule.

better. When the ratio of 2,4-D and NAA was 3:4 to 1:2 in medium, the induction frequency reached 52.3–75.6%. The highest frequency was more than 95%. If 2,4-D concentration was too high (over 4.5 μM), the induction frequency was lower.

CALLUS SUBCULTURE. Improved callus growth was obtained by transferring callus to subculture medium containing high concentrations of KIN and CH in addition to high concentration of auxin. On this medium the transferred callus began to proliferate and differentiate after about 10 to 15 days. Calli increased their size by one-third to one-half during 2 months of culture. New calli were mostly light in color or still yellow. Calli should be transferred into differentiation medium after 40 to 50 days of subculture. Few embryos were obtained during subculture.

CONDITIONS FOR EMBRYO FORMATION. Calli may differentiate into embryos on differentiation media, with or without KIN, and with reduced auxin concentration. As shown in Table 2, 55.1% embryos were produced on medium without KIN; only 16.9% were produced in medium with KIN. In addition, 8.9% embryos were obtained during subculture and 19.1% were obtained in medium supplemented with ADE or NAA. Therefore, different differentiation media could stimulate callus to differentiate into embryos, but medium without KIN gave the best results. When embryos were subcultured, supplementation of medium with KIN was necessary.

A majority of callus-derived embryos clustered together. Each cluster contained 2 to 22 embryos. At most, four clusters grew from one piece of callus. Embryos were smooth, white, and of different sizes.

EMBRYO PROLIFERATION. To increase the number of embryos, several media for embryo proliferation were developed. The results are

TABLE 2. Results of Various Media for Induction of Embryos

Medium*	Major Supplements	Embryos	
		No.	%
Differentiation medium	2,4-D 2.3 μM, CH 1 g/l, 176 mM sucrose	98	55.1
	2,4-D 2.3 μM, KIN 2.3 μM, CH 1 g/l, 176 mM sucrose	30	16.9
	NAA 270 μM, ADE 296 μM, CH 1 g/l, active carbon 0.5 mg/l	14	7.9
	NAA 270 μM, CH 1 g/l, active carbon 0.5 mg/l	20	1.2
Multiplication medium	2,4-D 4.5 μM, NAA 11 μM, KIN 4.6 μM, CH 1.0 g/l	16	8.9

*The modified MS medium was used as basic medium.

shown in Table 3. To achieve better embryo proliferation, NAA, ATP, royal jelly, and other substances were added to those media which had already yielded good results. The results in Table 4 show that good embryo proliferation was obtained. The number of regenerated embryos increased as the duration of culture was prolonged. High proliferation frequency was achieved in media with 2.2 μM BA or 2.3 μM ZEA. A supplement of royal jelly, 3 mg/l, in the medium yielded even better results than those supplemented with NAA and ATP. It was observed that proliferation frequency of embryos was reduced with increasing number of passages on subculture medium. Some embryos even became flat. Further study of this phenomenon is needed.

REGENERATION OF PLANTLETS. For embryo differentiation into plantlets, concentration of macroelements in the modified MS medium was reduced, microelements of H medium were used, and medium was supplemented with a low concentration of auxin.

With adequate light, embryos elongated and green buds developed. The buds grew into plantlets after a long period of culture. Intact plantlets were often regenerated directly from embryos since buds and roots developed simultaneously. It was not difficult to induce plantlets from most of the well-developed embryos (Fig. 2).

Culture of Roots and Leaves

Oil palm plantlets induced from roots have been obtained in England and France. The work presented here was based on root culture done by Ong (1975, 1976) and leaf culture by Ahee et al. (1981).

CALLUS INDUCTION FROM ROOTS AND LEAVES. Sterilized root tips were inoculated into modified MS medium with 0.05 μM NAA and

TABLE 3. Effects of Different Kinds of Growth Regulators on Proliferation of Embryos

			After 20 Days in Culture		After 48 Days in Culture		After 60 Days in Culture	
Growth Regulator (μM)	No. of Tubes	No. of Embryos Transferred	No. of Embryos	Proliferation Percentage	No. of Embryos	Proliferation Percentage	No. of Embryos	Proliferation Percentage
KIN 0.93	14	210	253	20.5	292	39.0	322	53.3
KIN 2.3	5	73	98	34.2	111	52.1	127	74.0
BA 2.2	6	48	62	29.2	88	83.3	109	127.1
ZEA 2.3	7	69	105	52.2	129	87.0	141	104.3

TABLE 4. Effect of Royal Jelly and Other Supplements on Proliferation of Embryos

Supplements in Medium (μM)	No. of Tubes	No. of Embryos Transferred	After 25 Days in Culture		After 38 Days in Culture		After 60 Days in Culture	
			No. of Embryos	Proliferation Percentage	No. of Embryos	Proliferation Percentage	No. of Embryos	Proliferation Percentage
BA 2.2	14	152	169	11.2	173	13.8	218	43.4
BA 2.2 NAA 1.1	13	155	176	13.5	188	21.3	226	45.8
BA 2.2 ATP 20 mg/l	9	141	166	17.7	172	22.0	198	40.4
BA 2.2 Royal jelly 3	8	94	123	30.9	129	37.2	163	73.4

2.3 μM KIN and cultured for 2 months. After several weeks in this medium, the root tips grew rapidly and lateral roots developed. If root tips were transferred into medium containing a moderate concentration of auxin, such as modified MS medium supplemented with 11–27 μM NAA, callus induction occurred (Ong, 1975, 1976). Modified MS medium was used to induce callus from leaves; auxin, nitrogen-containing compounds, sugar, and ascorbic acid were added. The first calli emerged after 60 days in culture. The number of explants yielding calli did not increase after 100 to 120 days in culture. Frequency of callus induction

Figure 2. Regenerated intact plantlet.

from different kinds of explants was 20–60%. The calli, composed of compactly combined cells, emerged from secondary leaf veins and grew along or around leaf veins in original explants. Pretreatment of leaves with a solution containing 175 mM sucrose and 0.25% active carbon or a solution of 0.83 mM cysteine reduced leaf browning. With this pretreatment, callus induction frequency reached 52.8% for sucrose carbon solution and 28.5% for cysteine pretreatment, respectively, after 100 days in culture. When medium supplemented with active carbon is employed for callus induction, auxin concentration should be raised to 10 or more times that of medium without active carbon.

CALLUS DIFFERENTIATION AND EMBRYO PROLIFERATION FROM LEAVES. Formation and differentiation of callus from leaves are closely related to the kind of explant, physiological state of the mother plants, leaf age, and site from which the leaves are taken. Callus was easily induced from leaves of plants growing in a nursery, but leaves from matured trees also produced callus, although the induction frequency was lower.

Two culture procedures were employed for callus differentiation. The first method was initiated in medium supplemented with active carbon, auxin, and cytokinin. In this medium, callus grew slowly, was brown, and had a nodelike appearance; embryos developed directly from these calli. A long culture time was necessary to obtain embryos with this method.

Another method was to transfer callus to medium containing auxin, in which secondary calli were induced after 6 months or longer in culture. Secondary calli which are fast-growing cultures are different from primary ones: light in color, loose in texture, and granular in appearance. They proliferated rapidly in an appropriate medium. Their weights doubled after 10 to 30 days in culture. The plant regenerating ability of these calli was maintained for at least 3 years if they were subcultured. They were considered to be embryonic cell lines which differentiated into embryos under embryonic culture conditions. A cluster of white tissues having polarity and distinct shape emerged from cultured calli. They turned green rapidly, separated from neighboring tissues, and possessed distinct epidermis and all the characteristics of somatic embryos.

Embryos obtained by these procedures proliferated while they developed. The coefficient of proliferation is 32% every 30 days. Low auxin concentration and low cytokinin concentration stimulate development of buds from embryos. Bud and root development occurs simultaneously in some embryos; most embryos require a 3-week culture period for root initiation. Some plantlets appear morphologically abnormal because of the affect of auxin during root initiation. These plantlets return to normal when auxin is removed. Embryos develop into buds frequently, but some embryos develop into globular structures instead. Buds may develop from these structures through successive subculture. The presence of NAA and IBA together with a cytokinin, such as BA and KIN, in low concentration stimulates development of embryos. Embryos turn brown if no auxin is added to the medium during culture. If auxin concentration is too high, the embryo may only develop roots but not buds. Typical needle-shaped leaves grow from buds in medium supplemented with GA.

FORMATION OF INTACT PLANTS. Although leaf-derived plantlets are regenerated through embryo proliferation, callus should be induced first. Until now, only some mature plants and nursery seedlings could produce callus. Some induced callus did not turn into fast-growing cultures.

STUDIES OF GENETIC STABILITY

Oil palm tissue culture from callus to plant is a long procedure. Genetic variation of calli may occur during culture. Therefore, studies of genetic stability are important for improvements of oil palm tissue culture.

Observations on Chromosome Numbers

Corley et al. (1977) observed root tip mitosis of oil palm clones and determined that all cells were diploid ($2n = 32$). Tan (1976) observed 6% abnormal mitoses in seedling root tips of a tenera variety. Although the rate of abnormal mitosis was low, chromosome number variations still occurred in clonal progenies.

Observations on Growth and Development

Corley et al. (1979) planted 60 plants of Clone JB32 in a nursery. They measured the relative growth rate (percentage each day) of leaves of these plants within 6 months and compared results with three other seedling population. The coefficient of variation in clonal plants was lower than in the control group. This indicated that genetic variations of the clones were small, an encouraging result.

Observations on Fruiting

Wooi et al. (1982) reported on 13,300 plants of clonal oil palm planted in 140 ha of experimental fields in Malaysia. The trees planted earlier had borne fruits. Variations between clones were distinct. However, variations among individuals within a clone were not significant. These results are evidence that clonal plants obtained by tissue culture are genetically stable and that tissue culture of oil palm can be applied in production.

PROSPECTS

Oil palm tissue culture may become commercially important, but further study of the reliability of tissue culture methods is required. The goal of the study should be to provide a means for in vitro propagation of any given improved plants. Various factors affecting cell division and differentiation of cells should be studied further, and the development of embryos and methods of regulation should be systematically analyzed. Only when cost is reduced and each stage is optimized can oil palm tissue culture be carried out on a large scale. Although oil palm tissue culture is still in the experimental stage and techniques are experimental, it can be concluded that it is feasible. Success with this system will provide experience for studies of tissue culture of other palm plants. Oil palm tissue culture techniques will soon be perfected and applied to plant production.

REFERENCES

Ahee, J., Arthuis, P., Cas, G., Duval, Y., Guenin, G., Hanower, J., Hanower, P., Lievoux, D., Lioret, C., Malaurie, B., Pannetier, C., Raillot, D., Varechon, C., and Zuckerman, L. 1981. Vegetative propagation of the oil palm in vitro by somatic embryogenesis. *Oleagineux* **36**:115–116.

Corley, R. H. V., Barrett, J. N., and Jones, L. H. 1977. Vegetative propagation of oil palm via tissue culture. In: *International Developments in Oil Palm*. pp. 1–8. Malaysian International Agriculture Oil Palm Conference, Incorporated Society of Planters, Kuala Lumpur, Malaysia.

Jones, L. H. 1974. Propagation of clonal oil palms by tissue culture. *Planter* **50**:374–381.

Ong, H. T. 1975. Callus formation from roots of the oil palm (*Elaeis guineensis*). *Proc. Natl. Plant Tissue Cult. Symp.*, (J. C. Rajarao and K. Paranjothy, eds.), Rubber Research Institute of Malaysia, Kuala Lumpur, 26–31.

———. 1976. Studies into tissue culture of oil palm. In: *International Developments in Oil Palm*. pp. 9–14. Malaysian International Agriculture Oil Palm Conference, Incorporated Society of Planters, Kuala Lumpur, Malaysia.

Paranjothy, K. 1982. Oil palm. In: *Handbook of Plant Cell Culture*. Vol. 3 (P. V. Ammirato, D. A. Evans, W. R. Sharp, and Yamada, Y., eds.), pp. 591–605. Macmillan, New York.

Rabechault, H. and Martin, J. P. 1976. Multiplication vegetative du palmier a huile (*Elaeis guineensis* Jacq.) a laide de cultures de tissus foliaries. *C. R. Acad. Sci.* **238**:1735–1737.

Staritsky, G. 1970. Tissue culture of the oil palm (*Elaeis guineensis* Jacq.) as a tool for its vegetative propagation. *Euphytica* **19**:288–292.

Wooi, K. C., Wong, C. Y., and Corley, R. H. V. 1982. Genetic stability of oil palm callus cultures. *Proc. Fifth Int. Congr. Plant Tissue Cult. 1982*, (A. Fujiwara, ed.) Japanese Association of Plant Tissue Culture, Tokyo, 749–750.

CHAPTER 33
Rubber Tree: Anther and Ovule Culture

Zhenghua Chen, Xuen Xu, and *Rensheng Pan*

INTRODUCTION

A complete procedure for haploid breeding has been established through the study of culture medium, culture conditions, optimal time of anther inoculation, subculture of callusing anthers, and development of pollen embryos, as well as transplantation techniques for pollen plantlets.

With the established anther culture techniques 100–140 embryos can be formed per 100 inoculated anthers, and the frequency of viable pollen plantlets has been increased to 10–30% (number of pollen plantlets per 100 anthers inoculated) in several clones (Chen et al., 1982a). More than 1000 pollen plantlets have been obtained from 10 rubber tree genotypes. Survival frequency of transplanted plants was 50–90% (Chen et al., 1981). There were 80 pollen-derived trees which could be cloned; from them several clones with high growth vigor and a clone with high latex productivity were selected. In 1984, haploid plantlets were obtained through unpollinated ovule culture of rubber tree (Chen et al., 1985).

Basic Medium

In 1977 there were two basic media used for callus induction of anthers (Chen et al., 1978): MS medium and MB medium, which contained macroelements of MS medium and microelements and organic growth substances of the medium designed by Bourgin and Nitsch for tobacco (1967). Since then we have carried out research on improvement of basic medium and optimal conditions for development of microspores into embryos.

The medium for rubber tree (RT medium) was designed by us as follows (Qian et al., 1982; Chen et al., 1982b; Chen et al., 1982b): macroelement: KNO_3 9.4 mM, NH_4NO_3 20.6 mM, KH_2PO_4 3.75 mM, $MgSO_2 \cdot 7H_2O$ 1.5 mM, $CaCl_2 \cdot 2H_2O$ 3.88 mM; microelements: $MnSO_4 \cdot 4H_2O$ 44.8 μM, H_3BO_3 200 μM, $ZnSO_4 \cdot 7H_2O$ 30 μM, $CuSO_4 \cdot 5H_2O$ 0.1 μM, $Na_2MoO_4 \cdot 2H_2O$ 1.0 μM; iron salt: Na_2 EDTA 100 μM, $FeSO_4 \cdot 7H_2O$ 100 μM; organic supplements: thiamine HCl 1.48 μM, pyridoxine HCl 2.43 μM, glycine 26.6 μM, nicotinic acid 4.06 μM, folic acid 2.27 μM, biotin 0.2 μM, inositol 555 μM.

Procedure of Haploid Induction

PROCEDURE OF ANTHER CULTURE. The induction of pollen plants was carried out in three steps. The first step was the inoculation of anthers on dedifferentiation medium to induce callus and microspore development. Dedifferentiation medium was rubber tree (RT) medium with KIN 4.6 μM, 2,4-D 4.5 μM, NAA 5.4 μM, CW 5%, sucrose 204 mM, and 0.6% agar, pH 5.8. The second step involved transferring the callusing anther to differentiation medium after 50 days in culture. At this point the small embryos grew and differentiated until they were visible to the naked eye, or the pollen calli differentiated into embryos visible to the naked eye. Differentiation medium was RT medium with KIN 2.3–4.6 μM, NAA 1.1 μM, GA 1.45 μM, sucrose 204–234 mM, agar 0.6% at pH 5.8. After this treatment the well-developed embryos were transferred to plantlet forming medium on which they developed into intact plantlets. Plant regeneration medium was MS medium (4/5 content of MS macroelements) with GA 2.9–5.8 μM, IAA 2.9–5.7 μM, 5-bromouracil 2.6–5.2 μM, sucrose 117–175 mM, and agar 0.6% at pH 5.8.

PROCEDURE OF UNPOLLINATED OVULE CULTURE. Unpollinated ovules were inoculated on dedifferentiation medium to induce haploid cells in the ovule to develop into embryos or calli. Dedifferentiation medium was MB or RT medium with KIN 4.6–6.9 μM (or BA 2.2–4.4 μM), 2,4-D 4.5 μM, NAA 5.4–11 μM, coconut water 5%, sucrose 204 mM, and agar 0.6% at pH 5.8. Calli or small embryos were transferred to differentiation medium after 45–65 days of culture on the dedifferentiation medium. The calli differentiated into embryos. Differentiation medium was RT medium with KIN 2.3–4.6 μM and agar-agar 0.6% at pH 5.8.

Mature embryos were transferred to plantlet forming medium on which they developed into intact plantlets. The plantlet forming medium is the same as that for anther culture.

Protocol

Selection and inoculation of anthers have been reported in detail elsewhere (Chen, 1983b). The following techniques were used for selection and inoculation of ovules:

1. Female flower buds were taken at the uninuclear stage of microspores in the male flower bud on the same inflorescence.
2. Buds were wrapped in gauze (3 × 3 cm) and dipped into 70% ethanol in a beaker for 30 secs.
3. Buds were immersed in 10% commercial bleach for 30 min then rinsed in sterile distilled water four to five times.
4. On a slide in a petri dish, ovules were excised from the ovaries in the flower buds; there are 3 ovules in each ovary. About 5 ovules could be inoculated in a small test tube (12 × 200 mm) and 10 ovules in a large one (20 × 200 mm) (Chen et al., 1985).

MORPHOGENESIS

Androgenesis

Observations were carried out on the developmental pathway of pollen grains in callusing anther. The following developmental pathways of the pollen grain were found:

1. Uninucleate pollen grains divided equally and then developed into multicellular masses. This corresponded to pathway B according to Sunderland et al. (1977). Most of the embryogenic microspores of *Hevea* developed into multicellular masses in this way.
2. Uninucleate pollen grains divided unequally and then formed a generative cell and a vegetative cell. The latter further developed into multicellular masses. A few of the embryogenic microspores developed in this way. This is called pathway A by Sunderland.
3. Uninucleate pollen grains developed into multicellular masses and gradually degenerated.

The data presented in Table 1 show that on the 20th day after inoculation one-third of pollen grains were exhausted. Most of pollen grains

TABLE 1. Development of Pollen Grains in Callusing Anthers

Days after Inoculation	No. of Callusing Anthers	Pollen Grains Total No.	Development of Pollen (%)						
			Uni- and Binuclear Pollen	Multinuclear Masses	Multicellular Masses		Embryos	Senescing Pollen
					In Exine	Released		
20	9	1071	33.5	13.1	12.1	1.9	0	39.3
25	5	591	19.8	10.3	12.7	7.1	3.0	47.0
30	8	1039	5.1	8.3	11.4	6.2	3.4	65.6

CHEN, C., CHEN, F., CHIEN, C., WANG, C., CHANG, S., HSU, H., OU, H., HO, Y., AND LU, T. 1979. A PROCESS OF DETAINING POLLEN PLANTS OF *HEVEA BRASILIENSIS* MUELL.-ARG. *SCI. SIN.* **22**:81–90.

were uni- or binuclear. There were also many multinuclear pollen grains and multicellular masses, and 1–2% of pollen had developed into multicellular masses with ruptured exine (outer wall). Twenty-five days after inoculation, some embryos appeared, the multicellular masses which broke through the wall increased (Fig. 1), and the uni- and binuclear pollen grains decreased. By the 30th day the multicellular masses and embryos were still increasing, while the uni- and binuclear pollen grains decreased greatly and some multicellular pollen grains had senesced. At this time some of the pollen multicellular masses continued to enlarge, broke through the anther wall, and finally formed embryos. In addition, some pollen multicellular masses degenerated because of poor nutrition.

Histological investigations of callusing anthers completed on paraffin serial sections showed that in 35-day-old callusing anthers embryos of different sizes remained separate in the anther sac and some large embryos had ruptured the dehiscent layer of the anther. The calli derived from pollen in the anther sac were distinct from somatic calli, and some

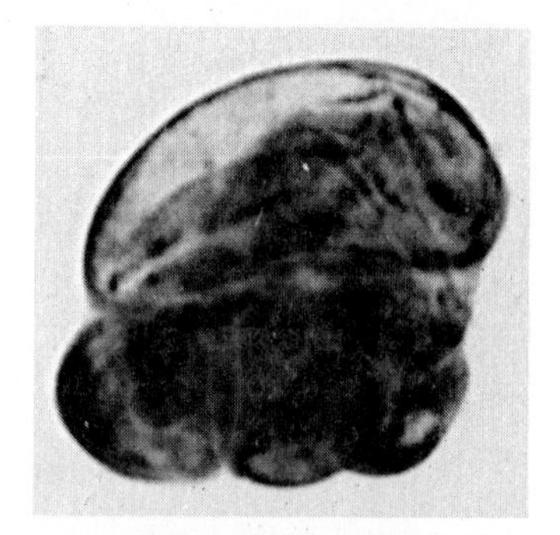

Figure 1. A multicellular mass emerging from a pollen grain.

of them were emerging through the anther wall. Only a few multicellular masses with intact exine were seen. At the 40th day after inoculation some large pollen calli appeared in the anther sac. At the same time embryos of different shapes and sizes could been seen. At the 45th–50th day, pollen calli proliferated rapidly. At the 50th day after inoculation callusing anthers were transferred to the second medium, where after about 10 days embryos were visible to the naked eye (Chen et al., 1979, 1983a).

Embryogenesis

There are three routes of embryo development:

1. A pollen grain or a haploid cell in the embryo sac develops directly into an embryo.
2. A pollen grain develops first into a callus and then differentiates into a clump of embryos.
3. The embryo derived from a pollen or haploid cell of the embryo sac produces several secondary embryos (Chen et al., 1986).

In (1) each embryo has its own genotype and the chromosome number is relatively stable. In (2) and (3) all the embryos originating from a single pollen grain or an ovule have the same genotype, but a few may show some variation in chromosome number.

Embryogenesis includes a young, mature embryo stage. Intact plantlets could be formed only when the mature embryo was transferred to plant forming medium. The mature embryo has root and cotyledon primordia and also primordia of shoot, true leaves, and terminal bud.

The main root and lateral root primordia in an embryo did not differentiate simultaneously. After transfer of the callusing anthers to differentiation medium, about 15 days later embryos visible to the naked eye emerged, although the organ primordia had not differentiated. After 30 days of culture the main root and cotyledon primordia formed, but the embryo was still small. After 30–60 days of culture the cotyledon primordia grew quickly and the lateral root primordia formed. At this time most of the embryos transferred to the plant forming medium could form a main root, lateral roots, and green cotyledons but did not organize shoots. About 60–90 days later the size of embryos markedly increased and the stem and terminal bud primordia formed. It is recommended that the embryos be transferred to the plant forming medium after about 90 days' culture on differentiation medium to increase induction of intact plantlets. During the 90 days of culture on differentiation medium it was necessary to subculture the embryos two to three times (Chen et al., 1982b).

The preceding methods and further improvement of culture media enhanced the induction frequency of intact plantlets (Fig. 2) to 28% (plantlets/100 embryos).

CYTOLOGICAL INVESTIGATION

Chromosome Number of Young Embryos and Root Tip Cells of Plantlets

Cytological investigation was carried out systematically on embryos and pollen plantlets of rubber tree. In 445 metaphases from 77 embryos 71.2% had 18 chromosomes, 19.3% had 27 chromosomes, 3.8% had 9 chromosomes, and only a few cells had more than 27 chromosomes (Table 2). In most cases the chromosome numbers were multiples of 9, i.e., $x = 9$, $2x = 18$, $3x = 27$, $4x = 36$, and $5x = 45$ (Chen et al., 1981). Root tip cells from 21 plantlets of rubber tree were observed. Among 175 metaphases 30.3% had 18 chromosomes (Fig. 3), 56.0% contained 27 chromosomes, 3.4% had 9 chromosomes, and 8.0% had aneuploid chromosomes. These results demonstrate that during the development of embryos to plantlets the number of cells with 27 chromosomes was in-

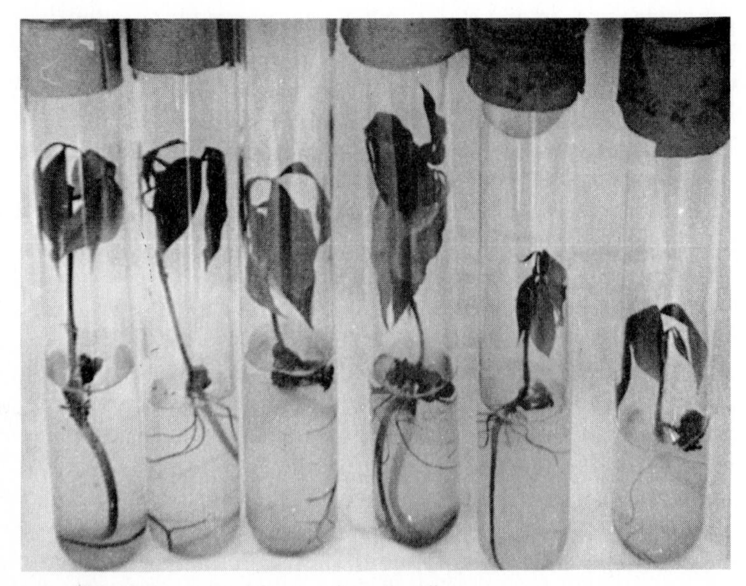

Figure 2. Plantlets derived from cultured pollen.

TABLE 2. Chromosome Numbers Observed in Mitotic Metaphases of Embryos and Root Tips of Plantlets

Tissue	No. Speci-men	No. Meta-phases	Chromosome Count								
			9	10–17	18	19–26	27	28–35	36	45	54
Embryos	77	445	17	1	317	4	86	4	10	3	3
Root tips	21	175	6	0	53	7	98	7	3	1	0

TABLE 3. Chromosome Counts in Leaf Cells of Embryo Sac Plantlets

Plantlet No.	No. of Metaphases Observed	Chromosome Count					
		9	18	19–26	27	28–35	36
1	37	3	15	1	11	3	4
2	61	0	23	6	24	5	3
Total	98	3	38	7	35	8	7

creased and that with 18 chromosomes decreased. These results confirm that the plantlets originated from pollen grains, but there were some aneuploid and polyploid cells in addition to the majority of metaphases with the haploid chromosome number.

The cytological investigation of young leaves of embryo sac plantlets showed that among 98 mitotic metaphases there were 41.8% with 9 and 18 chromosomes, 35.7% with 27 chromosomes, and 7.1% with 36 chromosomes. There were also 15.3% cells with an aneuploid chromosome number (Table 3). This result indicated that the plantlets originated from haploid cells in the embryo sac but became mixoploid. The variation of chromosome number in embryo sac plantlets was similar to that in pollen plantlets (Table 3).

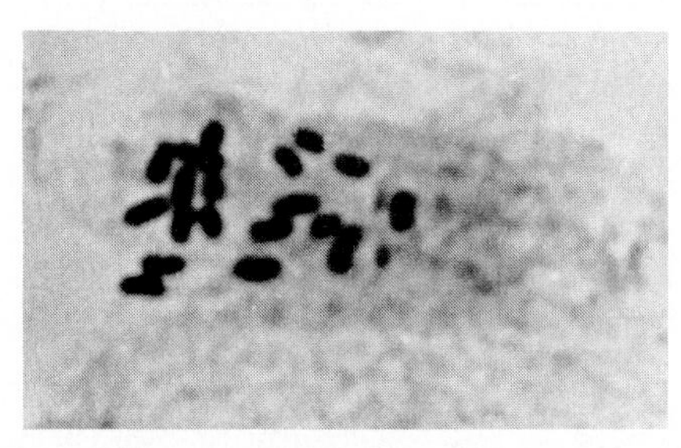

Figure 3. A multipolar mitotic metaphase.

Chromosome Count of Transplanted Pollen Trees

Systematic observations were carried out on transplanted pollen rubber trees. The chromosome counts of young leaves on pollen trees at different ages revealed that all of them were mixoploid.

CHROMOSOME COUNTS OF YOUNG LEAVES AT DIFFERENT TREE HEIGHTS. Chromosome counts of leaf cells taken from a pollen tree at a height of 50 cm revealed considerable variation in chromosome number, most of the cells having counts between 18 and 27 chromosomes (Table 4). However, the same tree at a height of 160 cm had many leaf cells with somatic chromosome numbers between 28 and 36 and relatively few cells with 18 to 27 chromosomes (Table 4). This showed an increase in diploidization during the growth of transplanted pollen trees.

In 1979, we observed the chromosome numbers of root tips and leaves from a transplanted tree. Leaves of this 50-cm-tall tree had 18–27 chromosomes. In 1980, when the tree reached 160 cm, most of the leaves had 28–36 chromosomes, but a few had 18–27 chromosomes (Table 3). This fact also revealed a successive increase in chromosome number during the growth period of the bud-grafted stock of pollen plants, i.e., continuation of the chromosome number change on the mother plant (Table 4) (Chen et al., 1982a, 1982c).

CHROMOSOME COUNT OF YOUNG LEAVES FROM DIFFERENT POLLEN TREES LESS THAN 50 CM. Chromosome counts of young leaves of pollen trees from clones Heiken No. 2 and Wuxeng 1–3 were completed. The results showed that the majority of metaphases had 18 chromosomes (54% and 53%, respectively) and part of the metaphases had 27 chromosomes (15% and 37%, respectively). Only a few metaphases had 36 chromosomes (0.6% and 7.9%, respectively). There were also some cells with aneuploid chromosome numbers. In most

TABLE 4. Chromosome Counts in Leaf Cells of a Pollen Plant and Its Clone

Material	Date of Obser-vation	Height (cm)	No. of Leaves Observed	No. of Metaphases Observed	Chromosome Count		
					9–17	18–27	28–36
Pollen	June	160	2	35	0	18	17
Plant	Mid-June	>160	3	14	1	6	7
	July	>160	4	18	0	1	17
Clone	May	60	2	11	3	8	0
	October	170	2	15	0	7	8
	November	>170	3	14	0	10	4
	December	>170	3	18	0	31	7

cases the chromosome numbers were multiples of 9. There was some difference in the proportion of cells with different chromosome number between different pollen trees (Table 5), though all of them were mixoploid with the haploid chromosome number predominating.

CHROMOSOME COUNTS OF YOUNG LEAVES IN POLLEN TREES ABOVE 150 CM IN HEIGHT. Chromosome counts of young leaves of the Heiken 2 clone (Fig. 4) were made. The results showed that the frequencies of cells with different chromosome number appeared to change continuously in seven pollen trees, with a tendency to regain diploidy in cells of the apical cone. In total about 30% cells had 18 chromosomes, 30% had 27 chromosomes, and 18% had 36 chromosomes. The proportions of the cells with different chromosome number between trees were quite different. For instance, pollen tree No. 2 had 38.7% cells with 18 chromosomes and only 7% cells with 36 chromosomes, whereas No. 5 pollen tree had 23.6% cells with 36 chromosomes and only a few cells with 18 chromosomes (Table 5).

Abnormal Mitosis

Multipolar mitoses were found in cytological observations not only in the embryos and root tips of plantlets but also in the young leaves of transplanted pollen trees. One such cell contained two poles on one side of the

TABLE 5. Chromosome Counts in Leaf Cells of Pollen Plants under 50 cm of Height

	Plant	Metaphases	Chromosome Number								
Clone	No.	Observed	9	10–17	18	19–26	27	28–35	36	45	54
Haiken-2	1	158	38	4	84	13	8	9	1	1	0
	2	80	10	4	41	6	15	1	2	1	0
	3	101	14	1	56	6	18	4	1	1	0
	4	63	10	1	29	1	16	3	0	2	1
	5	45	7	1	24	1	8	1	0	2	1
	6	21	2	0	17	1	1	0	0	0	0
	7	137	10	0	72	28	25	2	0	0	0
Total		595	91	11	323	46	91	20	4	7	2
Wuxing	1	13	3		8		2		0	1	
I-3	2	82	0		32		40		10		
	3	38	0		27		10		1		
	4	18	0		9		5		4		
	5	44	0		24		18		1		
	6	32	0		21		9		2		
Total		227	3		121		84		18	1	

Figure 4. A pollen rubber tree.

equatorial plate, one with 9 chromosomes and the other with 18 chromosomes while another cell had one pole with 9 chromosomes. In such a case, daughter cells with 27 and 9 chromosomes might be formed.

Critical Variables

The results can be summarized as follows:

MIXOPLOID CHROMOSOME NUMBER. The phenomenon of mixoploidy was found in cytological observations of pollen rubber trees.

Since the pollen rubber tree and perennial crops in general have a long growth period, the considerable part of abnormalities in chromosome number could be accumulated in the shoot apical cone, whereas in herbaceous plants most chromosome aberrations in somatic cells are usually eliminated in the sexual process.

THE TENDENCY TO SPONTANEOUS CHROMOSOME DOUBLING. Cells with chromosome numbers that were multiples of 9 were found not only in the embryos and root tip cells of plantlets but also in the young leaves of transplanted pollen trees. Cells with 27 chromosomes may have a great competitive advantage over those with 9 chromosomes, because the former were found more often than the latter during cytological observation.

The return of the chromosome number to the $2n$ level may be advantageous to the pollen plant, for the stable increase of diploid cells suggested that they may also have a selective advantage over those of other ploidies.

POSSIBLE MECHANISM OF VARIATION IN CHROMOSOME NUMBER. Endomitosis, fusion of nuclei, endoreduplication, and multipolar mitoses of cells are frequently observed in certain in vitro cultures (Crob, 1978). Multipolar mitosis could account for the formation of cells with chromosome number aberrations in anther culture of rubber tree (Chen et al., 1985).

Rubber tree is an amphidiploid (Ong, 1975). The basic chromosome number is 9. Most pollen trees which have predominantly diploid cells in the apical cone might be homozygous diploid, but some of them might be nonhomozygous. Haploid cells of rubber tree have both A and B genomes each with 9 chromosomes, the diploid cells might have three genotypes, i.e. AAAB, ABBB, AABB. As a result of repeated multipolar divisions, among them, only cells with AABB genomes are homozygous cells.

POTENTIAL OF BUD GRAFTING. As buds from pollen trees can be propagated, it should be possible to derive diploid, euploid, and aneuploid plants by bud grafting. These chromosomal variants may be useful in breeding work with the rubber tree. The buds and young leaves at the early developmental stages of pollen trees are mixoploid. If we can establish the techniques of induction of homozygous diploid plants by means of cell suspension culture and chromosome doubling with colchicine, we can obtain pure lines with various chromosome numbers. This approach will lead to a very interesting field of chromosome engineering.

ACCELERATION OF NATURAL PROCESSES. Artificial measures may accelerate the natural process of chromosome doubling of pollen plants and induce pure lines in the early phases of rubber tree culture. The ploidy of plantlets should be checked in advance in the process of in vitro culture, and artificial chromosome doubling should be carried out only when the plantlets are shown to be real haploids. The method is to treat the plant apex with 0.2–0.4% colchicine solution once in the morning for 10 days (Lu et al., 1980).

STUDY ON POLLEN TREE H_1 AND ITS CLONES

Pollen plantlets transplanted in the field grew quite slowly. Pollen plants H_1 were only 30 cm (average) in height half a year after transplantation, whereas the seedlings derived from the same inflorescence were 150 cm in height at the same time (Fig. 5). As a rule, the growth rate of pollen tree H_1 was 10–12 months slower than that of seedlings because the time for induction of pollen plantlets was much longer than that of seeds. In addition, the pollen plantlets lacked endosperm, and their roots were not as well developed as those of seedlings. It was quite difficult to choose a suitable control for comparison with pollen trees. We designed an accurate method for the study of transplanted pollen trees. Our experiments showed that the comparison of clones of pollen trees H_1 with those of

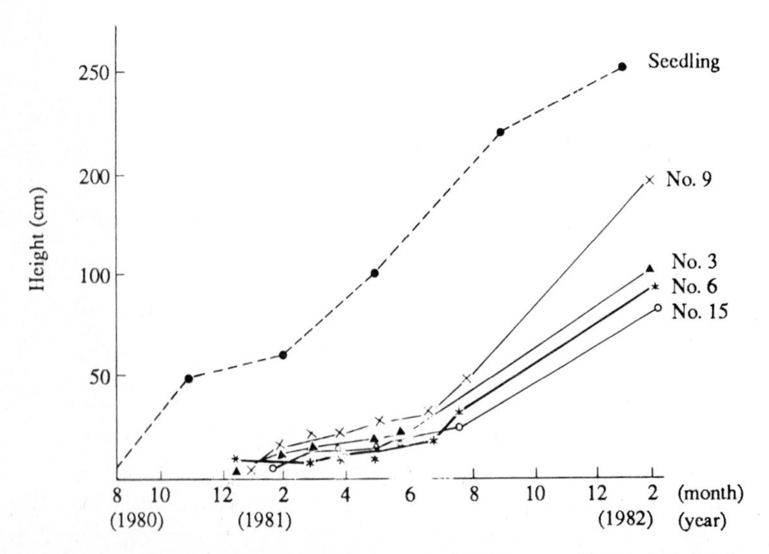

Figure 5. Height of pollen plants in comparison to seedlings derived from the same inflorescence.

TABLE 6. Height of Pollen Trees H_1 and Mean Girth of Their Clones of Rubber Tree

Clone No.	Height of 1-year-old H_1 (cm)	1983				1985			
		No. of trees Measured	Girth (mm) ($\bar{X} \pm SE$)	Comparison of Pollen Tree Clone with Control		No. of trees Measured	Girth (mm) ($\bar{X} \pm SE$)	Comparison of Pollen Tree Clone with Control	
				t	p			t	p
No. 3	98	24	95.38 ± 1.76	6.43	<0.001	24	238.6 ± 4.02	5.83	<0.01
No. 6	90	30	88.97 ± 1.57	4.55	<0.001	30	223.1 ± 3.42	3.62	<0.01
No. 9	146	36	88.92 ± 1.57	4.46	<0.001	39	254.4 ± 3.92	8.85	<0.01
No. 15	80	18	87.50 ± 2.33	3.41	<0.01	18	213.6 ± 1.48	2.28	<0.05
Donor (Control)	—	24	76.46 ± 2.26	—	—	27	203.4 ± 4.23	—	—

trees from seed appeared to be an ideal method for the evaluation of pollen trees. Since this method minimizes the variation in cultivation techniques and the differences between clones caused by environmental factors, the genetic differences between genotypes of pollen plants can be seen quite easily.

Four vigorously growing pollen trees of clone H_1 were selected on the basis of periodic measurement and then cloned by budding at the same time as their donor trees. The mean girth was measured 2 and 4 years after budding; the vigor of four clones of the pollen trees markedly surpassed that of the donor clone ($p < 0.01$). These results revealed that it was possible to select the most promising clones of pollen tree H_1 (Table 6). The results also showed that pollen tree No. 9 was significantly higher than all others in the first year, but its mean girth was not the largest one in the second. Pollen tree No. 3 was not higher than others in the first year, but the mean girth of its clone was the largest in year 2 ($p < 0.01$). However, mean girth of pollen tree No. 9 significantly surpassed those of other clones in the fourth year of growth ($p < 0.01$) (Table 6). It seemed to us that the environmental factors that change each year can induce various responses in different pollen tree clones (Chen, 1986).

One of the pollen trees transplanted in 1979 is now 7 years old. Its girth at a height of 50 cm was 60 cm. Tapping was completed once every 3 days; it indicated that latex yield of this tree was about 30% higher than that of the parent clone.

KEY REFERENCES

Chen, C., Chen, F., Chien, C., Wang, C., Chang, S., Hsu, H., Ou, H., Ho, Y., and Lu, T. 1979. A process of obtaining pollen plants of *Hevea brasiliensis* Muell.-Arg. *Sci. Sin.* **22**:81–90.

Chen, Z., Qian, C., Cen, M., Xu, X., and Xiao, Y. 1982a. Recent advances in anther culture of *Hevea brasiliensis* Muell.-Arg. *Theor. Appl. Genet.* **62**:103–108.

——— 1983a. Microscopic observation of *Hevea brasiliensis* cultures. In: *Cell and Tissue Culture Techniques for Cereal Crop Improvement*. pp. 47–54. Science Press, Beijing.

———, Xu, X., Liao, X., Pang, R., and Li, W. 1985. Production of haploid plantlets from unpollinated ovule in *Hevea brasiliensis*. Muell.-Arg. *Annu. Rep. Inst. Genet. Acad. Sin. (1984)*. 27.

——— 1986. Induction of Androgenesis in Woody Plants. In: *Haploids of Higher Plants In Vitro*. (H. Han and Y. Hongyuan, eds.), pp. 42–66. Springer-Verlag, Berlin.

REFERENCES

Bourgin, J. P. and Nitsch, J. P. 1967. Obtention de *Nicotiana* haploides a partir d'etamines cultivées in vitro. *Ann. Physiol. Veg.* **9**:377–382.

Cen, M., Chen, Z., Qian, C., Wang, C., He, Y., and Xiao, Y. 1981. Investigations of ploidy in the process of anther culture of *Hevea brasiliensis. Acta Genet. Sin.* **8**(2):169–174 (in Chinese, English summary).

Chen, C., Chen, F., Chien, C., Wang, C., Chang, S., Hsu, H., Ou, H., He, Y., and Lu, T. 1978. Induction of pollen plants of *Hevea brasiliensis* Muell.-Arg. *Acta Genet. Sin.* **5**:99–107 (in Chinese, English summary).

Chen, Z., Qian, C., Cen, M., Xu, X., Xiao, Y., Lin, M., Deng, Z., Wu, S., and Huang, N. 1981. Recent advances in anther culture of rubber tree and sugar cane. *Ann. Rep. Inst. Genet. Acad. Sin. (1980)*. 103–104.

———, Qian, C., Xu, X., and Deng, Z., 1982b. Anther culture techniques of rubber tree and sugar cane. In: *Plant Tissue Culture 1982*. (A. Fujiwara, ed.), pp. 533–534. Japanese Association of Plant Tissue Culture, Tokyo.

———, Xu, X., Zhang, S., Pan, M., Ke, B., and Zhang, L. 1982c. Cytological investigation of transplanted pollen plants *Hevea brasiliensis* Muell.-Arg. *Ann. Rep. Inst. Genet. Acad. Sin. (1981)*. 89–90.

———. 1983b. Rubber (*Hevea*) In: *Handbook of Plant Cell Culture*. Vol. II (W. R. Sharp, D. A. Evans, P. V. Ammirato, and Y. Yamada eds.), pp. 546–571. Macmillan, New York.

Lu, Y., Fung, X., Lu, Y., and Li, W. 1980. Induction of polyploid plants in *Hevea brasiliensis. Hereditas* **2**:23–26.

Ong, S. H. 1975. Chromosome morphology at the pachytene stage in *Hevea brasiliensis*: A preliminary report. In: *Int. Rubber Conf.* pp. 1–19.

Qian, C., Chen, Z., Cen, M., Lin, M., and Xu, X. 1982. Improvement of dedifferentiation medium in anther culture of *Hevea brasiliensis* Muell.-Arg. *Annu. Rep. Inst. Genet. Acad. Sin. (1981)*. p. 92–94.

Sunderland, N. and Dunwell, J. M. 1977. Anther and pollen culture. In: *Plant Tissue and Cell Culture*. (H. F. Street, ed.), pp. 223–265. Blackwell, Oxford.

CHAPTER 34

Guayule

Huizhu Xu and *Minzhi Qian**

INTRODUCTION

Guayule, *Parthenium argentatum* A. Gray, is a small perennial evergreen shrub belonging to the family Compositae. It is a native of northern Mexico and southwest Texas in the United States. *P. argentatum* is the only species of this genus that is known to produce a rubber substance.

Economic Importance (Rollins et al., 1977)

The rubber substance of *P. argentatum* is produced in the parenchyma cells. The cells containing rubber are distributed in the outer tissue of stems, branches, and roots as well as in the older cells of xylem and pith. The rubber content in a mature plant of wild *P. argentatum* accounts for about 10% of the total dry weight of the plant. A few varieties have been found to contain up to 26% rubber. The cultivars in the United States may contain up to 20% rubber after a 4-year growth period.

P. argentatum also contains resin, which accounts for 10–15% of the plant weight. The resin includes terpene, sesquiterpene, diterpene, glyceride, and hydrocarbons of low molecular weight, such as polyisoprene,

*The authors thank Shurong Hong, Ehui Fu, and Zhen Li for their participation in the work. Many thanks also go to Yimin Qin and Yinyin Liu, who supplied the photographs.

which are important raw materials in the chemical industry. The fiber, which remains after the extraction of rubber and resin, can be made into paper pulp.

The leaves of *P. argentatum* contain 25% dry weight as cuticle wax. The wax has a melting point of 76°C, higher than that of any other natural wax, and possesses excellent transparency. This is particularly important in the hard wax market. Furthermore, a kind of volatile oil possessing fragrance and bright color can be extracted from the resin and leaf blades by means of vapor distillation. This may also be of industrial value. Because of its tolerance of drought, poor soil, and high and low temperatures, this plant can be grown on soil where most other crops cannot be cultivated. Thus it bears particular economic significance for the development of remote and arid areas.

Geographical Distribution

Although *P. argentatum* has not been studied for very long, it has been cultivated successfully in Spain, Turkey, Israel, Argentina, southwest Australia, and a small part of the USSR. Most of the progress in breeding and cultivation techniques of *P. argentatum* has been made in the United States.

In China *P. argentatum* seeds were introduced from the USSR as early as the 1950s. Since 1978, they have been imported from Mexico and the United States. At present, cultivation is still at an introductory trial stage.

Research History

In their work on tissue culture of *P. argentatum*, Arreguin and Bonner (1950) studied the influence of the leaf extracts and other biochemicals on rubber formation in aseptic tissue culture. No intact plant was propagated from tissue culture. Radin et al. (1982) attempted to produce somatic embryos and haploid plants by tissue culture. Dastoor et al. (1981) published the effects of triethylamine derivatives on callus and bud induction. Staba and Nygaard (1983) reported that a complete plant was obtained from cultured seeds of nine strains and established that it was possible to propagate *P. argentatum* in vitro.

In 1981 the Wuhan Institute of Botany, Chinese Academy of Sciences, imported a small quantity of *P. argentatum*. Tissue culture methods were employed in 1983 for rapid propagation.

Intact plantlets were induced from the shoot tip and the young leaf explants after changing the type and concentration of hormones in the growth medium. During an 8-month experimental period, a single bud produced more than 2800 plants (which survived transplantation to the seedbed). A series of technical procedures from in vitro culture to transplantation was established, providing an effective method for accelerating propagation of good varieties and elite individual plants.

CULTURE TECHNIQUES

Sterilization of Explants and Inoculation

PREPARATION OF EXPLANTS. Two plantlets 15 cm high were taken into the culture room at the end of February and used for explants. The temperature was 26°C, and to accelerate growth supplementary fluorescent light of 1000 to 2000 lux was provided for 10 hr/day. Because of the higher temperature and weak light in the room, the newly sprouted buds were more slender than those cultivated under natural conditions outdoors. Experiments confirmed that these materials were easy to culture and had a low rate of contamination.

METHOD OF STERILIZATION. New shoots were cut, soaked briefly in 70% ethanol, and placed in a solution of 0.1% mercuric chloride for 3 min, then rinsed thoroughly with aseptic water three or four times.

INOCULATION TECHNIQUES. After sterilization, leaf blades were cut into sections 0.5×0.5 cm^2. Shoots were cut into segments 0.5 cm long at the same time and inoculated on the culture medium.

Culture Procedures and Media

INDUCTION OF BUDS. When the shoot tip was inoculated on MS medium with BA (2.2 μM) new shoots sprouted rapidly. The axillary buds grew out about 20 days after inoculation. Growing buds were transferred onto MS medium containing IAA and BA. Buds should grow well, forming a shoot cluster in a radial shape. (Fig. 1)

Leaf explants were inoculated on MS medium with IAA and BA. The buds differentiated directly a month later without subculture. Bud formation was closely related to the condition of the mother plant and the concentration of hormones in the medium. If the mother plant was maintained in the culture room for a long time, the newly sprouted slender leaves could rapidly become albino and lose the ability to differentiate. No albinism occurred when leaf explants were taken directly from an outdoor plant, but differentiation was very difficult. If the concentration of hormones in the culture medium was doubled, no albinism occurred;

Figure 1. The shoot cluster in radial shape.

in this case, the buds differentiated and shoot clusters formed. Large shoots could be cut for rooting.

ROOT DIFFERENTIATION. Shoots grown from shoot tip culture could root on MS medium supplemented with 11 μM IAA, generally in about 10 days. The roots grew up to 1 to 2 cm long under suitable temperature and light conditions, and at this time transplantation could be carried out (Table 1). When the culture temperature was over 40°C, the rooting rate decreased, especially on H medium, where deformed transparent plantlets with undeveloped roots often appeared. These deformed shoots rarely survived transplantation.

Though MS medium containing a low concentration of BA could also promote rooting, the roots grew very slowly, suggesting that this medium was unsuitable for rooting. Plantlets obtained from the leaf explants rooted on this medium, but the growth of buds was inhibited to a certain degree during rooting. The complete plant could then be transplanted, but the survival rate was low.

SUBCULTURE OF BUD CLUSTERS. When the large shoots were cut off, the remaining shoots could be excised for subculture. New clusters

TABLE 1. Effect of Different Media on Rooting Rate

	H + 9.0 μM IAA			MS + 2.5 μM IBA		
Experiment	No. of Inoculated Shoots	No. of Rooting Shoots	Rooting Rate (%)	No. of Inoculated Shoots	No. of Rooting Shoots	Rooting Rate (%)
I	105	98	93	115	104	90
II	69	69	100	75	74	99
III	25	24	96	50	48	96
IV	123	93	76	456	419	92

of buds formed in a week under suitable conditions. Subculture was carried out once every 7 to 10 days to propagate new buds. If deformed plantlets were observed during subculture, levels of the hormones in the subculture medium were reduced. Alternatively, medium containing a low concentration of BA could be used to return the plantlet to normal growth.

INCUBATION OF WELL-DEVELOPED PLANTLETS. Subculture was made at room temperature from July to September. The shoots grew well, and the developing ability of axillary buds was extremely strong, as evidenced by the fact that about 30 large shoots for rooting were obtained from each subculture in a 50-ml flask. During the rooting period the plantlets were exposed to natural variation of temperature (20–40°C) and were treated with intense light (3000–6000 lux for 10 hr/day). Plantlets were tall with thick stems, healthy roots, and large green leaves. Generally, after the shoots had been cultured on the rooting medium for 10 days, they were ready to be transplanted. Survival rates after transplantation of plantlets under intense light were much higher than those of controls (Table 2).

TRANSPLANTATION, CUTTING, AND PLANTING

Transplantation of Cultured Plantlets

If moisture levels were appropriate, the survival rate of transplanted plantlets was increased with increasing light intensity (Table 2). Multiple tests indicated that the plantlets of *P. argentatum* could be adapted to light at 7000 to 50,000 lux and could tolerate temperatures up to 40–50°C. Plantlets transplanted to the field directly during the hot season had thick and strong stems, earlier lignification, and thick, fresh leaves of deep color. Plantlets transplanted during the last 10 days of September even survived the rainy season during the first half of October, when it

TABLE 2. Influence of Strong Light Treatment During the Rooting Period on the Survival Rate of Transplants

		Lighting intensity during rooting period (lux)					
Treatment	No. of Plantlet	2000	5000	7000	5000	Total	Total Survival Rate (%)
Under intensive light	Transplanted	10	10	10	10	40	60
	Survived	1	6	7	10	24	
Control	Transplanted	10	10	10	—	30	23
	Survived	1	2	4	—	7	

rained for 20 days with a total rainfall of 380 mm. It was convenient to transplant to an outdoor sand bed, where the survival rate was high. A typical sand bed was 2 m long, 80 cm wide, and 30 cm high and was placed at the edge of the field which was exposed to the sun and was on the sheltered side. The sandbed was watered in preparation for transplants. Plantlets were removed from flasks and washed to remove medium. Plantlets were transplanted into the sand in rows 8 to 10 cm apart spaced 3 to 4 cm apart within each row. Beds were thoroughly watered and covered immediately with a plastic film. Further watering was carried out according to the weather conditions to maintain the humidity in the bed. The sand bed may be covered with a reed curtain during the first 3 days if the transplanted plantlets are growing in a high temperature and under the direct sunlight in order to reduce the amount of water needed. If the transplanting was done on an overcast or rainy day, aeration of the seedbed had to be ensured. Otherwise, the plantlets could readily become infected and the survival rate of transplantation would be decreased.

Cutting of Cultured Plantlets

Compared with the seedlings, the cultured shoots of *P. argentatum* possessed extraordinary rooting ability. When the plantlets were transplanted, any surplus branches could be excised and put directly into the sandbed. Rooting occurred within 8 to 10 days during high temperature season if proper procedures were taken to maintain appropriate humidity (Fig. 2). In our experiment, more than 500 shoots were cut when 1400 plantlets were transplanted. This can significantly increase the plantlet propagation coefficient while further decreasing the culture cost.

Figure 2. The rooted plantlets from the cuttings of test tube shoots.

Setting

When the plantlets had grown 15 cm high, they were transferred to the field. This setting was best done on a cloudy day because during a sunny day, watering must be frequent or suitable shade provided. During the growing period, the plants may flower within 3 months of transfer to the seedbed. Young plants transplanted in autumn could be set in the same year. During winter the seedbed was covered with a large plastic film. These transplanted plantlets grew 40 to 50 cm high in the field during the first year.

PROSPECTS

The rapid propagation of nursery stock of *P. argentatum* can be achieved by using tissue culture of shoot tips. Under the culture conditions we established, buds proliferated rapidly with a high propagation coefficient and fast, vigorous rooting. With the help of the plantlet cutting technique, a large number of plants was obtained in a short period. We transplanted more than 1300 plantlets that came from the same shoot tip during 1 week of September. We therefore concluded that shoot tip culture provides a means for industrial production of nursery stock. Furthermore, this tissue culture method can be used to select and breed improved strains. We bred two lines of *P. argentatum* with different agricultural characteristics using this technique. More attention should be directed to culture studies of different varieties and of different explants from plants at different growth phases so as to improve the technique of selecting and breeding *P. argentatum*.

REFERENCES

Arreguin, B. and Bonner, J. A. 1950. The biochemistry of rubber formation in guayule II. Rubber formation in aseptic tissue cultures. *Arch. Biochem.* **26**:178–186.

Dastoor, M. N., Schubert, W. W., and Petersen, G. R. 1981. Preliminary results of in vitro propagation of guayule. *J. Agric. Food Chem.* **29**:680–690.

Radin, D. N., Behl, H. M., Proksch, P. and Rodriguez, E. 1982. Rubber and other hydrocarbons produced in tissue cultures of guayule (*Parthenium argentatum*). *Plant Sci. Lett.* **26**:301–310.

Rollins, D. H., Allen, P. J., Benedict, H. M., Bland, D. H., Bonner, J. F., Bovey, F. A., Lopez, E. C., Dregne, H. E., Merson, R. E., Federer, W. T., Ferstel, T. C., Gentry, H. S., Haagen-Smit, A. J., Kelley, O. J., Marwel, C. S., Massengale, M. A., McGinnies, W. G. 1977. The Plant; Economics; Research Needs. In: *Guayule: An Alternative Source of Natural Rubber*. pp. 24–28, 54–69. National Academy of Sciences, Washington, D.C.

Staba, E. J., and Nygaard, B. G. 1983. In vitro culture of guayule. *Z. Pflanzenphysiol.* **109**:371–378.

CHAPTER 35
Tea: Anther Culture

Zhenguang Chen and *Huihua Liao**

INTRODUCTION

Cultivation

Tea, *Camellia sinensis* O. Ktze of the Theaceae family, originated in China. Tea has been consumed for 2000 years and is one of the standard beverages in the world. Chinese tea is famous in the world market for its unique quality, remarkable characteristics, and numerous varieties, such as green tea, black tea, and Oolong.

Geographical Distribution

Tea plants originally grew in the regions of broadleaved evergreen rain forests and seasonal rain forests located in the peripheral tropical zone of China and tropical zone of South Asia. At present, large areas of perennial tea plants are still found in the provinces of Yunnan and Guizhou in China. Ecological conditions in these regions are favorable for the growth of tea plants which have characteristics suited to these areas, including shade tolerance and resistance to heat and humidity. There are numerous varieties that have been developed from the long cultivation,

*English translation by Xiaxian Zhou.

breeding, selection, and improvement of tea processing. Tea plants were first introduced into India, Burma, Vietnam, and Japan in the 9th century; into the USSR and Great Britain in the 17th centry; and into Sri Lanka in the 19th century. In the early 20th century tea plants were introduced into East Africa.

Problems of Tea Plant Breeding

Tea plant breeding focuses on the selection of varieties for high productivity, good quality, stress tolerance, disease resistance, and ease of mechanical harvesting, so as to increase the productivity of tea leaves. Because of variations in natural conditions, it is important to breed early-, medium-, and late-maturing varieties with a combination of different traits for producing high-quality tea. Further study will be necessary for the collection, storage, and utilization of resources of tea plant varieties as well as new techniques in tea plant breeding.

Literature Review

There is very little information on tissue culture of tea plants. Kiyo Katsuo in Japan was the first to study anther culture of tea plants in 1968 and to achieve root differentiation, in 1972. Hoken Toi in 1981 also studied root differentiation in anther culture of tea plants. The highest frequency of root differentiation was achieved with the plant hormone combination of NAA 10 μM and KIN 10 μM. The present authors achieved root induction in cultures derived from anthers of tea in 1974, but no stem or leaf organs were differentiated from callus (Chen, 1979, 1982). In 1980, the authors obtained intact plantlets with roots, stem, and leaves from anther culture of tea for the first time. Transplantation experiments were carried out successfully in 1981. The plantlets were weak with very few roots and had a low survival rate when transplanted into soil. At present, at least 20 plants have been transplanted successfully using this method.

ANTHER CULTURE TECHNIQUES

Culture Procedure and Media

INDUCTION OF ANTHER-DERIVED CALLUS. In tea, a large amount of somatic callus may form from the anther wall. Callus from the filament makes the induction of haploid plantlets difficult. Therefore, it is important to control callus formation at the anther wall to promote the development of microspores. In our recent experiments, it has been found that somatic callus formation from the anther wall is closely related to medium and culture conditions. Results of experiments on various basal media revealed that SJ-1 basal medium gave the highest induction frequency of somatic callus of anther wall (35.6%). The next best were media H, modified MS, and Blaydes, having induction frequencies of 28.5%, 24.7%, and 13.6%, respectively (Institute of Botany, Chinese Academy of Sciences, 1972). Medium N_6 (Zhu et al., 1975) gave the lowest callus induction frequency, 7.8%, making it the most appropriate for induction of haploid plantlets from anther culture of tea.

Experiments on the effect of various plant hormone combinations on induction of anther-derived callus showed that callus induction increased with increasing concentration of 2,4-D when the level of KIN was held constant. Callus induction frequencies were 30.4%, 35.6%, and 45% when the concentrations of 2,4-D were 0.9, 2.3, and 4.5 μM, respectively. A concentration of 0.9–2.3 μM was therefore used to control excessive proliferation of somatic cells.

Experiments with different sucrose concentrations demonstrated that sucrose concentration of 146 mM gave the highest induction of anther-derived callus; 490 calli were obtained from 600 anthers for an induction frequency 81.7%. As sucrose concentration was increased to 8–10% in an experiment with more than 600 anthers, the induction frequencies decreased from 47.8 to 42.3%. Callus induction frequency was reduced to 4.4%, and some anthers turned brown and died when the sucrose concentration was 15%. In an experiment with 520 anthers, it was concluded that the optimum sucrose concentration in tea plant anther culture was about 10%. At this concentration, proliferation of somatic cell tissue was suppressed and microspores proliferated.

INDUCTION OF BUD DIFFERENTIATION. Smooth, nodulelike callus formed at the point of breakage of the anther was transferred to N_6 medium supplemented with 9.3 μM KIN, 2.3 μM 2,4-D, 0.95 mM serine, and 5.5 mM glutamine. Some individual calli turned green under illumination and were transferred for differentiation to N_6 medium with 9.1 μM ZEA, 86.5 μM ADE, and 10.0 mg/l lactalbumin hydrolysate. Mini leaf buds were differentiated after 6 to 10 weeks' culture. The calli with buds were cut into small pieces, and when these pieces were cultured separately, new buds continued to differentiate.

PLANTLET FORMATION. Calli with buds were further cultured in the preceding medium. Plantlets differentiated in medium with ZEA and ADE. However, roots could not be easily formed, despite the ease with which roots were formed in anther-derived callus. However, roots were developed and plantlets with root, stem, and leaves formed when small rootless plantlets were transferred to N_6 medium with 0.57 μM IAA.

Main Manipulation Techniques and Culture Conditions

SELECTION OF MATERIALS. Cultivars Fu Yun No. 6, Fu Yun No. 7, Purple leaf Len, and Zhenghe Bing White Lea were used as starting materials. However, plantlets were induced only from the cultivar Fu Yun No. 7. Tea plants have a long flowering period, so normally developed, healthy flower buds were selected in November. Anthers at mid- and late-uninucleate stage of androgenesis were taken for inoculation. Variation in color and size of flower buds existed because of the effects of climate and bud position. Also, the anthers' developmental stage varied because of buds that contained large amounts of anthers. In general, optimal anther color was orange.

STERILIZATION. Medium was autoclaved at 15 lb pressure for 20 mins; flower buds were rinsed in 70% alcohol and sterilized in 0.1% mercuric chloride for 10 min and then washed five times with sterile water. The anthers were isolated and inoculated after the petals were removed aseptically.

CULTURE CONDITIONS. Callus was induced in the dark. Organ differentiation was carried out in light at 20–25°C for 12 hr/day. Callus grew rapidly with loose texture when the temperature was higher than 25°C, but callus grew slowly and had compact texture and organs differentiated better at 20–25°C. However, at 20–25°C, 24-hr illumination per day resulted in no callus formation. This demonstrated the strong inhibitory effect of light on induction of anther wall callus. From 400 anthers inoculated in the dark at 20–25°C, 210 calli with good texture were obtained, an induction frequency of 52%. When illumination was less than 24 hr/day, there was no callus formation from 600 inoculated anthers.

CYTOLOGICAL OBSERVATION. Root tips of surviving plants transplanted into soil were fixed with Carnoy's fluid, stained with iron haematoxylin, and smeared by routine method or sectioned by standard paraffin method. Chromosome examination demonstrated that 3 of 12 plants were haploids (n = 15). The remaining 9 plants were aneuploids; most of their cells had a chromosome number of $2n$ = 18–22. No diploid plants were observed. Variation of chromosome numbers of anther-derived tea plants should be studied further.

TRANSPLANTATION OF PLANTLETS

After transplantation of the plantlet from tissue culture to soil, young plantlets often died because of sudden changes in environmental conditions. The most serious problem was fungal disease infection.

Experimental results indicated that fungal infection at the root neck of the plantlet could be prevented by smearing the plantlet with wettable 50% carbendazim powder solution at the time of transplanting, enhancing the survival rate of transplanted plantlets. In addition, plantlets were adaptable to natural environment and grew well for 20 to 30 days after being maintained with appropriate moisture, temperature, and light intensity.

REFERENCES

Chen, Z. 1979. An experiment on anther culture of tea plants. *Fujian Nongye Keji* **4**:39–40 (in Chinese).

——— and Liao, H. 1982. Obtaining plantlets through anther culture of tea plants. *Zhongguo Chaye* **4**:6–7 (in Chinese).

Institute of Botany, Chinese Academy of Sciences. 1972. Haploid plant culture and its significance in breeding. In: *Danbeiti Yuzhong Zilian*. Vol. 1. pp. 1–18. Science Press, Beijing (in Chinese).

Zhu, Z., Wang, J., Sun, J., Yin, G., Xu, Z., Zhu, Z., and Bi, F. 1975. Establishment of a better medium for anther culture in *Oryza sativa* through comparative experiment of nitrogen sources. *Sci. Sin*. 5:484–490 (in Chinese).

CHAPTER 36
Deodara Cedar

*Min Liu**

INTRODUCTION

Geographical Distribution and Economic Importance

Deodara cedar, *Cedrus deodara* (Roxb.) Loud., belongs to the family Pinaceae. *Cedrus deodara* Loud. together with *Sciadopitys verticillata* Sieb et Zucc. and *Araucaria cunninghamii* Sweet are the three best known ornamental tree species in the world.

Cedrus deodara Loud. originated in mountainous areas 1200 to 3000 m above sea level in India and Afghanistan. China has cultivated *Cedrus deodara* for more than 70 years. It is cultivated in the provinces of Liaoning, Shandong, Jiangsu, Zhejiang, Anhui, Jiangxi, Fujian, Hubei, Hunan, Henan, Shanxi, Sichuan, and Yunnan, and the cities of Beijing and Shanghai.

There are four species in *Cedrus* Loud., which are distributed in North Africa, western Asia, and the western Himalaya Mountains.

Cedrus deodara, an evergreen tree that reaches a height of 60 m, is a good timber tree and an ornamental species having a beautiful appearance. Its timber is moderately hard, fragrant, and durable and it is good building material for construction. The timber can also be used for ex-

*English translation by Xiaxian Zhou.

tracted aromatic oil. *Cedrus deodara* is an ideal tree for streets in large cities because of its freedom from bacterial diseases (Delectis Florae Reipublicae Popularis Sinicae, Agendae Academia Sinicae Edita, 1978).

Botanical Traits and Significance of Stem Tip Culture

Winter buds of *Cedrus deodara* are small and have few bud scales. There are bud scales at the basal sections of the branches. Leaves of the long branches are spirally arranged, stretching out radially. Leaves on the short branches are fascicular. The cone flower is monosexual and hermaphroditic and stands straight at the top of short branches. The male cone flower contains numerous stamens growing in spiral shape, with two anthers and very short filaments. The female cone flower is light purple in color, with many nucellar scales of spiral shape. A short bract scale is on the dorsal surface of the nucellar scales. The cone matures in over 1 year with two seeds.

Cedrus deodara blossoms and sets seeds when 30 years old. Natural pollination is difficult for *Cedrus*; however, mature seeds may be achieved with artificial pollination. At present, cuttings are mainly used for propagation. It is also acceptable to sow fewer seeds to obtain seedlings. Plantlet proliferation may be promoted by tissue culture (Luo Shiwei, 1978).

The author began to propagate *Cedrus deodara* by tissue culture in 1982. This opened new areas for investigation since negative results had been previously reported (Winton, 1978; Yang Naibo, 1982).

TISSUE CULTURE TECHNIQUE

Culture Procedure and Medium

BUD INDUCTION. Young stem segments taken from 1-year-old seedlings of *Cedrus* were used as explants. MS medium was used with either 9.3 μM KIN, 2.7 μM NAA, and 1.15 μM 2,4-D or 8.8 μM BA and 2.7 μM NAA for callus induction from stem segment and bud differentiation. Dedifferentiation of *Cedrus* took longer than that of herbaceous plants and some other perennial crops. It took 60 to 80 days from inoculation to callus formation. Two types of callus formed: one was loose and of granular shape and formed from the cut end of the stem segment; the other grew out of the stem segment, swelling the segment and breaking through the epidermis. The differentiation ability of the former callus was weak; the latter was easy to differentiate within a short period. The stem segments began to swell 12 days after inoculation and turned brown. The thickened brown stem segments were transferred to induction medium for subculture. Clustered *Cedrus* plantlets were differentiated 2 months later; each cluster contained 3 to 10 plantlets. The inoculated explants turned dark brown and dried out gradually. A differentiated internode was formed and swelled only where the plantlet was differentiated.

When stem segments were taken from the adult trees for inoculation, brown material often appeared around the stem segment 2 weeks after inoculation. This occurred because *Cedrus* branches are rich in pine resin and other chemicals. Toxic material was produced by explant metabolism, resulting in the death of the explants. It was necessary to replace the medium frequently. If young stem segments of seedlings were used as material, results were better. Vitamin C and active charcoal could be added to the medium to reduce oxidation and absorb toxic material (Cells Laboratory, Shanghai Plant Physiology Institute, Chinese Academy of Sciences, 1978).

ROOT INDUCTION. *Cedrus* cultured shoots were cut when 1–1.5 cm in height and planted in rooting medium. Rooting medium was half-strength MS medium + 2.7 μM NAA + 4.9 μM IBA. Two or three rootlets were produced in 1 to 2 months.

Protocol

1. Branch stem was excised from the 1-year-old *Cedrus* seedling and rinsed with running water for 2–4 h.
2. The excised branch stem was soaked in 75% alcohol for 30 min and soaked immediately in 0.1% mercuric chloride solution for 10 min, washed with sterile water four times, and then blotted with filter paper.
3. The stem segments were cut into 0.5-cm pieces and inserted into the culture medium at a 60 degree angle to the surface of the medium. It was important that the lower part of the growing stem section be inserted into the medium; otherwise differentiation could be prevented.
4. Culture temperature: 25 ± 2°C; light illumination: 10 hr/day; light intensity: 1500 to 2000 lux. After rooting of the culture occurred, illumination was switched to natural diffuse light and the temperature reduced to 20 ± 2°C.

Culture Time and Induction Frequency

It took 120 to 150 days from stem segment inoculation to plantlet formation and 30 to 60 days from plantlet formation to rooting formation. Thus, it was 150 to 210 days from explant inoculation to transplantation of the plantlet into soil. This time was much longer than for other perennial crop species.

Cedrus stem segments of 1-year-old seedlings yielded the highest induction frequency (70%). When stem segments of adult *Cedrus* trees were used for inoculation, frequency of plantlet formation decreased to 10%.

TRANSPLANTATION OF PLANTLET

Plantlets were transplanted to the soil as soon as roots formed in the rooting medium and covered with glass for 1 week in order to preserve moisture. Another rooting method, called horticulturally young branch cutting, is as follows: differentiated *Cedrus* plantlet is treated with auxin for 2 hours and planted in unmixed soil (composed of vermiculite and humus soil, 1:1). As soon as the roots form, it is transplanted into the soil.

The plantlet should not be kept in rooting medium for too long while it is forming roots. Roots will be slim and weak and have difficulty surviving at the time of transplanting. The plantlet should be transplanted into soil at the moment when the root is just visible to the naked eye.

REFERENCES

Cells Laboratory, Shanghai Institute of Plant Physiology, Chinese Academy of Sciences. 1978. Organ Culture. In: *Plant Tissue and Cell Culture*. pp. 190–216. Shanghai Science and Technology Press, Shanghai (in Chinese).

Delectis Florae Reipublicae Popularis Sinicae, Agendae Academiae Sinicae Edita. 1978. Pinaceae Lindl.; *Cedrus* Trew. In: *Flora Reipublicae Popularis Sinicae*. Volume 7. pp. 200–203. Science Press, Beijing (in Chinese).

Luo, S. 1978. Advances and application of plant tissue and cell culture. *Acta Phytophysiol. Sin*. **4**:91–112 (in Chinese).

Yang, N. 1982. List of test tube plants (continued). *Plant Physiol. Commun*. **5**:59–71 (in Chinese).

Winton, L. L. 1978. Morphogenesis in clonal propagation of woody plants. In: *Frontiers of Plant Tissue Culture*. (T. A. Thorpe, ed.), pp. 419–426. International Association of Plant Tissue Culture, Calgary.

CHAPTER 37
Xanthoceras sorbifolia

*Yongming Wang**

ECONOMIC IMPORTANCE AND HISTORY OF *XANTHOCERAS SORBIFOLIA* CULTIVATION

Xanthoceras sorbifolia Bge is a deciduous tree, belonging to the family Sapindaceae. It is the only species in the genus *Xanthoceras*, which is native to China (Fig. 1).

Xanthoceras sorbifolia is a hardy, adaptable oil-bearing tree species, with seeds that are an economically important product. The kernel contains 62.8% fat, 26–29% crude protein, and 9% sugars. The kernel oil is light yellow, has a pleasant odor, and is a high-quality edible oil and raw material for the chemical industry (Department of Forestry, Inner Mongolia Institute of Agriculture and Animal Husbandry, 1977). The fatty acids in the kernel are 35–43.7% oleic acid and 34.8–52.0% linoleic acid, which can be extracted and used in the manufacture of drugs for heart and circulatory diseases. The pericarp contains about 12% aldehyde.

Xanthoceras sorbifolia has only recently been introduced to intensive cultivation. Large-scale cultivation began in Inner Mongolia Autonomous Region in the late 1950s. Since then, the cultivation area in Inner Mongolia has been enlarged and it has been introduced and cultivated in Xinjiang Uygur Autonomous Region, Jiling Province, and Liaoning Province. In the middle and lower reaches of the Min and Yalong rivers of

*English translation by Jifang Huang.

Figure 1. Fruit-bearing shoot of *Xanthoceras sorbifolia*.

Sichuan Province, *Xanthoceras sorbifolia* was planted on a trial basis for afforestation in arid valleys, where it grew and set seed normally.

In the early 19th century, *Xanthoceras sorbifolia* was introduced into the United States of America, England, and Russia from China as an ornamental tree in gardens and parks. Around 1974, *Xanthoceras sorbifolia* was introduced into the Democratic People's Republic of Korea, where it is used as an oil-bearing tree.

GEOGRAPHICAL DISTRIBUTION

The natural distribution of *Xanthoceras sorbifolia* is concentrated in the Zhiwuling and Han River Valley of Gansu Province, Yan'an District of Shaanxi Province, the eastern slope of Luliang Mountain of Shanxi Province, and the mountain regions of western Henan Province. There are small areas of secondary forests distributed in the Mengda forest zone of Xunhua Xian of Qinghai Province, the southern foot of Daqing Mountain and Zhaowuda League of Inner Mongolia, and the Chengde and Zhangjiakou districts of Hebei Province.

Its distribution indicates that *Xanthoceras sorbifolia* requires an average yearly temperature of 4.1–14.2°C, precipitation of 140.7–984.3 mm, and total day length of 2341–3166 hr. The tree can normally grow and set seeds in the regions from Qinling Mountains and Huaihe River (around 33 degrees north latitude) northward up to 46 degrees north latitude. It undergoes slight freezing on the branch tips but is able to set seeds in Kedong Xian of Heilong-jiang Province and Aletai District of Xinjiang Uygur Autonomous Region, around 48 degrees north latitude, where the lowest temperature is −40°C. It was also planted in the United States of America, where the lowest temperature is −10° to −5°C.

In the Union of Soviet Socialist Republics, the farthest northeast regions where *Xanthoceras sorbifolia* can be cultivated successfully are Dnieperpetrovsk and Rostov. There are naturally scattered trees of *Xanthoceras sorbifolia* in the arid regions of Ningxia Hui Autonomous Region of China, where the yearly precipitation is only 148.2 mm.

Xanthoceras sorbifolia can normally grow and develop on many kinds of soil. There are natural forests and successfully introduced cultivated forests in the slightly acidic meadow soils of the mountainous forests along the Min River of Sichuan Province, in the poor soil of the mountain districts of western Henan Province, in the soil of the Altai mountainous districts of Xinjiang, in the drift sand of the suburban district of Chifeng City, and in the dark loose soil or mountainous poor soil of hilly lands. In the mountain regions of Inner Mongolia and Hebei Province, there is also natural growth of *Xanthoceras sorbifolia* trees. The pH of the soil in those regions is usually alkaline and the top soil is only 0.1–1 m. In Bayanzhou'er League of Inner Mongolia and Akesu District inhabited by the Uygur nationality in Xinjiang, *Xanthoceras sorbifolia* trees over 10 years old can only be seen in places where the pH of the soil is 8–8.5 (Department of Forestry, Inner Mongolia Institute of Agriculture and Animal Husbandry, 1977).

PROBLEMS IN PRODUCTION

In China, in regions where wild *Xanthoceras sorbifolia* grows and in areas where it has been cultivated, most populations are very heterogeneous, resulting in great variation in yield between individual trees and in very low yield per unit area. The first seed selection and improvement programs for *Xanthoceras sorbifolia* were carried out in the early 1970s in Zhaowuda League Institute of Forestry Sciences of Liaoning Province (presently under the Inner Mongolia Autonomous Region), the Institute of Afforestation for Arid Region of Liaoning Province, and the Inner Mongolia College of Agriculture and Animal Husbandry. Batches of improved individuals have been obtained in several regions. However, the regions best suited for particular clones have not yet been identified, the grafting and cutting techniques have not been standardized, and the improved trees have been propagated very slowly. In order to enhance the yield per unit area and to make full use of this oil-bearing source, *Xanthoceras sorbifolia* breeding programs should promote vegetative propagation of improved trees and establish regional tests with good clones.

TISSUE CULTURE TECHNOLOGY

The tissue culture techniques for *X. sorbifolia* have been established in China in recent years and are described here (Wang and Chen, 1981; Wang et al., 1982; Zhang et al., 1980).

Culture Procedures and Media

PROTOCOL

Inoculation of young stem segments bearing axillary buds

Induction of axillary bud differentiation

Clustered green shoot development

Induction of green shoots for root regeneration

Growing of clustered green shoots with roots

Development of root system

Plantlets developed for transplantation

MEDIA

For induction of green shoots: MS basic medium with BA 2.2–4.4 μM, CH 250–300 mg/l, sucrose 73.3–88 mM.

For rooting induction: ½ macro modified MS medium supplemented with IBA 4.9 μM and sucrose 44 mM.

For promoting development of root system: modified MS medium with 44 mM sucrose.

White cone-shaped root tips emerged from the center of the lower cut surface or near the cut surface of stem after 7–15 days of culture of green shoots in the root regenerating medium. Green shoots with cone-shaped root tips were transferred immediately to medium without hormones to promote growth of the root system. After about 25 days, healthy plantlets developed and were ready for transplantation. If plantlets were not transplanted at this time callus formation sometimes occurred on the bases of the young roots, greatly lowering the survival rate after transplantation.

Plantlets were subcultured to fresh medium at intervals of 25–30 days to proliferate green shoots. The amount of BA was 2.2–4.4 μM for inducing green shoots; it should be reduced to 2.2 μM for shoot subculture. With excessive doses of BA, white frostlike calli always developed on the petiole base, resulting in leaf abscision and etiolation.

Addition of casein hydrolysate and an increase in the amount of organic components could prolong the useful life of the medium for subculture and increase shoot differentiation.

The pH for the medium is 5.8–6.2, with agar concentration of 7–9.5 g/l.

Essentials in Culture Technique

SELECTION AND STERILIZATION OF EXPLANT. *Materials*. Young stem or semilignified branches with axillary buds were taken in sunny days in May to June and cut into 2–8-cm segments.

Sterilization. The prepared stem segments were soaked in 75% alcohol for 30 sec and transferred into a 0.1% solution of mercuric chloride. The mercuric chloride soaking time was altered, according to the maturity and condition of the materials; about 4–6 min was usually suitable. Excessive sterilization time injured young stem segments and axillary buds, causing browning and failure of normal growth and differentiation. However, insufficient sterilization resulted in contamination of the culture materials. After sterilization, the cultures were rinsed four times with sterilized water.

Cutting segments and inoculation. The sterilized culture materials were cut into 1–2-cm segments bearing axillary buds and placed in aseptic petri dishes. The water on the surface of the segments was blotted off with filter paper, and four to five segments were inoculated per 100-ml culture flask.

CULTURE CONDITIONS. The culture room temperature was 23 ± 3°C, and the illumination was 12–14 hr at 2000–3500 lux, provided by four 40-W fluorescent tubes.

The length of the photoperiod had an important influence on the growth of green plantlets of *Xanthoceras sorbifolia*. Ten hours or less light for 1 month retarded growth of some of the green plantlets or even stopped growth. This phenomenon could be prevented by changing the photoperiod to 12–16 hr.

A temperature range of 18–26°C in the culture room maintained normal growth and rooting of the green shoots, but temperatures over 28°C caused etiolation of the green plantlets, reduction in root regeneration, and decrease in the number of lateral roots.

Morphogenesis and Propagation of Green Plantlets

After inoculation, the axillary buds on the cultured young stem segments began to swell and became greenish. The swollen axillary buds were grown to a height of 1–2 cm in culture flasks. Green shoots emerged at this point, and the buds were excised and subcultured. The axillary buds sprouted up as clustered leaflike shoots after about 1 month in culture. As the culture developed, the young leaves gradually hardened and the leaf blades produced greenish pinheadlike outgrowths. In the following 2–3 months, solitary adventitious shoots or clustered shoots developed from the outgrowths (Fig. 2). When the shoots grew 2–3 cm in height after 4–5 months' culture, they were cut off from their bases to regenerate roots. The remaining clusters of adventitious buds were cut into several pieces and used in continuous subculturing, from which new adventitious clustered shoots were obtained.

The compound leaves of green *Xanthoceras sorbifolia* plantlets, especially those at the plantlet base, had a very strong potential for regenerating adventitious buds. As soon as the leaves contacted the subculture medium, they hardened, arched, or curled and turned dark brown or light green, and solitary adventitious shoots or clustered shoots emerged.

After many successive subcultures, the bases of the most clustered plantlets swelled to form masses of callus with compact texture, pale green or light brown on the outer layer and white inside, from which clustered green shoots differentiated. Subculture of compound leaves of green plantlets and clustered buds could be used for continuous propagation of *Xanthoceras sorbifolia*. According to the Laboratory of Tissue Culture of Anshan Institute of Forestry Sciences, the propagation rate for *Xanthoceras sorbifolia* was observed to be about 1:3 (Table 1). Cultures that have been subcultured for 4 years have maintained their vigorous capacity to differentiate up to the present.

TRANSPLANTATION TECHNIQUES AND CLONAL GROWTH

Transplantation Techniques

Cultured healthy plantlets with lateral roots were selected for transplantation. The tops of the culture tubes were removed to expose the plantlets to ambient conditions for 1 or 2 days. The plantlets were taken out, and the medium adhering to the root and root collar was removed. The plantlets were then transferred into pots of 3–4 in diameter. The soil

Figure 2. Clusters of shoots developing from axillary buds on cultured young stem segments.

TABLE 1. Results of Propagation of Green Shoots of *Xanthoceras sorbifolia**

Date of Transfer (Day/Month)	27/8	26/9	13/10	29/10	29/11	25/12
Passage number	2	3	4	5	6	7
Number of flasks	3	29	80	212	237	603
Number of shoots used for root regeneration	1	1	16	16	83	233

*From the fifth to the seventh subculture, some of the cultured plantlets were discarded during each subculture because of limitations of size.

field in the pots was composed of equal volumes of nursery bed soil plus field soil and sand.

Increasing Transplantation Survival Rates

Xanthoceras sorbifolia trees have large fleshy roots which are susceptible to root rot. The survival rate of transplanted plants could be increased by the following methods:

1. Before mixing soils for potting, the river sand used was rinsed with clean water. The sand mixtures were sterilized.
2. Before use with soil the earthen pots were immersed in water for 1 to 2 days prior to transplantation to wet the soil thoroughly.
3. When planted, deep planting and shallow covering procedures were applied. In deep planting, the plantlet root system was deeply embedded in the pot and spread out along the bottom so that the plantlet could absorb water from the bottom layer of soil even though the upper layer of soil was dry. In shallow covering, the plantlet root collar was located at the soil surface level or a little higher to keep dry soil around the root collar.
4. No watering was needed until the top soil in the pots dried out. This occurred about 3–5 days after transplantation, and the pots were then thoroughly irrigated.

These procedures resulted in a stable survival rate of about 90%. Because the plantlets grew very quickly, their root systems filled the pots in 2 to 3 months. The root system should be fully extended in the soil when plantlets are set in the field; otherwise, the roots curl in the upper layer of soil and have difficulty penetrating deeply into the lower stratum of soil, and the plantlets would lodge.

Figure 3. Stout plantlets in nursery.

Growth of *Xanthoceras sorbifolia* Clones

Of 309 plantlets which were planted in the field in May and June 1982, 281 clones survived and grew well (Fig. 3). The initial average height was 7 cm and the maximum 15 cm (Fig. 3). In September of the same year, the average height of the plantlets was 36 cm and the maximum was 94 cm. The average diameter at ground level was 0.52 cm and the maximum was 0.89 cm. In October 1983, the maximum height was 127 cm, with an average of 61.2 cm, and the average diameter at ground level was 1.23 cm. In May 1985, the maximum height was 239 cm and the maximum diameter at ground level was 3.33 cm.

In 1983, individual adult trees flowered and some adult trees flowered and set fruit in 1984. Most of the adult trees flowered and set fruit in 1985. No abnormal trees were found.

REFERENCES

Department of Forestry, Inner Mongolia Institute of Agriculture and Animal Husbandry. 1977. Economic value and natural distribution of *Xanthoceras sorbifolia* In: *Xantheroceras sorbifolia*. pp. 1–13. Inner Mongolia People's Press, Huhehaote.

Wang, Y. and Chen, Y. 1981. Preliminary study on tissue culture in *Xanthoceras sorbifolia*. *Corr. Sci. Technol. Forest*. **7**:7–9 (in Chinese).

———, Zhao, J., and Chen, Y. 1982. Reports of research on propagation of *Xanthoceras sorbifolia* by tissue culture. *Anshan Sci. Technol*. **4**:1–5 (in Chinese).

Zhang, G., Xu, X., and Zhao, Z. 1980. Preliminary achievement on induction and transplantation of plantlets from tissues of young stem in *Xanthoceras sorbifolia*. *Corr. Sci. Technol. Forest*. **7**:4–5 (in Chinese).

Species Index

Acacia, 34, 226
A. auriculaeformis, 34, 47
Acer, 226
A. pseudoplatanus, 83, 88, 94
A. saccharum, 34
Actinidia, 407, 411, 413–414, 416
A. arguta, 409, 415
A. chinensis, 95, 267, 407–409, 412, 415
A. eriantha, 409
Aesculus, 63, 226
A. hippocastanum, 63, 64, 67, 68, 69
Agrobacterium tumefaciens, 13, 113
Alchornea, 34
Alnus, 226
A. glutinosa, 95, 105
A. incanus, 105
A. indica, 95
Amygdalus davidiana, 284
Anacardiaceae, 225
Angelica sinensis, 194
Apium graveolens, 323
Arabidopsis thaliana + *Brassica* sp., 12
Araucaria cunninghamia, 34, 47, 129, 480
Aretium lappa, 137
Atriplex canescens, 34
Atropa belladonna, 12

Bacillus, 319
Betula, 226
B. alleghaniensis, 34
Betula (*Cont.*):
B. pendula, 94
B. platyphylla, 34, 43, 95
Bougainvillea spectabilis, 129, 132
Brassica, 12
B. napus, 64, 65
Broussonetia, 226

Camellia chrysantha, 129
C. japonica, 34, 63, 129, 132, 133, 135
C. oleifera, 429
C. sinensis, 64, 475
Carica papaya, 35, 77, 86
Carya, 226
Castanea, 226
C. sativa, 31, 35
Catalpa, 226
C. bungei, 35, 47
Catharanthus roseus, 135
Cathaya argyrophylla, 35
Cattleya, 33
Cedrus, 482–483
C. deodara, 129, 133, 480, 481
Celastrus, 226
C. orbiculatus, 35
Cerasus tomentosa, 284
Chaenomeles japonica, 133
Chosenia, 146, 147
C. macrolepsis, 146, 147, 148, 151, 155, 156

Cinnamomum, 226
C. albosericeum, 35
Citrus, 26, 27, 47, 63, 67, 84, 95, 102, 105, 108, 128, 257, 267, 365, 369–370, 372–374
C. aurantium, 88, 94, 367
C. ichangensis, 366
C. grandis, 35, 36, 48, 368
C. lemon, 94
C. limon, 368
C. microcarpa, 35, 63, 65, 129, 132, 367–369
C. nobilis, 366
C. paradisi, 94
C. × *poncirus,* 36
C. reticulata, 94, 365, 366, 368
C. sinensis, 35, 87, 88, 93, 94, 365, 368
C. trifoliata, 367
Codiaeum variegatum, 36, 129
Coffea sp., 95, 100, 105, 108
C. arabica, 32, 36, 84
C. canephora, 36
Copaifera multijuga, 94
Corylus, 226
Coryneum populinum, 171
Cotoneaster dammeri, 133
Crataegus, 133
C. braehyacantha, 133
C. pinnatifida, 329–335, 340–343, 346
Cryptomeria fortunei, 48
Cryptomoris japonica, 36
Cucumis sativus, 136
Cunninghamia lancelolata, 36
Cupressaceae, 128
Cycadaceae, 128
Cycas revoluta, 128
Cydonia oblonga, 266
Cymbidium goeringii, 128
Cyphomandra betacea, 36

Dahlia pinnata, 137
Daphne odora, 36, 122, 129, 132, 135
Datura innoxia, 88
Davidia involucrata, 37
Dendranthema morifolium, 128, 136, 137
Dianthus caryophyllus, 27, 48, 128, 136, 137
Dimocarpus longana, 385–386
Dryobalanops aromatica, 377

Elaegnus angustifolia, 421–422, 424, 427, 429
Elaeis guineensis, 37, 48, 323, 440
Ericaceae, 128
Eriobotrya cavaleriei, 398
E. deflexa, 398
E. fragrans, 398
E. japonica, 397–398
E. prinoides, 398
Eucalyptus, 48, 199, 226
E. alba, 201
E. botryodea, 37
E. botryoides, 204
E. camaldulensis, 200, 201
E. cinerea, 201, 202, 204
E. citriodora, 37, 42, 200, 201, 204
E. delegatensis, 200
E. diversifolia, 200
E. exserta, 200, 204
E. ficifolia, 201, 202
E. globulus, 200
E. grandis, 200
E. grandis × *E. exserta,* 202
E. gunnii, 204
E. maculata, 200
E. maideni, 200, 204
E. microcorys, 200
E. pilularis, 200
E. regnans, 201, 204
E. resinifera, 200
E. robusta, 200
E. saligna, 200, 204
E. saligna × *E. exserta,* 202
E. tereticornis, 200, 202
E. trabutii, 204
E. viminalis, 200
Eucommia, 122
E. ulmoides, 117, 122
Euphoria, 63, 257, 386
E. longana, 63, 257
Euphorbia pulcherrima, 129

Fagus, 226
F. sylvatica, 95, 96, 100
Ficus carica, 129, 132, 135
Forsythia suspensa, 122
Fortunella crassifilia, 132
F. crassifolia, 129
F. margarita, 37
Fragaria ananassa, 136, 137
Fuchsia, 33

Gerbera jamesonii, 128
Gingkgo biloba, 37, 122, 129

Gladiolus gandavensis, 128
Gleditsia, 226
Glycine max + *Nicotiana tabacum,* 12
Gmelina arborea, 37
Gomphicna globosa, 137
Gramineae, 128
Grevillea rasmarinfolia, 37

Hamamelis vernalis, 129
Helicobasidium, 171
Hevea, 63, 94, 95, 99, 455
  H. brasiliensis, 62, 65, 262
Hibiscus rosa-sinensis, 37, 129, 135
Hordeum vulgare, 10

Idesia polycarpa, 78, 79, 81
Iris tectoru, 137

Jasminum sambac, 135
Juglans regia, 432

Lagerstoemia indica, 37
Larix decidua, 94
  L. potaninii, 354
Leguminosae, 128
Leucaena leucocephala, 38, 94
Liliaceae, 128
Lilium, 137
Linum usitatissimum, 11
Liquidambar, 226
  L. styraciflua, 77, 86
Liriodendron, 226
Litchi, 63, 377
  L. chinensis, 65, 257, 376, 386
  L. philippinensis, 377
Lolium perenne, 10
  L. multiflorum, 10
Loropetalum, 226
  L. chinense, 38, 129
Lycium, 63, 122
  L. barbarum, 64, 66, 68, 69, 118, 119, 120
  L. chinense, 64
  L. chinensis, 95, 99, 100, 105, 108, 117, 118, 120, 184
  L. halimifolum, 62
Lycopersicon esculentum, 12

Magnoliaceae, 128
Malus, 26, 38, 63, 133, 233, 234, 239, 243, 245, 267
  M. asiatica, 234, 246

Malus (*Cont.*):
  M. baccata, 247
  M. domestica, 38
  M. micromalus, 234
  M. prunifolia, 63–67, 73, 234, 246, 256, 260, 261, 262
  M. pumila, 38, 95, 100, 105, 133, 136, 234, 235, 390
  M. spectabilis, 247
  M. sylvestris, 38
Manihot esculenta, 27
  M. utilissima, 38, 49, 51
Maytenus, 122
  M. hookari, 39, 121
Michelia alba, 39, 129
  M. macclurei, 39
Momordica grosvenori, 39
Morus alba, 39
Murraya exotica, 39

Nandina domestica, 129, 132
Nephelium longana, 386
Nicotiana glauca, 112
Nicotiana tabacum, 11, 12, 48, 64, 65, 81, 108, 137
  N. tabacum + *Atropa belladonna,* 12

Olea europaea, 39
  O. europaea sativa, 156
Oryza sativa, 64, 65

Paeonia suffruticos, 129, 135
Palmaceae, 128, 440
Palmae, 84
Papavar somniferum, 43
Parchira, 226
Parthenium argentatum, 468, 472–474
Passiflora, 226
Paulownia, 226
  P. elongata × *P. tomentosa,* 39
  P. fortunei, 39
  P. taiwaniana, 94
  P. taiwaniana + *Populus euramericana,* 95
Penicillium notatum, 12
Petunia × *hybrida,* 112
Phellodendron, 217, 226
  P. amurense, 150, 216, 217, 219, 222, 223
Picea abies, 47, 153
Pinaceae, 128
Pinus, 87

Pinus (*Cont.*):
P. caribaea, 39
P. contorta, 95, 105
P. echinata, 94
P. pinaster, 95, 98, 100, 101, 105
P. strobus, 129
P. teada, 27, 94
Poncirus, 63
P. trifoliata, 64, 68, 69, 70, 257
Potentilla, 133
Populus, 63, 67, 146, 161, 174, 183, 199, 226
P. adenopoda, 146, 148, 151, 153
P. alba, 148, 151, 157, 164, 194, 198
P. alba × *P. bolleana*, 146, 148, 151
P. alba × (*P. davidiana* × *P. Simonii*) × *P. tomentosa,* 146–148
P. alba × *P. laurifolia,* 192, 196, 197
P. alba × *P. simonii,* 162
P. albus, 101, 103, 108, 146
P. balleana, 164
P. balsamifera, 159, 179
P. berolinensis, 161–167, 174, 176–180
P. berolinensis × *P. pyramidalis,* 162
P. canadensis, 148
P. canadensis × *P. koreana,* 162, 166
P. canescens, 146, 148
P. davidiana, 48, 155, 164
P. davidiana × *P. alba,* 148, 155
P. davidiana × *P. bolleana,* 146, 148, 151
P. deltoides, 161, 179
P. diversifolia, 146, 148, 149, 151
P. euphratica, 162
P. × *euramericana,* 94
P. euroamericana, 146, 148
P. grandidentata, 146, 148, 149, 151, 153, 157
P. harbinensis, 161
P. harbinensis × *P. pyramidalis,* 162, 166
P. hopeiensis, 146, 148, 150, 151, 155, 157
P. japonica, 151
P. jrtyschensis, 191, 192
P. nigra, 62, 146, 148, 162
P. maximowiczii, 161
P. pekinensis, 162, 166
P. pseudosimonii, 161, 162, 166
P. pseudo-simonii × *P. pyramidalis,* 161, 162, 166
Populus (*Cont.*):
P. pyramidalis, 161
P. siebldii × *P. grandidentata,* 161
P. × *simonigra,* 185
P. simonii, 97, 161, 162, 191, 192, 196
P. simonii × *P. diversifolia,* 159
P. simonii × *P. italica* × *P. rassica,* 192
P. simonii × *P. nigra,* 155, 161, 162, 164–169, 171, 174–178, 180
P. simonii × *P. pyramidalis,* 164, 165, 166
P. tomentosa, 146, 148, 150, 151, 157
P. tomentosa × *P. bolleana,* 146
P. tremula, 95, 146, 148, 157
P. tremuloides, 94, 95, 96, 100, 146, 148, 174
P. trichocapa, 174
P. ussuriensis, 162
Prunus sp., 278, 281
P. amygdalus, 39
P. armeniaca, 39, 40
P. avium, 40
P. cerasifera, 40, 129, 132
P. cistena, 40
P. compressa, 279
P. communis, 280
P. davidiana, 48, 279, 284
P. densa, 279
P. domestica, 139
P. duplex, 279
P. fergarensis, 280
P. instititia, 40
P. mira, 279
P. nectarina, 279
P. pendula, 279
P. persica, 40, 136
P. pseudocerasus, 48, 136
P. tenella, 280
P. tomentosa, 133, 284
Pseudotsuga douglasii, 47
P. menziesii, 94, 95
Punica granatum, 40, 129
Pyracantha coccinea, 133
Pyrus sp., 136, 264, 267, 276
P. bretschneideri redh., 265
P. cammunis, 264, 265
P. communis, 40, 136
P. pashia, 266
P. persica, 266
P. pyrafolia, 265
P. salicina, 266

Black currant, shoot tip culture (*Cont.*):
culture procedures and media, 352–353
donor plant selection, 352
genetic stability of test tube plantlets, 362
inoculation of both dormant and germinating buds, 356–357
investigation of transplanted cultured plantlets, 361–362
media for differentiation and subculture, 352–353
prospects, 362–363
regulation of BA and GA concentrations, 359
root induction, 353
rooting outside test tube, 360–361
selection of large explants for inoculation, 355
selection of shoot tips, 355–357
shoot morphogenesis, 353–354
sterilization and culture conditions, 353
superiority of BA to other cytokinins, 357
superiority of light culture to dark culture in media with BA and GA, 358–359
terminal buds as explants, 355
time period for subculture and proliferation, 360
transplantation techniques, 360–361

Cell growth rate, determination of, 80–81
Cell suspension culture, 76–89
in *Carica papaya,* 86
in *Citrus sinensis,* 87
in *Liquidambar Styraciflua,* 86
morphogenesis of embryos, 83–86
bioreactors for large-scale suspensions, 85
conditions for embryo formation, 84–85
modes of morphogenesis, 83–84
somatic embryo delivery systems, 86
synchronization of embryogenesis, 85
mutants isolated from, 87–89
buffering growth of cells, 89
mutagenesis, 88
plant regeneration and field evaluation, 89
preparation of materials, 88
preparation of selective media, 89

Cell suspension culture, mutants isolated from (*Cont.*):
screening for disease-resistant mutants, 89
selection and preparation of toxin, 89
in *Pinus,* 87
in *Santalum album,* 86–87
techniques, 77–83
cell growth, 80–81
culture methods, 79–80
density of cultured cells, 78–79
induction and subculture of callus, 77
isolation of single cells, 78
selection of cultured cells, 81–83
synchronization of cell division, 83
Cell totipotency, 6–9
accomplishment of, 7–9
cell types, 7
Chosenia tissue culture, 148
characteristics of morphogenesis, 156
culture requirements, 155
history, 147
Chromosome number variation and somaclonal and gametoclonal variation, 11
Citrus:
anther culture, 365–374
chromosome number examination, 374
histological observation of development of microspores, 373–374
organ differentiation, 371
procedure and media, 367
successful induction of pollen-derived embryos, 368–371
effect of low-temperature treatment of flower bud on embryo formation, 368–369
optimal medium and plant hormone combinations, 369, 370
screening of materials capable of induction, 368
selection of optimal culture conditions for development of microspores, 370–371
techniques, 367
transplanting techniques for cultured plantlets, 372–373
geographical distribution, 365–366
history of culture, 366–367
importance of cultivation, 365
problems in breeding, 366

Cork tree, 216–224
characteristics of morphogenesis and stability of characteristics, 221–223
culture techniques, 217–221
appropriate hormone compositions, 219–220
culture procedures, 218–219
donor age and plantlet regeneration, 221
increased induction frequency using male materials, 220–221
selection of explants, 217–218
economic importance, 216
geographical distribution, 216
problems in seeding production, 216–217
prospects, 223–224
Crabapple:
anther culture, 256–263
androgenesis, 261
callus induction, 258–259
chromosome number in plantlet root tip cells, 261–262
culture procedures and media, 257–260
history, 257
induction of callus differentiation, 259–260
induction of healthy shoots, 260
inoculation techniques and culture conditions, 260
origin of plantlets, 261–262
prospects, 262–263
root induction, 260
selection of culture materials, 257–258
transplantation of plantlets, 260–261
problems in breeding, 256
Crossing-over and somaclonal and gametoclonal variation, 10
Cryoprotectants, 50–51
Culture methods:
feeder layer, 80
solid plate, 79–80
static culture with shallow liquid layer, 79
suspended drop (microdrop), 80
Cytokinin-auxin ratio and adventitious shoot development, 151, 152
Cytological techniques, 53–55

Deodara cedar, 480–483
botanical traits and significance of stem tip culture, 481
geographical distribution and economic importance, 480–481
Deodara cedar (*Cont.*):
tissue culture, 481–483
bud induction, 483–484
culture time and induction frequency, 484–485
root induction, 484
transplantation of plantlet, 483
Embryo culture, 45–46
apple shoot tip, 245–254
hawthorn, 346–350
peach, 285–290
pear, 267–269
Endosperm culture:
kiwi fruit, 413–414
of *Lycium,* 119
pear, 269–274
Eucalyptus, 199–215
classification, 200
culture techniques, 202–211
afforestation using cultured plantlets as nursery stock, 212–214
E. ficifolia culture medium, 202–203
induction of shoots from embryonic cell aggregates, 204–211
callus induction, 205
characteristics of embryo development, 209–211
differentiation of embryonic cell aggregates, 205–206
plantlet development, 206–208
subculture of embryonic cell aggregates, 208–209
plantlet differentiation from callus of adventitious buds, 203–204
callus induction, 203–204
root induction and plantlet formation, 204
shoot regeneration, 204
prospects for, 215
transplantation of plantlets, 211–212
economic importance, 199–200
history of research, 201–202
problems in production, 200–201
Explants, 23–27
agents used for disinfection, 29
browning, 31–33
factors affecting, 31–32
methods to overcome, 32–33
cells with low mitotic activity, 24
characteristics of juvenility and maturity, 23–24
disinfection, 30

Explants (*Cont.*):
factors determining juvenility and maturity, 24
maintaining genetic stability, 25
rejuvenation, 24–25
selection of, 27
types of, 26

Ficoll solution, 82

Gametes and cell totipotency, 7
Gametoclonal variation, sources of, 9–11
Gene number changes and somaclonal and gametoclonal variation, 10–11
Genetic variants, biotechnological methods for obtaining, 9–13
Grape anther culture, 300–310
chromosome number in anther- or pollen-derived plantlets, 309
culture procedure and media, 302–303
differentiation and proliferation of embryos, 303
economic significance, 300
effect of genotype on induction frequency, 305–308
explant culture conditions, 304
explant selection, 303
explant sterilization and inoculation, 304
history, 301–302
induction of callusing anthers, 302–303
media and supplements, 304–305
medium for platelet induction, 303
prospects, 309–310
requirements for grape breeding, 301
techniques, 302–307
transplantation of anther- or pollen-derived plantlets, 307, 309
Grape micropropagation, 312–327
characteristics and genetic stability of plantlets, 326
future prospects, 326–327
history, 314
protocol of in vitro propagation, 315–324
shoot multiplication, 315–323
combination of hormones and, 320–321
critical variables, 319–323
culture process and media, 316, 317
genotypic differences and, 319–320
germ plasm storage and, 321–323
sterilization and inoculation of explants, 316, 318–319
Grape micropropagation (*Cont.*):
significance of, 312–314
acceleration of propagation, 312–313
advantages for germ plasm preservation, 313
use in pathology research and disease control, 313–314
use in plant physiology research, 313
virus elimination, 313
somatic embryogenesis, 323–324
transplantation technique, 324–325
Guayule, 468–474
culture techniques, 470–472
culture procedures and media, 470–472
sterilization of explants and inoculation, 470
cutting of cultured plantlets, 473
economic importance, 468–469
geographical distribution, 469
prospects for tissue culture, 474
research history, 469–470
setting of plantlets, 474
transplantation of cultured plantlets, 472–473

Haploid induction, 62–73
achievements in, 62–64
perspectives, 73
variables in, 64–72
anther culture procedures, 68–70
basal medium, 67
characteristics of pollen plantlets, 72
differentiation in embryos and shoots, 66–67
embryo differentiation and plantlet formation, 65–66
hormones, 67
importance of haploids and homozygous diploids, 71–72
mixoploidy, 70–71
organic supplements, 67–68
stages of embryogenesis, 64–65
Hawthorn, 329–344
development of superior clones, 331–334
culture procedure, 331–332
medium, 332
morphogenesis, 332–334
economic importance, 329–330
embryo culture, 346–350
adult bud sterilization, 349
culture temperature and illumination, 349

Hawthorn, embryo culture (*Cont.*):
procedures and media, 346–348
seed sterilization, 348–349
transplantation of plantlets, 349–350
flower bud differentiation, 342–343
culture techniques, 342
morphogenesis, 342–343
future prospects, 343–344
geographical distribution, 330
history of culture, 331
problems in production, 330
root initiation, 335–339
comparison of rooting methods, 338–339
rooting in test tubes, 335–338
rooting in vessels, 338
shoot proliferation, 334–335
culture procedure, 334
differentiation and growth of plantlets, 334–335
transplantation, 339–340
factors in, 339–340
transplanting techniques, 339
young embryo culture, 340–342
culture procedure, 340–341
differentiation of green plantlets, 342
embryo development, 341–342
embryo induction, 342
incubation conditions, 341
manipulation techniques, 340–341
media, 341
morphogenesis, 341–342
Histological techniques, 52–53

Karyotypic variants, and somaclonal and gametoclonal variation, 10
Kiwi fruit:
economic importance, 408
problems in production, 408
research history, 408–409
taxonomy and geographical distribution, 407
tissue culture techniques:
culture conditions, 411
culture of cotyledon, leaf, and embryo, 415–416
culture of shoot tips and stem segments with axillary buds, 412–413
culture procedures and media, 409–411
endosperm culture, 413–414
explant dedifferentiation and bud induction, 409–410
Kiwi fruit, tissue culture techniques (*Cont.*):
inoculation, 411
regeneration of plants from stem segment callus, 409
root induction, 410–411
sterilization of material, 411
subculture, 410
transplantation of plantlets, 411–412

Litchi:
anther culture, 376–384
characteristics of morphogenesis, 383
cytological examination, 379–380
of plantlets, 383–384
essentials for successful, 380–383
experimental materials and inoculation techniques, 379
future prospects, 384
regenerating intact plants, 381–383
selecting appropriate medium and supplements, 380–381
steps in, 378–379
economic importance, 376
geographical distribution, 377
problems in breeding and cultivation, 377–378
taxonomy, 377
Longan:
anther culture, 389–395
chromosomal observations, 395
morphogenesis, 393–395
selection of media, 389–393
culture procedure, 387–389
economic importance, 385
geographical distribution, 385–386
present status and value of tissue culture, 386–387
taxonomy, 386
Loquat:
economic importance, 397
geographical distributions, 398
history of culture, 399
plantlet transplantation, 404–405
problems in breeding, 399
taxonomy, 398
tissue culture techniques, 399–404
culture conditions, 402
embryo formation in callus, 401
induction of bud germination and leaf expansion, 400–401
material, 399–400

Loquat, tissue culture techniques (*Cont.*):
procedure, 402
rooting, 401
shoot formation, 401
shoot tip culture, 402–404

Media, 46–48
B5, 47
BL, 47
BM, 47
ER, 47
H, 47
HB, 48
HE, 48
LS, 46–47
modified Nitsch (1951), 48
MS, 46
in anther culture, 67
N6, 47
Nitsch (1969), 47
for plant regeneration in ornamental perennial plants, 132
for protoplast culture, 102, 104–106
SH, 47
White (WH), 48
WS, 48
Medicinal perennial crops:
plant propagation of stem segments of *Maytenus hookeri,* 121
regeneration of plantlets from Eucommiaceae hypocotyls, 122
tissue culture of, 116–122
problems in production of, 116–117
tissue culture of *Lycium,* 117–121
anthers and unpollinated ovaries, 118–121
cell culture, 119–120
endosperm culture, 119
leaves and shoot tip, 117–118
protoplast culture, 120–121
Meristematic cells and cell totipotency, 7
Mitotic index, determination of, 80
Mixoploidy, 70–71
Mutation(s):
induction and selection of, 12
spontaneous, and somaclonal and gametoclonal variation, 11

Oil palm, 440–451
biological characteristics, 441
history and economic importance of cultivation, 440–441
Oil palm (*Cont.*):
problems of production, 441–442
prospects for tissue culture, 451
review of oil palm tissue culture, 442–443
studies of genetic stability, 450–451
tissue culture techniques in China, 443–450
callus differentiation and embryo proliferation from leaves, 449–450
callus formation and hormone regulation, 445–446
callus induction from roots and leaves, 447–449
callus subculture, 446
conditions for embryo formation, 446
culture condition, 445
culture procedure, 443
embryo proliferation, 446–448
formation of intact plants, 450
heat treatment of seed embryo, 445
media, 443, 444
regeneration of plantlets, 447, 448
sterilization and inoculation, 443–444
Oleaster, 421–430
economic significance, 421
geographical distribution, 422
tissue culture:
culture procedures and media, 422–423
embryonic cell masses and embryos, 427–430
optimal growth regulator combination, 423–427
Ornamental perennial plants:
advantages of tissue culture, 128
cultivation, 127
drawbacks of seed propagation, 128
identification of virus-free plants, 139–140
prevention of reinfection of virus-free plants, 140
significance of regenerating virus-free plants, 135–136
taxonomy, 127–128
tissue culture and virus elimination of, 127–141
tissue culture techniques, 130–135
culture materials, 130
culture procedure, 130–131
incubation conditions, 134–135
media requirements at each stage, 131–133
sterilization and inoculation procedures, 133–134

Ornamental perennial plants (*Cont.*):
  transplantation and genetic stability of plantlets, 140–141
  virus elimination, 135–141
    heat treatment, 136–137
    by micrografting of shoot tips, 138–139
    by shoot tip culture, 137–138
  viruses affecting, 135
Ovary culture, poplar, 183–189

Peach, 278–298
  botanical taxonomy, 279–280
  cultivation and economic importance, history of, 278–279
  culture, history of, 282–284
  ecological classification, 280–281
  embryo culture, 285–290
    process and medium, 285–288
    steps in, 288–290
  future prospects for culture, 298
  problems for breeding, 281–282
  stem apex culture, 290–297
    media, 290–294
    root induction media, 294–295
    steps in, 296–297
    subculture media, 295–296
  use of embryo and shoot tip cultures, 284–285
Pear, 264–276
  classification, 265–266
  economic importance, 264–265
  embryo culture in vitro, 267–269
    comparison of embryo cultured plantlets and seedlings from matured seeds, 269
    culture methods and media, 267–268
    transplantation of plantlets, 268–269
  endosperm culture in vitro, 269–274
    culture conditions, 272
    culture methods and media, 269–271
    cytological observation, 273–274
    mixoploidy in endosperm plantlets, 274
    morphogenesis, 272–273
    sterilization and inoculation, 271
    transplantation, 272
  geographical distribution, 265
  history of culture, 266–267
  production goals, 266
  shoot tip culture, 274–275
    culture methods and media, 274–275
    transplantation and field setting of plantlets, 275–276
Perennial crop(s):
  advantages of biotechnology in, 6
  aromatic (*see* Aromatic perennial crops)
  breeding techniques, 5
  cell suspension culture and mutant screening in, 76–89
  disadvantages of biotechnology in, 5–6
  economic importance, 3
  haploid induction in, 62–73
  improvement, potential of biotechnology in, 3–19
  medicinal (*see* Medicinal perennial crops)
  protoplast culture and fusion in, 92–113
  quality, 4
  stress tolerance, 4
  tissue culture techniques in, 22–55
  yield, 4
Perennial ornamental plants (*see* Ornamental perennial plants)
Plating rate, determination of, 81
Poplar:
  anther culture, 161–181
    chromosome inspection of plantlets cultured in flask, 175, 176
    chromosome inspection of plants growing in nursery, 176–179
    chromosome variation and spontaneous doubling, 174–180
    culture method, 163–164
    culture procedures, 162–163
    effect of development stage on pollen dedifferentiation, 172, 174
    enhancing frequency of callus differentiation into plantlets, 169–170
    future prospects, 181
    mechanism of variation in chromosome number, 179–180
    rooting of plantlets, 170–171
    species and hybrids used, 161–162
    spontaneous chromosome doubling, 179
    successful haploid induction, 164–168
    transplantation of plantlets into pots, 171
    types of microspore development, 172, 173
  ovary culture, 183–189
    culture procedures and media, 184–185
    history, 183–184
    hormone selection for shoot differentiation, 187–188

Poplar, ovary culture (*Cont.*):
manipulation procedures, 185–186
plantlet transplantation and examination of chromosome number, 188–189
restraining growth of somatic callus, 186–187
stimulating development of haploid cells in embryo sac, 187
successful plant regeneration from embryo sac, 186–189
rapid propagation, 191–198
culture procedures and media, 192
embryo morphogenesis, 193–194
field observation of clonal progeny, 197
future prospects, 197–198
induction of large numbers of embryos, 194–195
manipulation techniques, 192–193
root initiation in bottle, 196–197
root initiation in pots, 195–196
tissue culture:
agar in medium, 153
basal medium, 151
characteristics of morphogenesis, 155–156
culture procedures, 147–149
culture requirements, 155
growth regulators, 151, 152
history, 146–147
light and temperature, 153–155
prospects of, 157, 159
selection of organs as explants, 149–150
sterilization of explants, 150–151
sucrose in medium, 151, 153
transplantation of test tube plantlets, 156–158
Protoplast culture and fusion, 12, 92–113
culture, 102, 104–108
culture conditions, 104, 106–108
medium, 102, 104–106
fusion, 108–113
electrofusion apparatus, 110
electrofusion procedure, 110–111
potential of, 92–93
selection of donor materials, 95–99
status of, 94
use of enzymes in isolation, 99–103

Rapid propagation, 191–198
Recessive variation, expression of and somaclonal and gametoclonal variation, 9–10
Recombinant DNA techniques for crop improvement, 12–13
Rubber tree:
anther and ovule culture, 453–466
androgenesis, 455–457
basic medium, 453–454
cytological investigation, 458–466
abnormal mitosis, 461–462
acceleration of natural processes, 464
chromosome count of transplanted pollen trees, 460–461
chromosome number of young embryos and root tip cells of plantlets, 458–459
mechanism of variation in chromosome number, 463
mixoploid chromosome number, 462–463
potential of bud grafting, 463
tendency to spontaneous chromosome doubling, 463
embryogenesis, 457–458
morphogenesis, 455–458
procedure of haploid induction, 454–455
protocol, 455
study on pollen tree H_1 and its clones, 464–466

Salicaceae:
economic importance, 146
geographical distribution, 145
taxonomy, 146
tissue culture in, 148
(*See also* Poplar; Willow tissue culture)
Shoot tip culture:
black currant, 351–353
ornamental perennial plants, 137–138
peach, 284–285
pear, 274–275
walnut, 435
Somaclonal variation, sources of, 9–11
Spontaneous mutation and somaclonal and gametoclonal variation, 11
Staghorn sumac, 225–228
callus induction, 226–227
future prospects, 228
induction of adventitious buds from callus, 227
possible uses of tissue culture techniques, 225–226
rooting and transplantation, 227–228
Stem apex culture of peach, 290–297

Tea:
  anther culture, 475–479
    culture conditions, 478
    culture procedure and media, 476–478
    cytological observation, 478
    selection of materials, 478
    sterilization, 478
  cultivation, 475
  geographical distribution, 475–476
  history of tissue culture, 476
  problems of tea plant breeding, 476
  transplantation of plantlets, 479
Tissue culture techniques, 22–55
  culture media, 46–48
  culture procedures, 33–46
  cytological techniques, 53–55
  embryo culture, 45–46
  establishment of aseptic tissue cultures, 28–31
    agents used for explant disinfection, 29
    application of antibiotics and fungicide, 30–31
    explant disinfection, 30
    methods of preventing contamination, 28–29
  explants used for, 23–27
    cells with low mitotic activity, 24
    characteristics of juvenility and maturity, 23–24
    factors determining juvenility and maturity, 24
    maintaining genetic stability, 25
    rejuvenation, 24–25
    selection of explants, 27
    types of explants, 26
  germ plasm preservation, 49–52
    freeze preservation, 50–52
    growth inhibition, 49–50
  histological observation, 52–53
  in vitro fertilization, 43–44
    of ovule carrying placentas, 44
    ovule, 44
    pistil, 43–44
  in vitro propagation, 33–43
    primary culture, 33–42
    root regeneration, 34–43
    shoot multiplication, 34–42
  methods to prevent browning of explants, 31–33
  for ornamental perennial plants, 130–135
Tree improvement:
  biotechnology and propagation of, 14–15
Tree improvement (*Cont.*):
  breeding systems in, 15–17
  program for via biotechnology, 15–19

Vegetative propagation, genetic gains of, 14, 15
Virus elimination from ornamental perennial plants, 135–141
Virus-free plants:
  identification of, 139–140
    antiserum identification, 139
    electron microscopy, 140
    indicator plant, 139
  prevention of reinfection of, 140

Walnut, 431–438
  embryo culture, 436–437
    material selection and pretreatment, 436
    medium, 436
    multiple bud induction, 436
    sterilization procedure, 436–437
  history and economic importance, 431
  methods to prevent browning, 437–438
  problems in breeding and cultivation, 432
  taxonomy, 432
  tissue culture:
    callus induction, 434–435
    future prospect, 438
    history, 432–434
    multiple bud formation, 435
    shoot tip culture, 435
Willow tissue culture, 148
  history, 147

Xanthoceras sorbifolia, 484–491
  economic importance and history of cultivation, 484–485
  geographical distribution, 485–486
  growth of clones, 491
  increasing transplantation survival rates, 490
  problems in production, 486
  tissue culture, 486–489
    culture conditions, 488
    culture protocol, 486
    media, 486
    morphogenesis and propagation of green plantlets, 488–489
    selection and sterilization of explant, 488
  transplantation techniques, 489–490

Zygotes, and cell totipotency, 7

About the Editors

ZHENGHUA CHEN is professor and director of the Laboratory for Cell Engineering of Economic Crops, Institute of Genetics, Chinese Academy of Sciences.

WILLIAM R. SHARP and DAVID A. EVANS are co-founders of the DNA Plant Technology Corporation in Cinnaminson, New Jersey.

PHILIP V. AMMIRATO is professor and former chairman of the biology department at Barnard College of Columbia University.

MARO R. SONDAHL is senior research director of tropical crops at DNA Plant Technology Corporation.